Innovations in Soft Computing and Information Technology

Jayeeta Chattopadhyay · Rahul Singh
Vandana Bhattacherjee
Editors

Innovations in Soft Computing and Information Technology

Proceedings of ICEMIT 2017, Volume 3

Springer

Editors
Jayeeta Chattopadhyay
Department of Chemistry
Amity University
Ranchi, Jharkhand, India

Vandana Bhattacherjee
Department of Computer Science
and Engineering
Birla Institute of Technology, Mesra
Ranchi, Jharkhand, India

Rahul Singh
Department of Mechanical Engineering
Amity University
Ranchi, Jharkhand, India

ISBN 978-981-13-3184-8 ISBN 978-981-13-3185-5 (eBook)
https://doi.org/10.1007/978-981-13-3185-5

Library of Congress Control Number: 2018960735

This Springer imprint is published by the registered company Springer Nature Singapore Pte Ltd.
The registered company address is: 152 Beach Road, #21-01/04 Gateway East, Singapore 189721, Singapore

Preface

The book entitled "Innovations in Soft Computing and Information Technology" will present the innovative scientific research works evaluated by technologists, academicians, research scholars, and students, which will be presented in International Conference on Energy, Materials and Information Technology, 2017, at Amity University Jharkhand, India. The widespread implementation of information technology has such an impact on today's society; the life cannot complete without its application. The fast-growing technology in this area requires more exposure to the researchers and technologists to apply this knowledge in new work. This book will include research works on system solutions based on soft computing techniques. It will cover innovative soft computing techniques and tools with advanced applications. The presentation of interdisciplinary problems and their solutions through information technology with innovative linkage with other disciplines is the critical focus of this book. It will also include cloud computing and WSN-related real-time research work. A cloud application leverages the cloud in software architecture, often eliminating the need to install and run the application on the customer's own computer, thus alleviating the burden of software maintenance, ongoing operation, and support. This book will cover all aspects of information technology through its state-of-the-art research work in the current time.

Ranchi, India

Jayeeta Chattopadhyay
Rahul Singh
Vandana Bhattacherjee

Acknowledgement

It is our pleasure to present this volume consisting of selected papers based on the oral presentation in International Conference on Energy, Materials and Information Technology at Amity University Jharkhand, Ranchi, India, on December 23–24, 2017. We would like to take this opportunity to thank all of the participants in the conference—invited speakers, presenters, and audience alike. Our special thanks go to all the keynote speakers present in the conference. We would like to thank SERB-DST for financial sponsorship for this conference. We express our heartfelt thanks to our Founder President, Dr. Ashok K. Chauhan, and Chancellor, Dr. Atul Chauhan, for showing us the right path. We would like to thank Prof. (Dr.) R. K. Jha, Vice Chancellor, Amity University Jharkhand, for his enormous support. We thank Prof. (Dr.) Ajit Kr. Pandey, Director, Amity University Jharkhand, for his valuable guidance. Here, we would like to give special thanks to Dr. Nitya Garg and Mr. Pravin Singh, without whom this proceedings would not have taken proper shape. We would like to thank all the faculty and staff of Amity University Jharkhand for their continuous support to make this conference successful. Our special thanks go to the students' volunteer team, who continuously helped us with their enormous support.

Contents

About the Editors

Jayeeta Chattopadhyay received her Master of Science in Chemistry from Devi Ahilya University, Indore (2003) and her Master of Technology in Fuels and Combustion from Birla Institute of Technology, Mesra, Ranchi (2005). She subsequently completed her Ph.D. in New Energy Engineering with the best doctoral thesis award from Seoul National University of Science and Technology, South Korea (2010). She received the prestigious Fast Track Young Scientist award from the Department of Science and Technology, Government of India (2010). She has published more than 20 research and review articles in high-impact peer-reviewed international journals, and holds one international patent. Her research interests include nano-structured materials for energy applications, and thermo-degradation of solid waste materials. She also works on oscillatory chemical reactions and pattern formation in reaction–diffusion systems. She is presently working at Amity University, Ranchi, Jharkhand, India.

Rahul Singh is an Assistant Professor at the Mechanical Engineering department of Amity University, Ranchi, India. He received his B.E. in Mechanical Engineering from BIT Durg and his M.E. in Machine Design from BIT Mesra, Ranchi. He is presently pursuing a Ph.D. in Mechanical Engineering at NIT Patna. Rahul's areas of interest are fluid mechanics, engineering mechanics, engineering thermodynamics, strength of materials, finite element methods, and machine design. He has authored 3 books.

Vandana Bhattacherjee is a Professor and Head of the Department of Computer Science & Engineering at Birla Institute of Technology, Mesra, Ranchi, India. Dr. Bhattacherjee has 16 years of research and 16 years of teaching experience. Her current research areas are object-oriented software metrics, soft computing, software cost estimation, process models, data mining and machine learning. She has been on the reviewer panel of the Journal of Computing and Information Technology, International Journal of Software Engineering and Knowledge Engineering, International Journal of Computers and Applications and Journal of Computer Science. She is a member of the IEEE Computer Society, Computer Society of India – Ranchi Chapter, and Indian Society for Technical Education. Dr. Bhattacherjee has several journal publications and 66 conference papers to her credit.

Sinusoids and Its Orthogonality

Abhinav Ranjan and Shraddha Prasad

1 Introduction

Any function having the following form is said to be sinusoidal:

$$x(t) = A\ \sin(\omega t + \phi)$$

where t is the independent or real variable, and the fixed parameters A, ω, and ϕ are all real constants.

One cause for the great significance of sinusoids is that they form a necessary base or core in physics. Another cause for its importance is that they are eigen-functions of linear systems. Now it can be said that they are of great importance in the detailed examination of filters such as reverberators, equalizers, some audio effects, etc [1–4].

All sinusoid functions can be expressed as the addition of a sine (phase zero) and a cosine (phase $\pi/2$) function. Note: If the sine part is said to be the "in-phase" component, then the cosine part will be called as the "phase-quadrature" component. The "phase quadrature" means "90° out of phase" or a relative phase shift of $\pm\pi/2$ [5].

A. Ranjan
Faculty of Science and Engineering, Jharkhand Rai University, Ranchi, India
e-mail: profaranjan@gmail.com

S. Prasad (✉)
Applied Science Department, Jharkhand Rai University, Ranchi, India
e-mail: shraddha.prasad@jru.edu.in

J. Chattopadhyay et al. (eds.), *Innovations in Soft Computing and Information Technology*, https://doi.org/10.1007/978-981-13-3185-5_1

1.1 Definition of DFT

The discrete Fourier transform (DFT) of a signal x can be defined by [6–10]

$$X(\omega_k) \triangleq \sum_{n=0}^{N-1} x(t_n)e^{-j\omega_k t_n}, \quad k = 0, 1, 2, \ldots, N-1$$

where "$\triangleq$" means "is defined as" or "equals by definition", and

$$\sum_{n=0}^{N-1} f(n) \triangleq f(0) + f(1) + \cdots + f(N-1)$$

$x(t_n) \triangleq$ Input signal amplitude (real or complex) at time t_n (s)

$t_n \triangleq nT = n$th Sampling instant (s), n an integer ≥ 0

$T \triangleq$ Sampling interval (s)

$X(\omega_k) \triangleq$ Spectrum of x (complex valued), at frequency ω_k

$\omega_k \triangleq k\Omega = k$th Frequency sample (radians per second)

$\Omega \triangleq \frac{2\pi}{NT} =$ Radian-frequency sampling interval (rad/s)

$f_s \triangleq 1/T =$ Sampling rate [samples/s, or Hertz (Hz)]

$N =$ Number of time samples = number of frequency samples (integer).
The sampling interval T is also called the sampling period.

1.2 Inverse DFT

The inverse DFT (i.e., IDFT) is given by [11–15]

$$x(t_n) = \frac{1}{N}\sum_{k=0}^{N-1} X(\omega_k)e^{j\omega_k t_n}, \quad n = 0, 1, 2, \ldots, N-1.$$

In digital signal processing, by setting $T = 1$ in the previous definition, the DFT and its inverse in the more pure form can be obtained below as:

$$X(k) \triangleq \sum_{n=0}^{N-1} x(n)e^{-j2\pi nk/N}, \quad k = 0, 1, 2, \ldots, N-1$$

$$x(n) = \frac{1}{N}\sum_{k=0}^{N-1} X(k)e^{j2\pi nk/N}, \quad n = 0, 1, 2, \ldots, N-1$$

where $x(n)$ denotes the input signal at time (sample) n, and $X(k)$ denotes the kth spectral sample.

1.3 DFT Sinusoids

From Euler's identity, $e^{j\theta} = \cos\theta + j\sin\theta$

So, the generation of sampled sinusoids done by the integral powers of the Nth roots of unity is shown in Fig. 1.

$(W_N^k)^n = e^{j2\pi kn/N} = e^{j\omega_k nT}$ are the sampled sinusoids used by the DFT. Now it can be said that by taking individually higher integer powers of the point W_N^k on the unity circle generates samples of the kth DFT sinusoid, giving $[W_N^k]^n =$ $n = 0, 1, 2, \ldots, N-1$. The kth sinusoid generator, i.e. W_N^k, is in succession the kth Nth root of unity (kth power of the primitive Nth root of unity W_N).

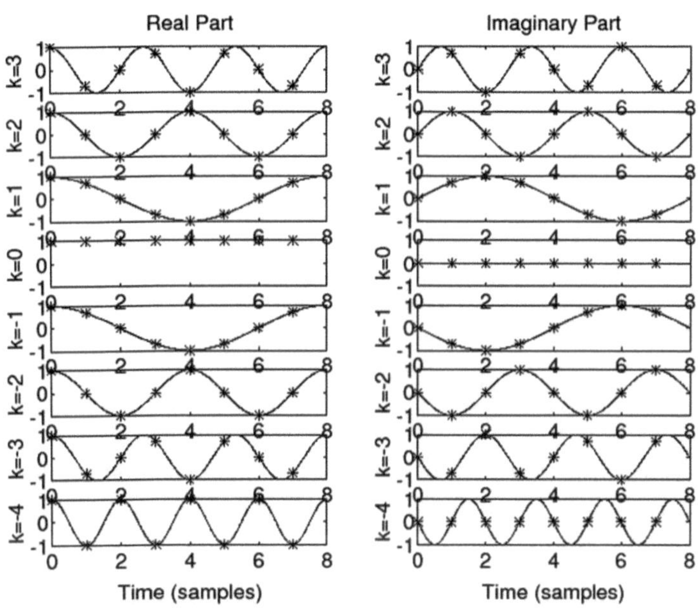

Fig. 1 Complex sinusoids used by the DFT for $N = 8$

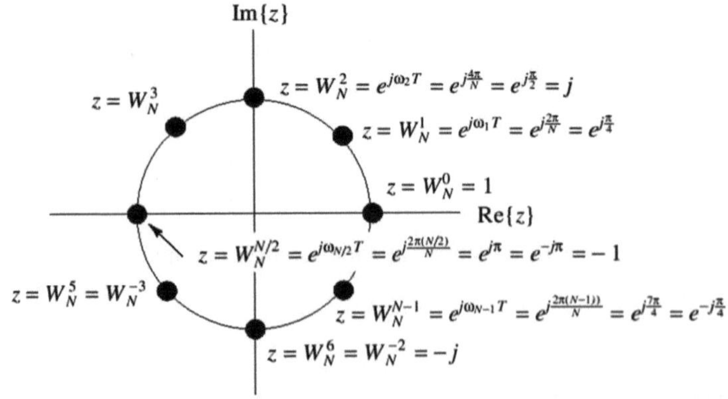

Fig. 2 *N*th roots of unity for $N = 8$

Note that in Fig. 1, the range of k is taken to be $[-N/2, N/2-1] = [-4, 3]$ instead of $[0, \mathrm{N}-1] = [0, 7]$. This is the most "physical" choice since it corresponds with our notion of "negative frequencies." However, we may add any integer multiple of N to k without changing the sinusoid indexed by k. In other words, $k \pm mN$ refers to the same sinusoid $\exp(j\omega_k nT)$ for all integers m.

The *N*th roots of unity are plotted for $N = 8$ as shown in Fig. 2.

2 Orthogonality of the DFT Sinusoids

Now it can be shown mathematically that the DFT sinusoids are totally orthogonal. Let

$$s_k(n) \triangleq \mathrm{e}^{j\omega_k nT} = \mathrm{e}^{j2\pi kn/N} = \left[W_N^k\right]^n, \quad n = 0, 1, 2, \ldots, N-1$$

Then, denote the *k*th DFT complex-sinusoid, for $k = 0 : N-1$. Then

$$\begin{aligned}\langle s_k, s_l \rangle &\triangleq \sum_{n=0}^{N-1} s_k(n)\overline{s_l(n)} = \sum_{n=0}^{N-1} \mathrm{e}^{j2\pi kn/N}\mathrm{e}^{-j2\pi\ \ln/N} \\ &= \sum_{n=0}^{N-1} \mathrm{e}^{j2\pi(k-l)n/N} = \frac{1-\mathrm{e}^{j2\pi(k-l)}}{1-\mathrm{e}^{j2\pi(k-l)/N}}\end{aligned}$$

where the final step made use of the closed-form expression for the sum of a geometric series. If $\triangleq k = l$, the numerator is zero while the denominator is nonzero. This proves $s_k \perp s_l, \quad k \neq l$.

Only by looking at zero-phase complex sinusoids and unit amplitude, as used by the DFT, it is easily verified that the (nonzero) amplitude and phase have no effect on its orthogonality.

3 Conclusions

At different frequencies, sinusoids are orthogonal if their durations are finite. The sampled sinusoids equivalent to the *N*th roots of unity are used by the DFT. Also by the reason that each sinusoid is of a different frequency and each is a harmonic of the sampling rate divided by *N*, the DFT sinusoids are orthogonal.

References

1. Lin Chin-Teng, Yu Yuan-Chu, and Van Lan-Da (2008) Cost-effective triple-mode reconfigurable pipeline FFT/IFFT/2-D DCT processor, Very Large Scale Integration (VLSI) Systems, IEEE Transactions
2. Yu Yuan-Chu and Yu Yuan-Tse (2013) Design of a high efficiency reconfigurable pipeline processor on next generation portable device
3. Yeh Wen-Chang and Jen Chein-Wei (2003) High-speed and low-power split-radix FFT, Signal Processing
4. Lin Y. T., Tsai P. Y., and Chiueh T. D. (2005) Low-power variable-length fast fourier transform processor
5. Rangarao K V, Mallik R K http://books.google.co.in/books?id=niiudDJK5zICeam
6. Dimitris G, Manolakis J and Proakis G. (2007) Digital Signal Processing, Efficient Computation of the DFT: Fast Fourier Transform Algorithms, Pearson Prentice Hall
7. Chang W. H. and Nguyen T. (2006) An OFDM-specified lossless FFT architecture, IEEE Trans. Circuits Syst. I; 53(6):1234–1243
8. Shousheng He and Torkelson M. (1996) A new approach to pipeline FFT processor. Proceedings of IPPS '96
9. James W, Cooley and John, Tukey W. (1965) An algorithm for the Machine calculation of complex fourier series Math. Comp.
10. Sidney C. B. (2012) Multidimensional index mapping. http://cnx.org/contents/3c48e4b5-0786-4d1f-bd30-a0cd860be3ab@12
11. Chen Yuan, Tsao Yu-Chi, Lin Yu-Wei, Lin Chin-Hung, and Lee Chen-Yi (2008) An indexed-scaling pipelined FFT processor for ofdm-based wpan applications
12. Magar S., Shen S., Luikuo G., Fleming M., and Aguilar R. (1988) An application specific DSP chip set for 100 mhz data rates Vol. 4
13. He S. and Torkelson M. (2012) Designing pipeline FFT processor for OFDM (de)modulation, in Proc. URSI Int. Symp. Signals, Syst., Electron, Vol. 28
14. Pratap R. (2006) Getting Started with MATLAB 7: A Quick Introduction for Scientists and Engineers
15. Manolakis D. G., Proakis J. G. (2007) Digital Signal Processing, Chapter Efficient Computation of the DFT: Fast Fourier Transform Algorithms

Abhinav Ranjan is presently working as Guest Faculty in Electronics and Communication Engineering Department at University College of Engineering and Technology, VBU Campus, Hazaribag. At present, he is pursuing as Research Scholar from Jharkhand Rai University, Ranchi in the Faculty of Science and Engineering under the supervision of Dr. Shraddha Prasad on the topic – "An Advanced Access to Design and Application of FFT/IFFT Processor based on Radix - 4^2 Algorithm". He has successfully completed M.Tech in the Faculty of Electronics and Communication at Monard University, Hapur, Uttar Pradesh in 2014. He has completed his B.Tech in the Faculty of Instrumentation and Control (Applied Electronics) from NM Institute of Engineering and Technology, Bhubaneswar, Odisha under Biju Pattnaik University of Technology in 2008.

Dr. Shraddha Prasad is presently working as Associate Professor in the Faculty of Science and Engineering. She is heading the Department of Applied Science at Jharkhand Rai University, Ranchi. She is also the Academic Coordinator for Engineering. She has done Ph.D from Birla Institute of Technology, Mesra, Ranchi on the topic - "Investigation of Optical Beam Propagation in Nonlinear Media: Spatial Soliton Formation and Their Interaction for All Optical Devices". M.Sc. in Physics from Ranchi University, Ranchi, B.Ed. from St. Xavier's College, Ranchi. She has completed 89th Orientation Programme with Grade-A, organized by UGC- Human Resource Development Centre Ranchi University, Ranchi. Dr. Shraddha Prasad has 6 research papers published in different reputed International journals with good impact factors e.g., Elsevier, Scopus, WoS etc. and numbers of research papers presented in different International Conference/Seminars within India. She is an Associate Member of "The Institution of Engineers". Dr. Shraddha Prasad has more than 11 years teaching experience in educational sector including 3 years research experience as a Project Fellow in U.G.C Project in Birla Institute of Technology, Mesra, Ranchi. Her areas of research are Nonlinear Optics, Solitons, All-optical Devices, Communication and material science.

A Case Study of Hindi–English Example-Based Machine Translation

Aditya Kumar Pathak, Priyankit Acharya and Rakesh Chandra Balabantaray

1 Introduction

Machine translation (MT) is a subtype of computational linguistics that uses software to translate text from one natural language (NL) into another natural language (NL). Machine translation performs simple substitution of words from one language to another language, but only substitution is not sufficient to give accurate translation of a text due to requirement of identification of whole phrases and their immediate counterparts. To solve this problem with corpus, statistical and neural techniques are the leading fields for better handling of differences in linguistic typology, isolation of anomalies, and translation systems. There are many types of machine translation systems; one of those is example-based machine translation or EBMT approach [1].

A deep learning approach-based machine translation (MT), i.e., neural MT, has made rapid progress in recent years. As per recent reports, Google translation services are currently using this technology in place of its previous statistical approach. In the competition, there are many providers including KantanMT, Omniscient Technologies, and SDL.

A. K. Pathak · R. C. Balabantaray (✉)
Department of CSE, IIIT, 751003 Bhubaneswar, India
e-mail: rakeshbray@gmail.com

P. Acharya
Department of IT, IIIT, 751003 Bhubaneswar, India

J. Chattopadhyay et al. (eds.), *Innovations in Soft Computing and Information Technology*, https://doi.org/10.1007/978-981-13-3185-5_2

1.1 Machine Translation

MT system along with simply substituting the words as per model also takes the application of complex linguistic knowledge, morphology, meaning, and grammar into consideration. The standard metric people are using for evaluation of MT systems is BLEU score.

Bilingual evaluation understudy (BLEU) is the algorithm to determine the quality of text translated by a machine translation. Quality is the comparison between machine-translated output to that of human-generated output; the closer machine translation is to human-generated translation, the better is the BLEU score. BLEU score is a n-gram overlap of machine translation to that of reference translation.

$$\text{BLEU} = \min\left[1, \frac{\text{output length}}{\text{reference length}}\right]\left(\sum_{i=1}^{n} \text{precision}\right)^{\frac{1}{n}} \tag{1}$$

The different classifications of machine translation-based systems are elaborated in the subsections mentioned below:

1.1.1 Direct Machine Translation (DMT)

Direct MT or DMT is among the most basic approach for machine translation. This approach provides word to word translation in direct manner according to a bilingual dictionary.

1.1.2 Rule-Based Machine Translation (RBMT)

RBMT consists of set of rules, known as grammar rules or bilingual lexicon; for processing these rules, software systems are needed.

1.1.3 Interlingua-Based Translation (IBT)

In this system, the source language (SL) is converted into interlingua (IL) form. The advantage of this approach is that the parser and analyzer of SL do not depend on generator for the target language (TL). So there is a requirement for complete resolution of ambiguity in SL text.

1.1.4 Knowledge-Based Machine Translation (KBMT)

This system requires a complete understanding of the source text on the basis of world knowledge and linguistic semantic knowledge about meaning of words and also about the combinations of words prior to the translation process into the target text. This system is implemented over interlingua architecture.

1.1.5 Statistical-Based Machine Translation (SMT)

SMT is a work on bilingual text corpora. The system is based on the probabilistic models in accordance with probability density function, indicated by $p(e|f)$. The best translation procedure is found out by picking the highest probability, as shown in Eq. (2).

$$e = \arg\max p(e|f) = \arg\max p(f|e)p(e) \quad (2)$$

1.1.6 Example-Based Machine Translation (EBMT)

The goal of this system is to reuse the examples of already existing translations. Bilingual corpus is used as its main knowledge base, and this bilingual corpus is translated by analogy.

1.1.7 Hybrid-Based Machine Translation (HBMT)

Hybrid-based approach is based on the combination of many machine translation techniques like DMT, SMT, EBMT.

1.1.8 Neural Machine Translation (NMT)

This system uses neural networks and deep learning approach. It is a type of KBMT system. NMT models require only a fraction of the memory needed by traditional SMT. Furthermore, unlike conventional translation systems, all parts of the NMT model are trained jointly, i.e., end to end, to maximize the translation performance.

2 Literature Survey

The Machine translation system for English to Indian language and vice versa had started long back mostly from 1990 onwards. The various systems and their approaches [1] are described in Table 1.

According to latest research, the performance report of baseline systems is translating Indian languages-based text (Bengali, Hindi, Malayalam, Punjabi, Tamil, Telugu, Gujarati, and Urdu) into English text with an average of 10% accuracy for all language pairs [2].

Table 1 Literature survey for machine translation

Year	A	B	C	D	E
1997	MANTRA MT [7]	Eng-Hin	T	General	It mainly focuses on Indian languages and is built over the concept of paninian grammar
1999	MANTRA Rajya Sabha [7, 8]	Eng-Hin	T	Office and admin documents	It uses TAG and LTAG to represent a grammar
2001	UNL-based English–Hindi MT system [3, 9]	Eng-Hin	I	General	It is easy to add new languages for translation in this system. This system is based on UNL grammar
2002	English–Hindi translation systems [10]	Eng-Hin	T	Weather narration	It works on creation of Hindi tree and uses translation module for processing and post-processing of English tree
2003	AnglaHindi [11]	Eng-Hin	I	General	The accuracy of this system is 90%; it uses different modules of Angla Bharti
2003	ANUBHARTI-1 [7]	Hin-Eng	H	General	It is a combination of corpus-based and example-based machine translation systems
2004	ANUBHARTI-2 [7]	Hin-Eng	H	General	It copies human learning process for storing knowledge from the past experience and uses it in the future
2004	Hinglish MT system [12]	Hin-Eng	E	General	It is based on ANUBHARTI-2 and ANGLABHARTI-2
2004–2006	MaTra: A fully automatic indicative English–Hindi MT [13, 14]	Eng-Hin	T	News, annual report, technical phrases	It works as MSIR and uses transfer frame-like structure to resolve ambiguities. It uses heuristic approach
2009	Hindi to English GB theory based translation system [15]	Hin-Eng	E	General	Government and binding (GB) theory is used by this system for translation. It consists of parsing as well as generating module
2014	Pure EBMT English to Hindi translation system [16]	Eng-Hin	E	Sentences are compared to extract the translation	Bi-text is used for translation. Various modules such as tagging matrix as well as similarity matrix are present

(continued)

Table 1 (continued)

Year	A	B	C	D	E
–	Interlingua-based English–Hindi machine translation [17]	Eng-Hin	I and T	General	Here information is represented sentence wise. L-UW handles lexical-semantic divergence dictionary, whereas analyzers handles Syntactic divergence

Nomenclature of Table
Y: Year
A: Machine translation systems
B: Language approach for translation
C: Approaches
D: Application
E: Observation
T: Transfer-based MT
I: Interlingua-based MT
E: Example-based MT
H: Hybrid-based MT

3 OpenNMT

OpenNMT is an open-source tool developed at MIT, based on neural machine translation system built upon the Torch/Py-Torch mathematical toolkit. The system is designed to be user-friendly and easily scalable while providing with good translation accuracy. This system provides general-purpose interface, which requires only source and target files with speed as well as memory optimizations. OpenNMT has an active open community welcoming industrial as well as academic contributions.

The schematic view of a neural machine translation is explained in Fig. 1. The red source words after being mapped to word vectors are fed into a recurrent neural

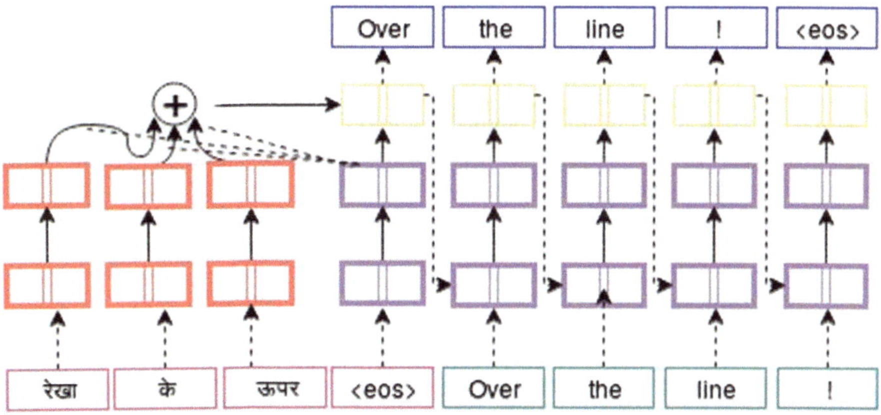

Fig. 1 Semantic view of NMT

network (RNN). After finding the <eos> symbol, i.e., end of sentence, the final time step initializes a target blue RNN. At each target time step, attention is applied over the source RNN and combined with the current hidden state to produce a prediction as mentioned in Eq. (3). This prediction is afterward fed back into the target RNN.

$$p(W_t|W_{1:t-1}, X) \tag{3}$$

3.1 Dataset

Our dataset consists of 1.4 million Hindi–English parallel corpus which is again the combination of all the underlying datasets. The various datasets considered for the task is mentioned in the following subsections.

3.1.1 CFILT

CFILT dataset [3] is English–Hindi parallel corpus by IIT Bombay. This corpus was used during 2016–17 in open shared task for Asian language translation for Hindi–English as well as English–Hindi language pairs, and for the Hindi–Japanese and Japanese–Hindi language pairs, it acted as a pivot. Size of the bi-text is 1.4 million, out of which we have used one million for our approach.

3.1.2 HindEncorp

HindEnCorp [4] is a parallel corpus of Hindi–English. HindEnCorp contains 273885 parallel sentences (3.8 million English and 3.9 million Hindi tokens). It is freely available for non-commercial research.

3.1.3 Tanzil Corp

Tanzil corp [5] is a OPUS (Open Source Parallel Corpus) release of English–Hindi parallel corpus. It consists of 191, 433 parallel sentences. Out of which we are taking 126, 115 sentences.

3.2 Methodology

The various processes of OpenNMT tool are elaborated in the following subsections.

3.2.1 Normalization

This process is used for uniform transformation on the source sequences to identify and protect some specific sequences into unique representation simpler for the translation process.

3.2.2 Tokenization

Given a n character sequence, tokenization is the task of splitting it up into pieces called, tokens. OpenNMT uses space-separated technique for tokenization.

3.2.3 Learning BPE

BPE or byte pair encoding is a text compression scheme which works on pattern substitution. On the basis of tokenized corpus, we are making a BPE model. For languages sharing an alphabet, learning BPE on two or more involved languages increases the consistency of segmentation and reduces the problem of insertion or deletion of characters when copying names [6].

3.2.4 Re-Tokenization

We are again tokenizing the previously tokenized sentences with respect to the learnt BPE model, and there is an introduction of case feature and joiner annotate. Case feature adds additional features to the encoder; an embedding will be optimized for each label and then fed as additional source input alongside the word it annotates. On the target side, these features will be predicted by the network. The decoder is then able to decode a sentence and annotate each word. On the other hand, by activating joiner annotate marker, the tokenization is reversible.

3.2.5 Preprocessing

Preprocessing is a data preparation process which passes over the data to generate word vocabularies and sequences of indices, which is to be used in training.

3.2.6 Training

In this process, we are taking default OpenNMT encoder and decoder, LSTM layers (default is 2), and RNN size (default is 500).

3.2.7 Translation

During translation, OpenNMT creates a binary file and translates source language to target language.

4 Results and Discussions

We are training the system with 1.4 million parallel corpus in OpenNMT model. The validation source and target have been taken as 20% of training corpus (i.e., 280k sentences), then for testing the model we have identified 10k sentences randomly from the corpus and calculated the BLEU score for the same in every epoch. The details of the various epochs are mentioned in Table 2 and Fig. 2.

Table 2 Epoch-wise results represented with BLEU score

Epoch	BLEU score	System information
Epoch-1	0.0860	CPU@ 3.40 GHz, Intel (R) core (TM) i7-6700
Epoch-2	0.1403	
Epoch-3	0.1403	
Epoch-4	0.1403	
Epoch-5	0.1860	
Epoch-6	0.1920	

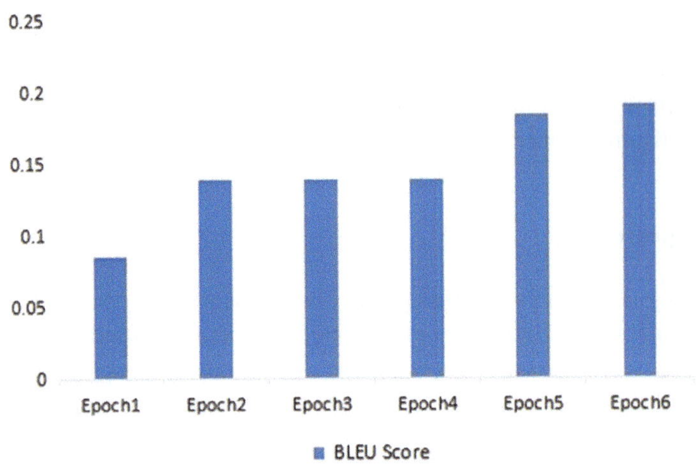

Fig. 2 Histogram representation of BLEU score of different epochs

5 Conclusions

We have taken training corpus from various sources consisting of different variety of sentences. The training phase of our system is done epoch wise, and it has been observed that the BLEU score increases with number of epochs; accuracy of the system obtained after six number of epochs is satisfactory compared to other machine translation systems. We are still trying to improve the BLEU score of Hindi to English translation system by applying some post-editing techniques after the model has generated the translation.

References

1. Amruta Godase and Sharvari Govilkar, Machine Translation development for indian language and its approaches, International Journal on Natural Language Computing (IJNLC) Vol. 4, No. 2, (2015) 55–74.
2. Nadeem Jadoon Khan, Waqas Anwar and Nadir Durrani, Machine Translation Approaches and Survey for Indian Languages, arXiv:1701.04290, (2017).
3. Anoop Kunchukuttan, Pratik Mehta and Pushpak Bhattacharyya, The IIT Bombay English-Hindi Parallel Corpus, (2016–2017).
4. Ondrej Bojar, Vojtech Diatka, Pavel Rychly, Pavel Straňák Vít Suchomel, Ales Tamchyna and Daniel Zeman, HindEnCorp Hindi-English and Hindi-only Corpus for Machine Translation, (2014).
5. Tanzil corpus, http://opus.lingfil.uu.se/Tanzil.php.
6. Rico Sennrich, Barry Haddow and Alexandra Birch, Neural Machine Translation of Rare Words with Subword Units, arXiv:1508.07909, (2015).
7. Sudip Naskar and Shivaji Bandyopadhyay, Use of Machine Translation in India: Current status, AAMT Journal, (2015) 25–31.
8. Mantra machine translation, https://cdac.in/index.aspx?id=mc_mat_mantra_rajbhasha.
9. Machine Translation Using UNL based Interlingua Approach, http://www.cfilt.iitb.ac.in/machine-translation/ eng-hindi-mt.
10. Lata Gore and Nishigandha Patil, English to Hindi Translation System, In proceedings of Symposium on Translation Support Systems, IIT Kanpur, (2002) 178–184.
11. R.M.K. Sinha and A. Jain, AnglaHindi: An English to Hindi Machine-Aided Translation System, International Conference AMTA (Association of Machine Translation in the Americas), (2002).
12. Machine Translation of Bilingual Hindi-English (Hinglish) Text, http://www.academia.edu/7986160/Machine_Translation_of_Bilingual_Hindi-English_Hinglish_Text.
13. Ananthakrishnan R, Kavitha M, Jayprasad J Hegde, Chandra Shekhar, Ritesh Shah, Sawani Bade and Sasikumar M., MaTra: A Practical Approach to Fully-Automatic Indicative English Hindi Machine Translation, In the proceedings of MSPIL-06, (2006).
14. http://www.ncst.ernet.in/matra/.
15. Choudhary, A. Singh, M., (2009), GB theory based Hindi to english translation system, Computer Science and Information Technology, 2nd IEEE International Conference, (2009), 293–297.
16. Ruchika A. Sinhal, Kapil O. Gupta, A Pure EBMT Approach for English to Hindi Sentence Translation System, I.J.Modern Education and Computer Science, (2014) 1–8, http://www.mecs-press.org/.

17. Shachi Dave, Jignashu Parikh And Pushpak Bhattacharyya, Interlingua based English-Hindi Machine Translation system and Language Divergence, (2003).
18. Guillaume Klein, Yoon Kim, Yuntian Deng, Jean Senellart and Alexander M. Rush, OpenNMT: Open-Source Toolkit for Neural Machine Translation, arXiv:1701.02810v2, (2017).
19. Graham Neubig, Neural Machine Translation and Sequence-to-sequence Models: A Tutorial, arXiv:1703.01619, (2017) 1–65.
20. Yongchao Deng, Jungi Kim, Guillaume Klein, Catherine Kobus, Natalia Segal, Christophe Servan, Bo Wang, Dakun Zhang, Josep Crego and Jean Senellart, SYSTRAN Purely Neural MT Engines for WMT2017, arXiv:1709.03814, (2017).

Aditya Kumar Pathak My area of research is Language Translation, Natural Language Processing. At present, I am working as a research Assistant at IIIT Delhi in the field of speech recognition using deep learning. I am enthusiastic about learning new thing.

Priyankit Acharya I am an experienced engineer with a demonstrated history of working in the computer software industry as well as research. I am a strong engineering professional skilled in programming, algorithmic application and Natural language processing.

Dr. Rakesh Chandra Balabantaray research interests lie in the areas of Intelligent Systems, Information Retrieval, and Natural Language Processing. He has published over 100 research papers in reputed journals and conferences. He has copyrights for two of the software he has developed. He was also the principal investigator of a project on Development of Cross Lingual Access sponsored by GOI.

Comparative Study of Different Machine Learning Models for Breast Cancer Diagnosis

Aman Kumar and M. Poonkodi

1 Introduction

The second driving reason of death among women is breast cancer, first being lung cancer [1]. In the wake of expanding at a disturbing rate for more than 20 years, breast cancer occurrence rates started diminishing in 2000, and dropped by around 7% out of 2003. Nonetheless, insights have demonstrated that close to 1.7 million new cases had been analyzed in 2012. While breast cancer is more predominant in the developed world, right around half of breast cancer cases and 58% of death due to it happen in developing nations [2]. Breast cancer survival rate is around 80% in North America and Japan and that of 60% in middle town nations. Low-income nations have a survival rate beneath 40%. The low survival rates in developing or under-developed nations are normally a consequence of the absence of essential identification programs.

Machine learning approaches in medical areas are developing because of improvement in efficiency of the strategies in classification and prediction frameworks, essentially in serving to therapeutic experts in foreseeing illness [3]. Notwithstanding its significance in discovering approaches to upgrade persistent outcomes, it helps in downsizing the cost of medications and encourages in improving clinical investigations. Machine learning and data mining rely on classification techniques that are the most crucial and significant tasks. A few experiments have been performed on medicinal datasets utilizing various classifiers.

Machine learning and data mining include the utilization of modern and refined information control tools to discover usable patterns and relations in datasets. In this paper, different classifiers like logistic regression, naïve Bayes, SVM, and so forth have been utilized on a moderate size dataset which contains 30 attributes and 569

A. Kumar (✉) · M. Poonkodi
Department of Computer Science and Engineering, SRM Institute of Science and Technology, 600026 Vadapalani, Chennai, India
e-mail: aman29091996@gmail.com

J. Chattopadhyay et al. (eds.), *Innovations in Soft Computing and Information Technology*, https://doi.org/10.1007/978-981-13-3185-5_3

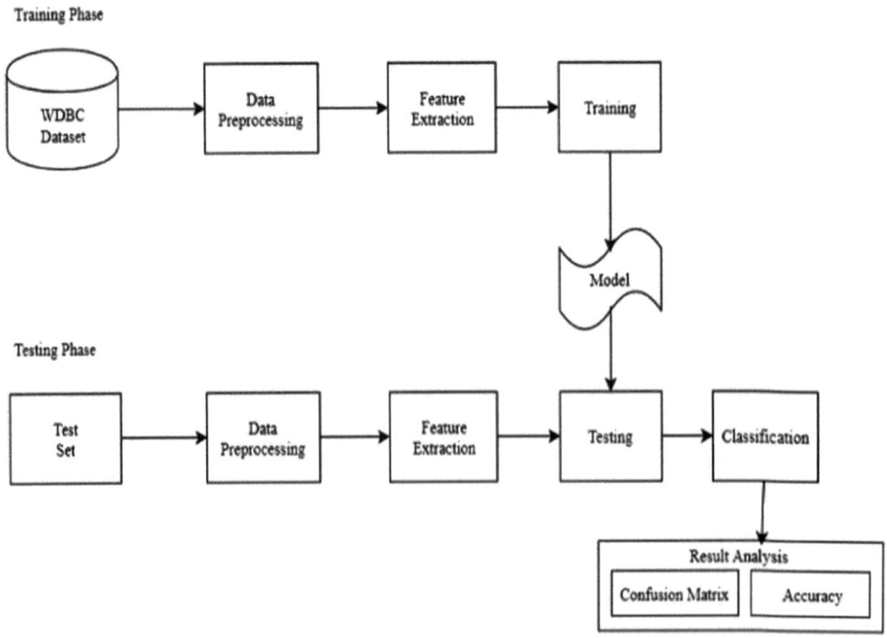

Fig. 1 Architecture diagram

records. Examples have been arranged in light of 'output' characteristic, i.e., 'B' for Benign and 'M' for Malignant. Point of this paper is to decide the superior classifier for breast cancer diagnosis in view of confusion matrix and accuracy. Figure 1 represents the design outline of the experiment.

This paper is ordered as follows. Section 2 gives the clarification about the dataset. Section 3 gives a concise clarification about the distinctive machine learning models and Sect. 4 comprises of the outcomes obtained. Closing articulations are given in Sect. 5.

2 Dataset Description

Features are determined from a digital image of a breast cancer sample obtained from fine-needle aspiration (FNA). The characteristics of the cell nuclei existing in the image are determined from these features.

Number of instances: 569

Number of attributes: 32 (ID, diagnosis, 30 real-valued input features)

Attribute information [4]

1. radius (mean distance from center to perimeter points)
2. texture
3. area
4. perimeter
5. smoothness (variation in radius)
6. compactness (perimeter^2/area − 1.0)
7. concavity (severity of contour concave portions)
8. concave points
9. symmetry
10. fractal dimension.

The mean, worst, and standard errors of each feature were computed for each image, which resulted in total 30 features.

3 Machine Learning Algorithms

3.1 Principal Component Analysis

Principal component analysis is the technique of extracting necessary features in the form of principal components from large number of features. It extracts features to catch the maximum amount of information conceivable. It is performed on correlation or variance matrix. Principal components can be characterized as the linear combinations of original variables and the eigenvectors. The principal component is given as

$$Z^1 = \Phi^{11}X^1 + \Phi^{21}X^2 + \Phi^{31}X^3 + \cdots + \Phi p^1 X \tag{1}$$

3.2 Linear Discriminant Analysis

LDA is utilized to change over a dataset into a low-dimensional dataset to abstain from overfitting and diminish computational expenses. LDA is alike PCA yet alongside finding the component axes that aim to maximize the information, it likewise decides the axes that maximize the separation between classifications. LDA extricates the features from a high-dimensional dataset and projects it into a low dimension such that the number of features is less than the number of classes. LDA processes the axes, i.e., linear discriminants keeping in consideration the class labels. Consider a n-dimensional dataset. LDA figures a d-dimensional vector for the different classes from the dataset. Eigenvectors are arranged in descending order of eigenvalues, and k eigenvectors are chosen to frame $d \times k$ matrix W. This $d \times k$ matrix is utilized to transform samples into new subspace.

$$Y = X \times W \tag{2}$$

where

X is the original sample space
Y is the remodeled sample space.

3.3 Logistic Regression

Logistic regression is a classification technique utilized when the reliant factors have two class labels. It is a special case of linear regression when the outcome is categorical. Instead of choosing parameters that minimize the sum of squared errors, logistic regression picks the parameters that improve the likelihood of sample values. It predicts the final product by fitting the factors to a logistic function otherwise called the sigmoid function.

$$Y = 1/(1 + e^{-x}) \tag{3}$$

Input esteems (x) are linearly combined on the basis of coefficient values to foresee the output value y.

3.4 KNN Classifier

K-nearest neighbor classifier is a nonparametric classifier that can be utilized for classification in addition to regression. KNN classifier focuses to discover the k-nearest points to the test point and decides the class of the test point utilizing the majority of vote technique [5]. KNN stores the whole dataset. To decide the k neighbors, different distance measurements are utilized.

$$\text{Euclidean Distance:}\, y = \sqrt{\sum\nolimits_{i=1}^{k} (x_i - y_i)^2} \tag{4}$$

$$\text{Manhattan Distance:}\, y = \sqrt{\sum\nolimits_{i=1}^{k} |x_i - y_i|} \tag{5}$$

$$\text{Minkowski Distance:}\, y = \sqrt[q]{\sum\nolimits_{i=1}^{k} (|x_i - y_i|)^q} \tag{6}$$

3.5 *Support Vector Machines*

Support vector machine is a supervised learning technique. In SVM, every data point is considered as a point in the space. Classification is done by deciding a hyperplane that isolates the classes [6]. Support vectors are the coordinates of the points. If more than one hyper planes are available, then hyper plane is chosen on the basis of margin. Margin is the most maximum separation between closest data point and hyperplane. SVM algorithm is employed using a kernel. Prediction is made utilizing the inner product of input and each support vector.

$$f(x) = B_0 + \mathrm{sum}(a_i * (x, x_i)) \tag{7}$$

3.5.1 Linear Kernel SVM

The dot product is called as the kernel.

$$K(x, x_i) = \mathrm{sum}(x * x_i) \tag{8}$$

3.5.2 Polynomial Kernel SVM

Instead of dot product, polynomial function can also be used.

$$K(x, x_i) = \mathrm{sum}(x * x_i)^\wedge d \tag{9}$$

3.5.3 Radial Kernel SVM

$$K(x, x_i) = \exp(-\mathrm{gamma} * \mathrm{sum}(x * x_i)^\wedge 2) \tag{10}$$

3.6 *Naive Bayes Classifier*

Naive Bayes is a classification technique based upon Bayes' Theorem (11). Naive Bayes considers every one of the attributes as autonomous; i.e., it considers that the presence of a specific feature in a class is not identified with the presence of different features [7]. Rather than prediction, naive Bayes classifier decides the likelihood of the data to be in a class. Probabilities utilized by naive Bayes are class probability and conditional probabilities. Class probability is the likelihood of each class, and conditional probability is the likelihood of each input value given each class value.

$$P(c/x) = \frac{P(x/c)P(c)}{P(x)} \tag{11}$$

3.7 Decision Tree

Decision tree is a supervised learning algorithm utilized for classification and regression. It splits the data recursively in view of the attributes until the point that a stop condition happens [8]. Branch of decision tree represents test condition, nodes describe the properties, and leaf nodes represent class labels [9]. Root node represents entire sample space. Disadvantage of decision tree is that it might over fit information and when working with numerical data it looses information.

3.8 Random Forest

Random forest is a mix of various decision trees over the same training set. It can be utilized for both regression and classification. It also performs dimension reduction, treats missing information, outlier values and other data preprocessing steps. In random forest, different trees are developed for a training set. Each tree gives a classification result. Unlike CART, it does not utilize greedy algorithm. Consider a training set of size N. It is separated into 'm' number of subsets. Decision tree is developed for every subset, and classification is resolved. Utilizing the majority of vote technique class label is determined.

3.9 Stacking

Stacking or stacked generalization is an ensembling technique, which consolidates at least two different models prepared on same training set. Predictions of various models are combined utilizing another model called as metamodel. A simple linear model, for example, logistic regression, is utilized to consolidate the submodels. It is important that the submodels that are being combined deliver uncorrelated predictions. The best models utilized for combination can be replaced by basic strategies, for example, majority vote, averaging or weighted average. Stacking consolidates the upsides of both bagging and boosting. It reduces the variance and increases accuracy.

4 Results

In the given paper, confusion matrix and accuracy have been used to compare the models. Confusion matrix consists of the information about actual and predicted results. A confusion matrix C gives the count of true negatives C_{00}, false negatives C_{10}, true positive $C_{11,}$ and false positive C_{01}. In Table 1, TN, FP, FN, and TP represent true negative, false positive, false negative, and true positive, respectively.

Accuracy is given by

$$\text{Accuracy} = \frac{\text{TN} + \text{TP}}{\text{TN} + \text{FP} + \text{FN} + \text{TP}} \tag{12}$$

Table 2 represents accuracy of classifiers with LDA, and Table 3 represents accuracy of classifiers with PCA. Kernel SVM with PCA has highest accuracy 98.24%. Table 4 represents fusion of classifiers by stacking with LDA, and Table 5 represents fusion of classifiers by stacking with PCA. KNN, SVM, naïve Bayes combined using logistic regression with PCA are better than other ensembling models with 97.36% accuracy.

Table 1 Confusion matrix

	No (predicted)	Yes (predicted)
No (actual)	TN	FP
Yes (actual)	FN	TP

Table 2 Accuracy of classifiers with LDA

Classifier	Accuracy
Logistic regression	0.96491228
KNN classifier	0.96491228
Decision tree	0.92982456
Naïve Bayes	0.96491228
SVM	0.95614035
Kernel SVM	0.96491228
Random forest	0.93859649

Table 3 Accuracy of classifiers with PCA

Classifier	Accuracy
Logistic regression	0.96491228
KNN classifier	0.95614035
Decision tree	0.88596491
Naïve Bayes	0.85087719
SVM	0.98245614
Kernel SVM	0.98245614
Random forest	0.92982456

Table 4 Accuracy of stacking with LDA

Classifier	Accuracy
Random forest, KNN, SVM	0.94736842
KNN, decision tree, SVM	0.92982456
KNN, SVM, naive Bayes	0.95614035
KNN, decision tree, kernel SVM	0.92982456
KNN, decision tree, naive Bayes	0.92982456
KNN, decision tree, random forest	0.93859649

Table 5 Accuracy of stacking with PCA

Classifier	Accuracy
Random forest, KNN, SVM	0.95614035
KNN, decision tree, SVM	0.92105263
KNN, SVM, naive Bayes	0.97368421
KNN, decision tree, kernel SVM	0.91228070
KNN, decision tree, naive Bayes	0.87719298
KNN, decision tree, random forest	0.94736842

5 Conclusion

In this paper, comparison between different supervised learning classification algorithms such as logistic regression, KNN classifier, SVM, naïve Bayes, decision trees has been done with different dimensionality reduction techniques. Combination of different algorithms is also used. The main objective of this paper is to compare different machine learning algorithms for breast cancer diagnosis on the basis of confusion matrix and accuracy. All algorithms have been programmed in Python 3 in Anaconda environment. Kernel SVM with PCA gave the superior accuracy of 98.24% as compared to others. Fusion of KNN, SVM, naïve Bayes by stacking, combined using logistic regression, gives an accuracy of 97.36%, higher than other ensembling models.

References

1. U.S. Cancer Statistics Working Group. United States Cancer Statistics: 1999–2008 Incidence and Mortality Web-based Report. Atlanta (GA): Department of Health and Human Services, Centers for Disease Control.
2. World Health Organization. Breast Cancer: Prevention and Control. http://www.who.int/cancer/detection/breastcancer/en/index1.html.
3. G.I. Salama, M.B. Abdelhalim and Abd-elghany Zeid. Experimental Comparison for Breast Cancer Diagnosis. Egypt: IEEE 2012.
4. UCI Machine Learning Repository. http://archive.ics.uci.edu/ml/datasets/Breast+Cancer+Wisconsin+%28Diagnostic%29.

5. Angeline Christobel Y, Dr. Sivaprakasam. “An Empirical Comparison of Data Mining Classification Methods”. International Journal of Computer Information Systems, Vol 3, No. 2. 2011.
6. Vapnik. V.N., “The Nature of Statistical Learning Theory”. 1st ed., Springer-Verlag. New York, 1995.
7. L. Yaun. “An Improved Naïve Bayes Text Classification Algorithm In Chinese Information Processing”. Science, pp. 267–269. 2010.
8. H. Jiawei and K. Micheline. “Data Mining- Concepts and Techniques”, Second Edition, Morgan Kaufmann, Elsevier Publishers, ISBN: 978-1-55860-901-3, 2008.
9. Zahra Nematzadeh, Roliana Ibrahim and Ali Selamat. Comparative Studies on Breast Cancer Classifications with K-Fold Cross Validations Using Machine Learning Techniques. Malaysia, UTM: IEEE 2015.

Aman Kumar He is currently pursuing B.Tech in Computer Science and Engineering from SRM Institute of Science and Technology, Chennai. He will graduate in May 2019.

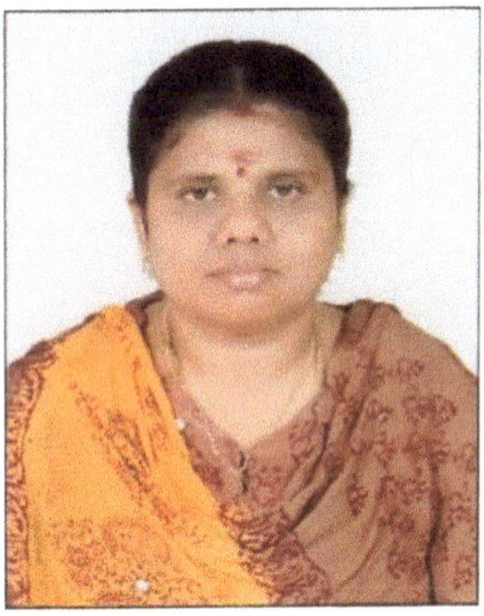

M. Poonkodi Has received M.Tech Degree from the department of Computer Science and Engineering, SRM Institute of Science and Technology, Chennai, India, in the year 2011. She is currently an Assistant Professor in the department of Computer Science and Engineering, SRM Institute of Science and Technology, Chennai, India. Presently being a research scholar in the same institution, her research interests include computer vision, pattern recognition, video analytics.

A PID Load Frequency Controller Design via Pole Placement and Approximate Frequency Response Matching

Anand Kumar, Md Nishat Anwar and Ashraf Raza

1 Introduction

If the load demand in power system suddenly varies, then the minimal frequency and tie-line power exchange also vary in that interconnected area, because of power imbalance between power generating unit and load demand of the system. Load frequency control problem is to retain nominal frequency and tie-line power exchange within quantified limits [1].

Various types of advance control techniques have been developed recently to solve the problem of frequency deviation error and also tie-line power exchange for power system such as internal model control (IMC) [2], H_∞ controller [3], DS controller [4–6], adaptive fuzzy control [7–9], optimal control [10, 11], robust control [12–18]. A comprehensive literature survey of LFC is explained [19–21]. In power system, LFC plays a vital role to provide reliable operation [22]. Anwar and Pan [23] proposed PID controller for LFC via a two-point frequency response matching. Settling time and peak value are still high in case of Anwar and Pan [23] and Gundes and Chow [24], whereas Tan [25] proposed anti-GRC configuration to overcome GRC problem and implement it for a single-area LFC and also extended for a two-area LFC and its response has been improved by the proposed design method.

A PID load frequency controller design via pole placement and approximate frequency response matching has been proposed in this paper. There is no restriction in complicatedness of power system model in the proposed method. Different performance parameters for nominal system as well as perturbed system are evaluated to check the satisfactory performance of the proposed method. The proposed method is used for a single-area power system and also extended for a

A. Kumar (✉) · M. N. Anwar · A. Raza
Department of Electrical Engineering, National Institute of Technology, Ashok Rajpath, Patna 800005, Bihar, India
e-mail: anandgya@gmail.com

J. Chattopadhyay et al. (eds.), *Innovations in Soft Computing and Information Technology*, https://doi.org/10.1007/978-981-13-3185-5_4

two-area power system with non-reheat turbine. All case studies in this paper are taken from the literature, and its response has been compared with well-known method that is explained in the literature.

The paper is arranged in four sections. Section 1 consists of introduction as discussed above. The detailed description of the system model for a single-area LFC and proposed design method is elucidated in Sect. 2. Simulation results of different case studies and its discussion are elucidated in Sect. 3, and finally in Sect. 4, conclusion of the paper is drawn.

2 System Model and Design Method

The power system dynamic modeling of the LFC is well discussed [1, 17, 22]. The single-area power system for LFC with small changes in load, which is expressed by linear model as shown in Fig. 1, and the term used in LFC are elucidated in Table 1.

Mainly, three components are used for LFC in power system of the dynamic models which are as follows:

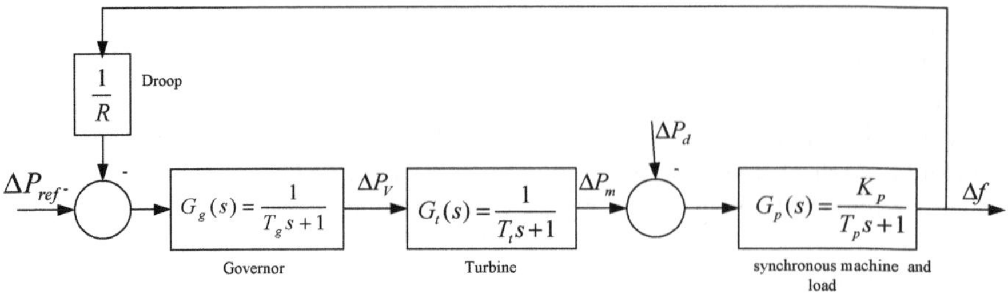

Fig. 1 Linear model of a single-area power system

Table 1 Nomenclature used to elucidate power system dynamics

ΔP_{d}	Load disturbance (p.u. MW)
ΔP_{ref}	Load reference setting
K_{p}	Power system gain
T_{p}	Power system time constant (s)
T_{t}	Turbine time constant (s)
T_{g}	Governor time constant (s)
R	Speed regulation due to governor action (Hz/p.u. MW)
$\Delta f(t)$	Incremental frequency deviation
$\Delta P_{\mathrm{m}}(t)$	Incremental change in generator output (Hz/p.u. MW)
ΔP_{V}	Incremental change in governor valve position

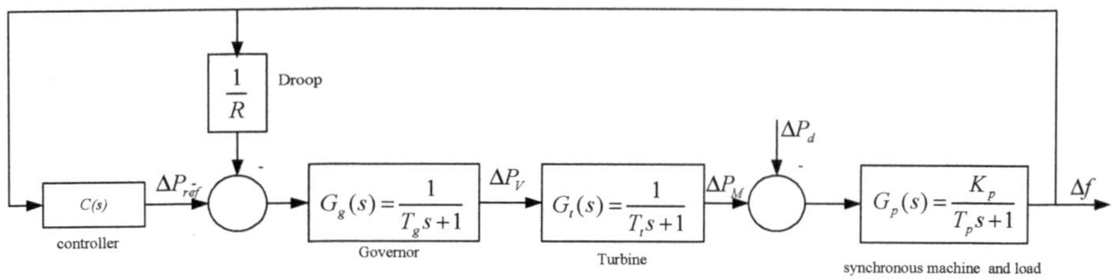

Fig. 2 Unity negative feedback configuration for a single-area LFC

$$\text{Governor dynamics} : G_{\rm g}(s) = \frac{1}{T_{\rm g}s + 1}$$

$$\text{Turbine dynamics} : G_{\rm t}(s) = \frac{1}{T_{\rm t}s + 1}$$

$$\text{Synchronous machine and load dynamics} : G_{\rm p}(s) = \frac{K_{\rm p}}{T_{\rm p}s + 1}$$

The PID controller is used for LFC of power system with unity negative feedback as shown in Fig. 2. The PID controller is chosen in "parallel form" as given by:

$$C(s) = K_C + \frac{K_I}{s} + K_D s \tag{1}$$

Consider $G_{\rm d,f}(s)$ as the closed-loop transfer function (CLTF) from $\Delta P_{\rm d}$ to Δf with a PID controller $C(s)$ as shown in Fig. 2, and it is given by

$$G_{d,f}(s) = \frac{G_{\rm p}(s)}{1 + G_{\rm p}(s)G_{\rm g}(s)G_{\rm t}(s)\left[C(s) + 1/R\right]} \tag{2}$$

A desired reference model $M_{d,f}(s)$ for a load frequency control is given by

$$M_{d,f}(s) = \frac{N_{\rm md}(s)}{D_{\rm md}(s)} = \frac{Ks}{(\lambda s + 1)^r} \tag{3}$$

where $\frac{N_{\rm md}(s)}{D_{\rm md}(s)}$ is a transfer function in rational form, r is the order of desired CLTF, the time constant λ of desired CLTF is the tuning parameter of the system, and $K = 1/K_I$ [26]. In the desired reference model, one zero is placed at origin, to confirm frequency deviation at steady state to zero.

The frequency points around ω that tends to zero will give good steady-state performance for most of the control system. Therefore, the selection of low-frequency points around 0.1% of bandwidth frequency for matching purpose is chosen. The variation of frequency points in this range will not affect the system performance. The objective of the proposed design technique is to equate the desired CLTF response with that of closed-loop disturbance response of the power system, i.e., represented in mathematical form as

$$M_{d,f}(s) \cong G_{d,f}(s) \tag{4}$$

To obtain such performance matching, the aspect we have considered is as follows:

1. For the desired transient response, the pole of the desired reference model at $s = -1/\lambda$ is taken as the pole of the closed-loop control system which provides the following characteristic equation.

$$1 + \left(C(s) + \frac{1}{R}\right) G_g(s) G_t(s) G_p(s) = 0 \text{ for } s = \frac{-1}{\lambda} \tag{5}$$

$$\text{Or, } C(s) = \frac{-1}{G_g(s) G_t(s) G_p(s)} - \frac{1}{R}\bigg|_{s=\frac{-1}{\lambda}} \tag{6}$$

$$\text{Or, } K_C - \lambda K_I - \frac{K_D}{\lambda} = \frac{-1}{G_g(s) G_t(s) G_p(s)} - \frac{1}{R}\bigg|_{s=\frac{-1}{\lambda}} \tag{7}$$

2. To get a better steady-state performance, the frequency response matching of the two system at low frequency point (i.e., ω = 0.001 rad/s) which gives the following equation,

$$G_{d,f}(j\omega) \cong M_{d,f}(j\omega) \tag{8}$$

$$\frac{G_p(j\omega)}{1 + \left(C(j\omega) + \frac{1}{R}\right) G_g(j\omega) G_t(j\omega) G_p(j\omega)} = M_{d,f}(j\omega) \tag{9}$$

$$C(j\omega) = \frac{1}{M_{d,f}(j\omega) G_g(j\omega) G_t(j\omega) G_p(j\omega)} - \frac{1}{G_g(j\omega) G_t(j\omega) G_p(j\omega)} - \frac{1}{R} \tag{10}$$

$$K_C + \frac{K_I}{j\omega} + K_D j\omega = \frac{1}{M_{d,f}(j\omega) G_g(j\omega) G_t(j\omega) G_p(j\omega)} - \frac{1}{G_g(j\omega) G_t(j\omega) G_p(j\omega)} - \frac{1}{R} \tag{11}$$

where

$$M_{d,f}(j\omega) = \frac{M_{DY}(j\omega)}{K_I}$$

Assuming

$$A = \frac{1}{M_{\mathrm{DY}}(j\omega)G_{\mathrm{g}}(j\omega)G_{\mathrm{t}}(j\omega)}$$

and

$$B = \left(\frac{1}{G_{\mathrm{g}}(j\omega)G_{\mathrm{t}}(j\omega)G_{\mathrm{p}}(j\omega)} + \frac{1}{R}\right)$$

Equation (11) may be written as

$$K_C + j\left(\frac{-K_I}{\omega} + K_D\omega\right) = (K_I\mathrm{Re}[A] - \mathrm{Re}[B]) + j(K_I\mathrm{Im}[A] - \mathrm{Im}[B]) \tag{12}$$

Separating the imaginary and real parts of Eq. (12) yields the following equations.

$$K_C - K_I\mathrm{Re}\,[A] = -\mathrm{Re}\,[B] \tag{13}$$

and

$$\left(\frac{-1}{\omega} - \mathrm{Im}\,[A]\right)K_I + K_D\omega = -\mathrm{Im}\,[B] \tag{14}$$

Combine Eqs. (7), (13), and (14) in matrix form as

$$\begin{bmatrix} 1 & -\lambda & -1/\lambda \\ 1 & -\mathrm{Re}\,[A] & 0 \\ 0 & \left(\frac{-1}{\omega} - \mathrm{Im}\,[A]\right) & \omega \end{bmatrix} \begin{bmatrix} K_C \\ K_I \\ K_D \end{bmatrix} = \begin{bmatrix} X \\ -\mathrm{Re}\,[B] \\ -\mathrm{Im}\,[B] \end{bmatrix} \tag{15}$$

The solution of Eq. (15) provides the parameter of PID controller (such as K_C, K_I, K_D) by pole placement and approximate frequency response matching done in MATLAB programming.

Maximum sensitivity (M_{s}):

The maximum sensitivity, (M_{s}), is used to estimate the robustness of a control system, which is expressed as

$$M_{\mathrm{s}} = \max_{0 \le \omega \le \infty} |1/[1 + C(j\omega)G(j\omega)]| \tag{16}$$

A smaller value of M_{s} is preferred which typically lies in the range of 1.2–2.0 [4].

3 Simulation Result

In this section, we analyzed various case studies for a single-area power system with non-reheat turbine (NRT) and also extended for a two-area power system model and its simulation is done in MATLAB and Simulink environment.

Case Study 1 Consider a single-area power system with NRT. The system parameters are as follows [23, 25, 27] $K_p = 120, T_p = 20, T_t = 0.3, T_g = 0.08, R = 2.4$.

Therefore, the desired closed-loop transfer function for LFC (at $r = 2$) is

$$M_{d,f}(s) = \frac{Ks}{(\lambda s + 1)^2}$$

where the steady-state frequency deviation returns back to zero due to one zero at origin in desired reference model. The frequency matching point is selected as $\omega = 0.001$ rad/s. At $\lambda = 0.2$, we get proposed PID load frequency controller as

$$C(s) = 1.2728 + \frac{2.1767}{s} + 0.3003s$$

To verify the better performance of the proposed PID load frequency controller, a step change in load $\Delta P_d = 0.01$ p.u is applied at $t = 1$ s and its response is compared with that of Tan [25], Anwar and Pan [23], and Padhan and Majhi [27] as shown in Fig. 3 and Table 2. The robustness of the proposed PID controller with +50% change in system parameter K_P and T_P has been elucidated, and its frequency deviation response is shown in Fig. 4 and Table 3. Figures 3 and 4 show better performance in both nominal and perturbed plants in terms of settling time, maximum sensitivity (M_s), ISE.

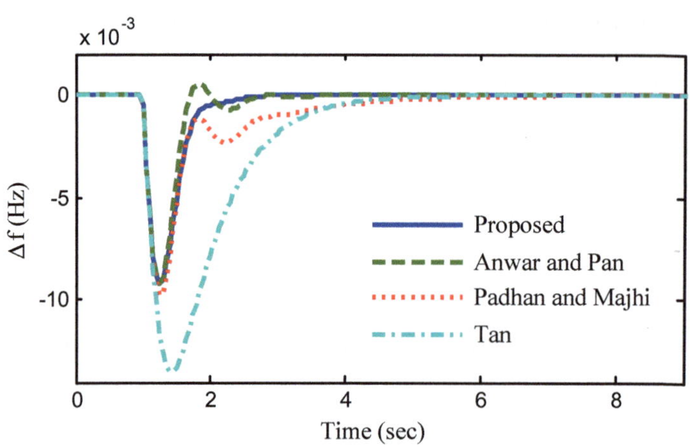

Fig. 3 Frequency deviation for Case Study 1 with nominal plant

Table 2 Comparison performance for Case Study 1 of nominal plant

Method	K_C	K_I	K_D	M_s	Peak value ($\times$ 10^{-3})	Settling time (s)	ISE ($\times$ 10^{-5})
Proposed (at λ = 0.2)	1.27	2.17	0.30	1.48	9.2	2.70	2.89
Anwar and Pan [23]	1.52	2.50	0.27	1.74	9.3	2.75	2.63
Padhan and Majhi [27]	1.49	1.30	0.23	1.77	9.8	5.40	3.62
Tan [25]	0.40	0.63	0.18	1.24	13.4	4.70	13.87

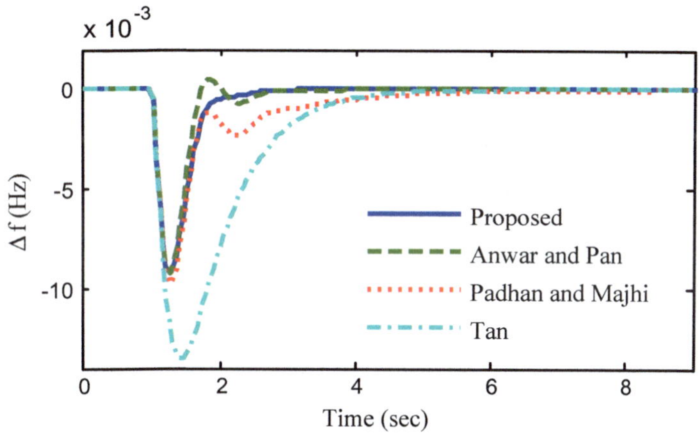

Fig. 4 Frequency deviation for Case Study 1 with +50% changes in K_P and T_P

Table 3 Comparison performance for Case Study 1 of perturbed plant

Method	K_C	K_I	K_D	M_s	Peak value ($\times$ 10^{-3})	Settling time (s)	ISE ($\times$ 10^{-5})
Proposed (at λ = 0.2)	1.27	2.17	0.30	1.48	9.2	2.75	2.86
Anwar and Pan [23]	1.52	2.50	0.27	1.74	9.27	2.80	2.60
Padhan and Majhi [27]	1.49	1.30	0.23	1.77	9.7	4.8	3.58
Tan [25]	0.40	0.63	0.18	1.24	13.4	5.6	13.67

Case Study 2 Consider a single-area power system with NRT. The system parameters are as follows [23, 24]

$$K_{\rm p} = 1, T_{\rm p} = 10, T_{\rm t} = 7, T_{\rm g} = 0.2, R = 0.05$$

Therefore, the desired closed-loop transfer function for LFC (at $r = 2$) is

$$M_{d,f}(s) = \frac{Ks}{(\lambda s + 1)^2}$$

The frequency matching point is selected as $\omega = 0.001$ rad/sec. At $\lambda = 0.4$, we get proposed PID load frequency controller as

$$C(s) = 1.58.44 + \frac{22.42}{s} + 146.98s$$

To verify better performance of the proposed PID load frequency controller, a step change in load $\Delta P_{\mathrm{d}} = 0.01$ p.u. is applied at $t = 1$ s and its response of the system is compared with that of Anwar and Pan [23], and Gundes and Chow [24] as shown in Fig. 5 and Table 4. The robustness of the proposed PID controller with +50% change in system parameter K_{P} and T_{P} has been elucidated, and its frequency deviation response is shown in Fig. 6 and Table 5. Figures 5 and 6 show better performance in both nominal and perturbed plants in terms of settling time, maximum sensitivity (M_{s}), ISE.

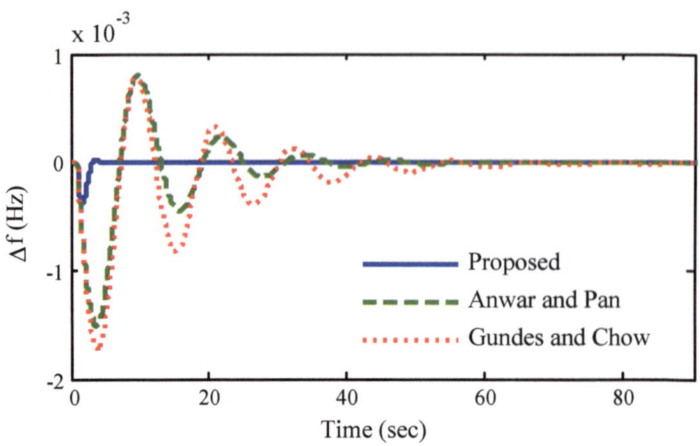

Fig. 5 Frequency deviation response for Case Study 2 with nominal plant

Table 4 Comparison performance for example 2 of nominal plant

Method	K_C	K_I	K_D	M_s	Peak value ($\times 10^{-4}$)	Settling time (s)	ISE ($\times 10^{-5}$)
Proposed (at $\lambda = 0.4$)	158.4	22.42	146.9	1.49	3.9	4.5	0.02
Anwar and Pan [23]	0.267	2.50	9.70	2.80	15.1	41.52	0.9
Gundes and Chow [24]	1.50	1.05	0.439	4.16	17.4	64.4	1.4

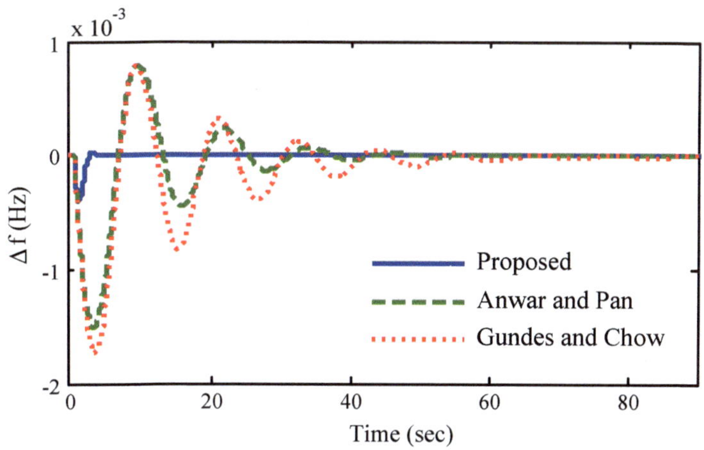

Fig. 6 Frequency deviation response for Case Study 2 with +50% changes in K_P and T_P

Table 5 Comparison performance for example 2 of perturbed plant

Method	K_C	K_I	K_D	M_s	Peak value ($\times 10^{-3}$)	Settling time (s)	ISE ($\times 10^{-5}$)
Proposed (at $\lambda = 0.4$)	158.4	22.42	146.9	1.49	0.4	4.4	0.014
Anwar and Pan [23]	0.267	2.50	9.70	2.80	1.58	54	1.17
Gundes and Chow [24]	1.50	1.05	0.439	4.16	1.82	62	1.8

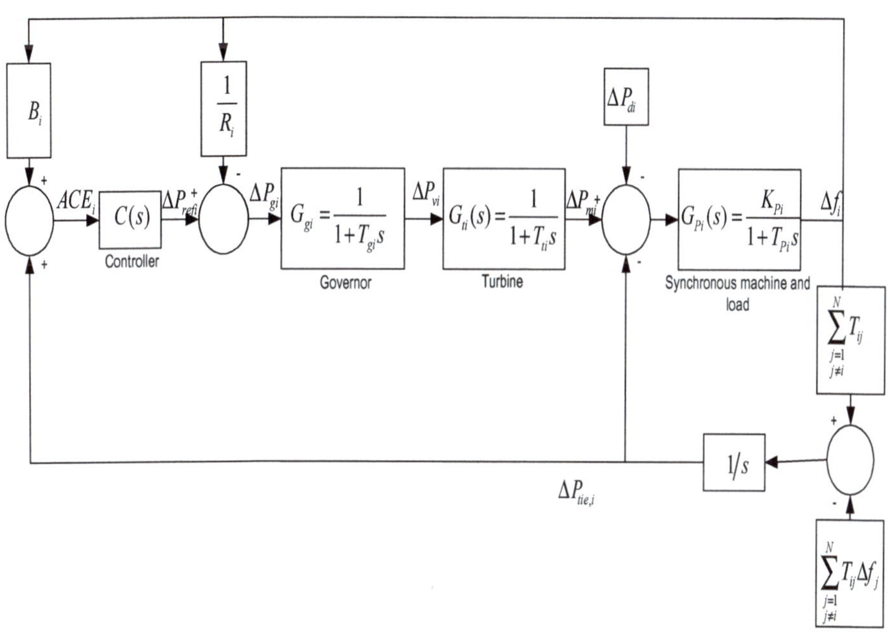

Fig. 7 Block diagram representation of control area *i*

3.1 The Generation Rate Constraint

The generation rate constraint (GRC) used as nonlinearity in load frequency control problem affects the performance of the system and sometimes causes instability [9]. To solve the limitation of GRC problem, an anti-GRC model is proposed by Tan [25]. An anti-GRC model is shown in Fig. 8. The proposed PID controller is used for anti-GRC model, and its parameters are taken from Case Study 1 [23, 25, 27]. The frequency deviation response occurs on the power system as shown in Fig. 9 along with that of Tan [25], and its GRC of 0.01 p.u/min, i.e., 0.0017 p.u/s, is considered. The response of mechanical power generated by turbine with the proposed PID controller along with that of Tan [25] is shown in Fig. 10 where K is used as compensator gain to reduce the error that occurs in between actual turbine

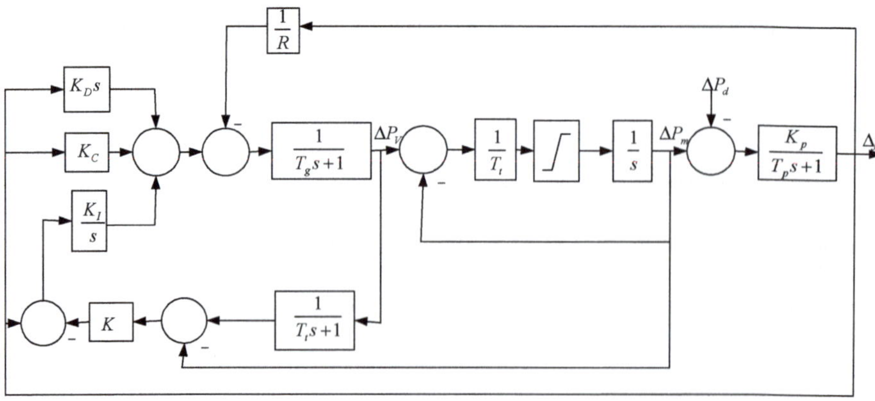

Fig. 8 Block diagram of anti-GRC model

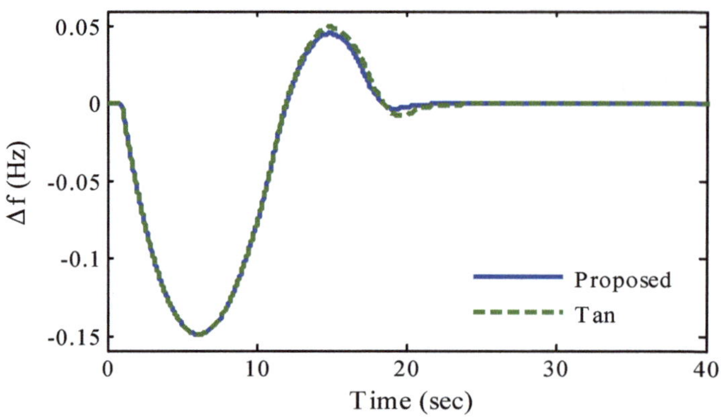

Fig. 9 Frequency deviation response with anti-GRC for Case Study 1

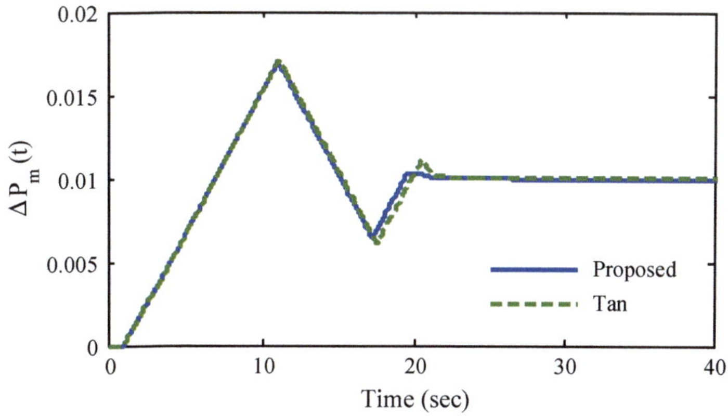

Fig. 10 Mechanical power generated by turbine with anti-GRC for Case Study 1

output and turbine output with GRC. Hit and trial error method is used to select better choice for K; here, we select $K = 2$.

3.2 Multi-area Power System

The proposed PID load frequency controller design method is extended for a two-area power system, to obtain frequency deviation response for each control area and its tie-line power exchange between interconnected areas. Figure 7 represents the basic block diagram of multi-area power system. The area control error (ACE) is composite parameter of incremental change in frequency deviation and tie-line power exchange between interconnected areas. Under steady-state condition, change in frequency deviation Δf and tie-line power exchange between interconnected areas becomes zero because ACE is used as actuating signal for adjustment of reference power set points and the ACE ith control area is expressed as

$$\mathrm{ACE}_i = \Delta P_{\mathrm{tie}i} + B_i \Delta f_i \tag{17}$$

where B_i is frequency bias factor, and total tie-line power exchange between ith control areas with other area is given as follows [17]

$$\Delta P_{\mathrm{tie}i} = \sum_{\substack{j=1 \\ j \neq i}}^{N} \Delta P_{\mathrm{tie}ij} = \frac{1}{s} \left[\sum_{\substack{j=i \\ j \neq i}}^{N} T_{ij} \Delta f_i - \sum_{\substack{j=1 \\ j \neq i}}^{N} T_{ij} \Delta f_j \right] \tag{18}$$

where T_{ij} is the synchronizing power coefficient. In multi-area power system, the plant model is given by

$$G_i = B_i \frac{G_{gi} G_{ti} G_{pi}}{1 + G_{gi} G_{ti} G_{pi}/R_i} \tag{19}$$

Assume there is no tie-line power exchange for tuning of decentralized PID controller. Multi-area load frequency controller design technique is as similar as a single-area power system; Except, each control area in power system turned independently with assumption of plant model as in Eq. (19).

Case Study 3 Consider a two-area power system with NRT discussed in [25]. Here, for each area similar system parameters are taken for simplicity in implementation.

$$\begin{aligned} &T_{g1} = T_{g2} = 0.08, T_{t1} = T_{t2} = 0.3, T_{p1} = T_{p2} = 20, K_{p1} = K_{p2} = 120, \\ &R_1 = R_2 = 2.4, B_1 = B_2 = 0.425, T_{12} = T_{21} = 0.545 \end{aligned}$$

Therefore, the desired closed-loop transfer function for LFC (at $r = 3$) is

$$M_{d,f}(s) = \frac{0.1757s}{(0.2s + 1)^3}$$

The frequency matching point is selected as $\omega = 0.001$ rad/s. At $\lambda = 0.2$, we get the proposed PID load frequency controller as

$$C(s) = 5.5298 + \frac{5.6906}{s} + 1.1579s$$

To verify the better performance of the proposed decentralized PID load frequency controller, a step change in load $\Delta P_d = 0.01$ p.u is applied at $t = 1$ s and its response of the system is compared with that of Tan [25] as shown in Figs. 11, 12, 13 and Tables 6, 7, 8. The robustness of proposed PID controller and its frequency deviation response is shown in Figs. 11 and 12. These figures show better performance of plant in terms of settling time, maximum sensitivity (M_s) and ISE. Figure 13 shows tie-line power interchange between two areas.

4 Conclusion

A PID load frequency controller design via pole placement and approximate frequency response matching has been proposed for a single-area power system and also extended for a two-area power system along with NRT. The characteristic equation of desired closed-loop transfer function and frequency response matching

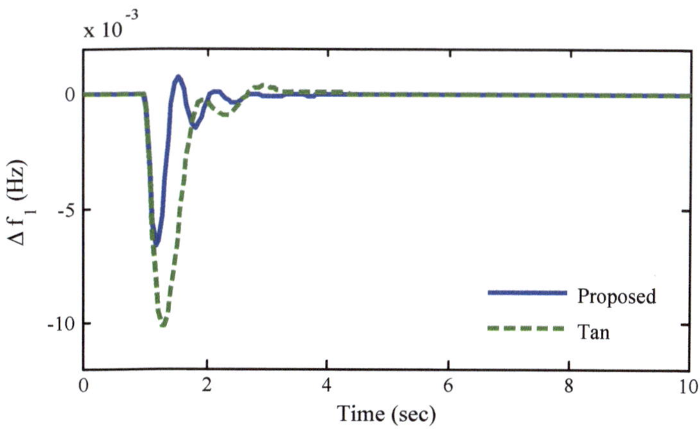

Fig. 11 Frequency deviation response of area 1 for Case Study 3

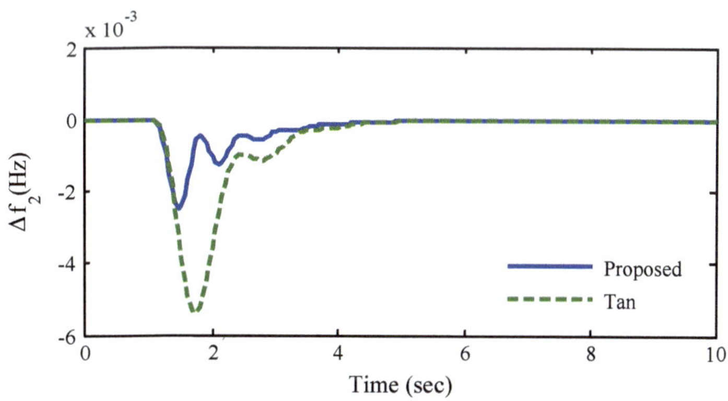

Fig. 12 Frequency deviation response of area 2 for Case Study 3

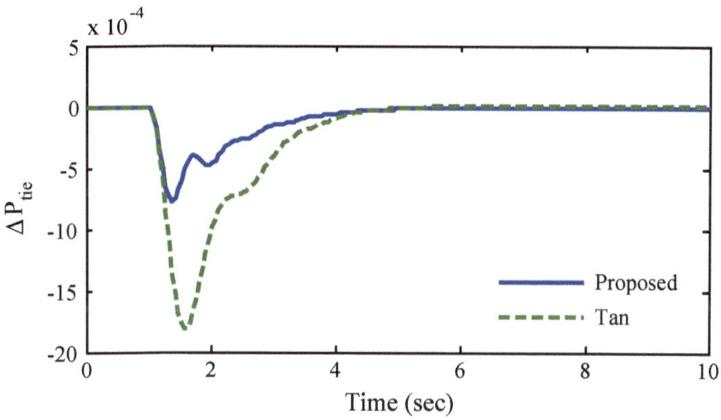

Fig. 13 Tie-line power interchange for Case Study 3

Table 6 Comparison performance of frequency deviation in area 1 for Case Study 3

Method	K_C	K_I	K_D	M_s	Peak value ($\times 10^{-3}$)	Settling time (s)	ISE ($\times 10^{-5}$)
Proposed	5.53	5.69	1.16	1.29	6.50	2.65	0.20
Tan [7]	1.57	2.39	0.52	1.60	10.1	3.35	1.36

Table 7 Comparison performance of frequency deviation in area 2 for Case Study 3

Method	K_C	K_I	K_D	M_s	Peak value ($\times 10^{-3}$)	Settling time (s)	ISE ($\times 10^{-5}$)
Proposed	5.53	5.69	1.16	1.29	2.50	3.20	0.005
Tan [7]	1.57	2.39	0.52	1.60	5.40	3.60	0.03

Table 8 Comparison performance of tie-line power for Case Study 3

Method	K_C	K_I	K_D	M_s	Peak value ($\times 10^{-3}$)	Settling time (s)	ISE ($\times 10^{-5}$)
Proposed	5.53	5.69	1.16	1.29	0.75	3.50	0.03
Tan [7]	1.57	2.39	0.52	1.60	1.80	4.10	0.22

of desired CLTF and plant disturbance CLTF forms linear algebraic equation which gives the PID controller parameters. Generation rate constraint is used as nonlinearity in the proposed design method, and its limitation is minimized by anti-GRC model. When a +50% changes in power system gain, power system time constant and its response is found to be similar as nominal plant; which shows the robustness of the controller. Simulation results of proposed design method show better performance in terms of maximum sensitivity, settling time, peak value, ISE than other well-known methods that were explained in the literature.

Acknowledgements The authors acknowledge the financial support from the Department of Science and Technology, India (under project grant no. ECR/16/001547) at NIT Patna, India.

References

1. Kundur P. (1994) New York: McGraw-Hill.
2. Saxena S, Hote YV (2013) IEEE Trans. Power Syst., vol. 28, no. 3, pp. 2749–2757.
3. Bevrani H, Hiyama T (2009) IEEE Trans. Energy Convers., vol. 24, no. 1, pp. 292–300.
4. Seborg DE, Edgar TF, Mellichamp D A. John Wiley & Sons: New York, 1989.
5. Astrom, K. J.; Hagglund, T. 2nd ed.; Instrument Society of America: Research Triangle Park, NC, 1995.
6. Chen D, Seborg DE. Ind Eng Chem Res, 41: 4807–22, 2002.
7. H. Bevrani and P. R. Daneshmand, IEEE Syst. J., vol. 6, no. 1, pp. 173–180, Mar. 2012.
8. I. Kocaarslan and E. Cam, Electr. Power Energy Syst., vol. 27, pp. 542–549, 2005.

9. Pan CT, Liaw CM. IEEE Trans Power Syst 4(1): 122–8. 1989.
10. Ali ES, Abd-Elazim SM. Electr Power Energy Syst,; 51: 224–31, 2013.
11. C.E. Fosha, O.I. Elgerd, IEEE Trans. PAS 89 563–567, (1970).
12. Khodabakhshian A, Edrisi M. Control Eng Pract; 16: 1069–80, 2008.
13. Wang Y, Zhou R, Wen C. "Robust load–frequency controller design for power system." IEE Proc Gen Transm Distrib Part C; 140(1): 11–6, 1993.
14. Khodabakhshian A, Hooshmand R Electr Power Energy Syst; 32: 375–82, 2010.
15. Morari, M.; Zafiriou, E. Prentice Hall: Englewood Cliffs, NJ, 1989.
16. Tan W; Xu Z Electric Power System Research; 79; 846–53, 2009.
17. Bevrani H. Springer; 2009.
18. H. Bevrani Y. Mitani. K, Tsuji. H. Bevrani, Energy convers 46, 1129–1146, 2005.
19. H. Shayeghi, H. A. Shayanfar, and A. Jalili, Energy Convers. Manag., vol. 50, pp. 344–353, 2009.
20. Ibraheem, P. Kumar, and D. P. Kothari, IEEE Trans. Power Syst., vol. 20, no. 1, pp. 346–357, Feb. 2005.
21. Ravi Shankar, S.R. Pradhan, Kalyan Chatterjee,*, Rajasi Mandal, Renewable and Sustainable Energy Reviews, 76, 1185–1207 (2017).
22. H. Saadat, McGraw-Hill, 1999.
23. Anwar, M. N., & Pan, S. Electric Power Energy Systems, 67, 560–569, 2015.
24. Gundes AN, Chow L. In: American control conference (ACC), Washington, DC, USA; June 17–19, 2013.
25. Tan W. Energy Convers Manage; 50: 1465–72, 2009.
26. Victor M, Alfaroa and Ramon vilanova, Journal of process control, vol. 22. pp. 359–374, (2012).
27. Padhan DG, Majhi S. ISA Trans; 52: 242–51, 2013.
28. Kothari DP, Nagrath IJ (2009) McGraw-Hill, 2009.

Anand Kumar Received Bachelor of Engineering degree in Electrical and Electronic Engineering from OIST, Bhopal, India, in 2013 and currently pursuing M.Tech-PhD dual degree from NIT, Patna in Department of Electrical Engineering since 2015. His research interest includes Load frequency control, Automatic voltage regulator, PID controller, PIDA controller.

Md Nishat Anwar He obtained his bachelor and master degree in Electrical Engineering from AMU, Aligarh, India. He obtained his Ph.D. from IIT (ISM), Dhanbad, India in 2015. He has teaching experience of 4 years and currently working as an Assistant Professor in the Department of Electrical Engineering, NIT Patna, India. His research interests include industrial control and automation, PID controller, model predictive control and has several masters and Ph.D. students working under him. His publication includes 4 papers in referred journals, 12 papers in international conferences and 5 papers in national conferences.

Ashraf Raza Received bachelor degree in Electrical Engineering from AMU, Aligarh, India, in 2010 and master degree from IIT, Roorkee, India, in 2012 in the specialization Electric drives and power electronics. He has 3 years teaching experience and currently pursuing Ph.D. from NIT Patna in Electrical Engineering Department. The research interest includes process control, PID control for time delay system, control of integrating and unstable process etc.

Processor Scheduling in High-Performance Computing Environment Using MPI

Annu Priya and Sudip Kumar Sahana

1 Introduction

The effectiveness of operating systems mainly depends on the scheduling algorithm used by the processor. Processor scheduling is one of the most commonly researched and challenging topic for the researcher in these days. Processor scheduling means that allocation of the resource time to the jobs that wait in the ready queue. Every job should get a chance to execute on the processor it is only possible when fairly and correctly scheduling algorithm applied. Processor scheduling classified into two parts: (i) Local processor scheduling and (ii) Global processor scheduling. Local processor scheduling uses single processor platform where the number of processes scheduled. By using the list scheduling, local processor scheduling constructs a schedule, one cycle at a time. The goal of the multiprocessor scheduling algorithm [1–4] is to meet deadlines, whereas the arrival of the tasks has implicit deadlines. Global processor scheduling into different families are biologically inspired and further, they grouped into several classes of heuristic approaches such as Genetic algorithm (GA), Ant Colony Optimization (ACO), etc. to compare processor scheduling there are number of performance has been done. The performance trials have been proposed for comparing processor scheduling techniques [5]. Processor scheduling includes various criteria such as waiting time, average waiting time, turnaround time, average turnaround time for the scheduling. For processor scheduling [6], there are a number of existing techniques available such as First-Come First-Serve (FCFS), Shortest Job-First (SJF), and Round-Robin (RR), etc., whereas there are several evolutionary computing approaches for

A. Priya (✉) · S. K. Sahana
Department of Computer Science Engineering, Birla Institute of Technology, Mesra, 835215 Ranchi, Jharkhand, India
e-mail: annu.priya12@yahoo.com

J. Chattopadhyay et al. (eds.), *Innovations in Soft Computing and Information Technology*, https://doi.org/10.1007/978-981-13-3185-5_5

processor scheduling such as Genetic algorithm (GA), Ant Colony Optimization (ACO), and particle swarm optimization (PSO), etc.

1.1 First-Come First-Serve (FCFS)

FCFS processor scheduling approach used to determine processor utilization and throughput. By using FCFS scheme performance improvement can be improved. The nodes that can't assign by user request FCFS determines them during calculation process. This algorithm is the easiest way of processor. Scheduling is the processes in the queue where these processes execute concurrently. The arrived new process always kept at the end of the queue so that process is also called the last process in the queue. The new arrive process that store in the queue after the last process automatically becomes the last process in the given queue. This algorithm is easy to implement, but sometimes processor usage can be wasted in a certain condition. The proposed FCFS technique is an easier method of scheduling to reduce the process starvation, and also ranking all types of processes correspondingly at the same time it takes care of whether each process is able to get a chance to get processor time and resources and also short ordering of processes so that they do not have to wait for higher burst time processes to get executed.

1.2 Shortest Job-First (SJF)

SJF is a non-preemptive scheduling algorithm in which smaller amount job executes first in waiting number of jobs in other words when a processor is available for the user, it assigns to the job that has smallest next processor burst. It allows the average minimum time for a set of jobs that is probably optimal. This technique factors short jobs at the expense of longer ones. The disadvantage with SJF is to determine the length of the next process. The advantage SJF scheduling algorithm when providing minimum average waiting time it is optimal in nature.

1.3 Genetic Algorithm (GA)

Genetic Algorithm (GA) [4, 7], firstly introduces by Holland in 1950. GA used to mimic the processes of natural evolution and selection. Beside GP (Genetic Programming), an evolutionary approach used to resolve optimization problem. Because of its robustness, genetic approach becomes familiar than other traditional scheduling approaches which is used to solve various optimization problems. Every individual processor search node is represented based on the order of the tasks. The genetic operator is used to represent on the precedence relations between the tasks, and these tasks assign to the given resources. GA differs from other existing traditional optimization approaches in the following ways:

GA uses coding constraint set rather than parameter.

i. Genetic algorithm search approach search nodes from population instead of a single one.
ii. Genetic algorithm applied probabilistic transition rules.

In GA, genes represented as a string of the search nodes. GA operators used to generate the new search nodes. Fitness function used by GA for calculating the search nodes, and a stochastic assignment provides to control the genetic operators.

A. Initial Population
In GA initially, population is set of task that randomly generated then after that randomly select of element of first array which range from 1 to total number of has given tasks. The selection of second array element is same as the first one but the range varies from 1 to total number of available processors. This operation is repetitive in nature until all the tasks have been assigned to the number of available processors.
B. Fitness Function
The objective of GA is to achieve is to schedule with minimum completion time for given scheduling problem. The fitness function is calculated by keeping the objective to minimizing Makespan is as follows:
$f(x) = 1/S$. Where fitness function = $f(x)$ scheduling, length = S, where scheduling length (S) is the maximum accomplishment time of a process in an individual chromosome. The aim of fitness function ($f(x)$) is to get a higher value for individual chromosome with minimum accomplishment time.
C. Genetic Operators
After applying crossover and mutation operators, next phase is selection which is applied on allocation of reproductive trails that is used for chromosomes according to their fitness function $f(x)$. For this procedure, the roulette wheel is used in which each chromosome of population has a slot in portion to its fitness function $f(x)$.

An algorithm for processor scheduling is working well when we have a small number of jobs available to schedule on a uniprocessor. There is chance to conflict when the number of jobs available in large size on multiprocessor. They may conflict when the number of jobs available in large size on multiprocessor (so on they scheduled the processors statically). A good scheduling technique is that which improves the performance criteria. The optimization, performance criteria are based on maximizing CPU utilization, maximizing throughput, minimizing turnaround time, minimizing waiting time, minimizing response time, minimizing the context switching. The main focus of this paper is on developing an algorithm for processor scheduling in HPC environment using MPI for low-dimensional mesh architectures. In Fig. 1, represent the layer architectural view of Open MPI which has mainly three layers: OPAL, ORTE, and OMPI.

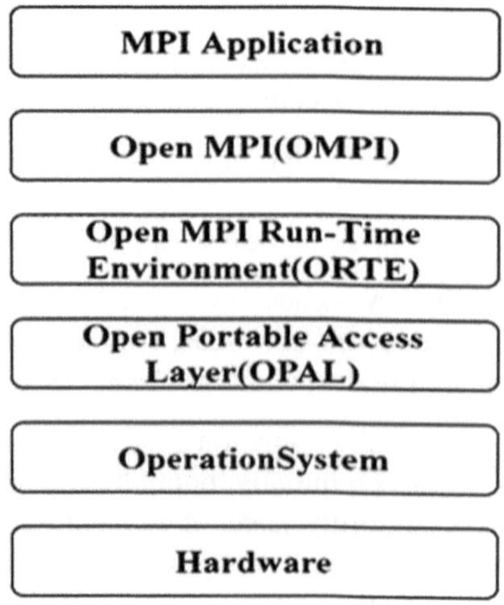

Fig. 1 Open MPI has three main abstraction project layers

i. Open Portable Access Layer (OPAL): this layer provides portability between different operating systems and basic services.
ii. Open MPI Run-Time Environment (ORTE): this layer supports to launch, monitor individual processes, and make set of specific processes into "jobs".
iii. Open MPI (OMPI): this layer supports to public MPI API and exposed to applications.

MPI is a message-passing application programmer interface. The goal of MPI is to achieve high performance also provide scalability and portability for the application. This MPI model supports in parallel programming used in the high-performance computing system (HPC) environment for parallel programming. It is a communication protocol computes programming that runs parallel on distributed memory systems. The MPI programming model architectures have various benefits because MPI supports memory locality. MPI has a specific set of routine calls that can easily callable from C, C++, Fortran, and another language that able to interface with such libraries, including C#, Java or Python. The rest of this paper is organized as following Sect. 1, deals with the introduction of processor scheduling whereas in Sect. 2 deals with the literature survey of the existing approaches for processor scheduling. Section 3, contains the research methodology for problem statements whereas in Sect. 4, contains the experimental results and discussion with other existing algorithms. Finally in Sect. 5, concludes the conclusion and the outline of the future scope of this work.

2 Literature Survey

The authors Xingwu et al. [8], proposed a novel scheduling system that uses simulated annealing as the optimization engine to maintain resource management on HPC systems. They demonstrate that simulations with workload on super-computers by using queue-based approach with FCFS with backfilling technique. This novel scheduling system reduced the job waiting time by 40% and also reduces the job response timing by 30%. This technique has the ability to run online also solve the difficult scheduling problem at each scheduling iteration within a second then constructs it practically for production HPC systems. Byung and Chita [9] addressed fast and efficient processor allocation approach which meets demand for messy interconnected multicomputer and also proposed algorithm that is based on stack allocation which locates a free sub-mesh for a job in a very fast manner by using simple coordination and spatial subtraction.

That prevents starvation and also reduces the response time significantly by minimizing the queuing delay for the jobs. Hsiu-Jy and Wei-Ming [10] proposed a novel scheduling method based on size variance among the jobs in queue.

That demonstrates self-adjustment between the two orders in real time. The benefit of this method is to reserve significantly diminishing chance of blocking situation. Shun-LI et al. [11] present model that is based on resource management and agent structure this function is based on local resource's adoption scheme of FCFS. This technique achieves the goals of task scheduling. William et al. [12] provide a job selection heuristic-based technique on resource balancing which supports for the construction of global K-resource scheduling algorithms. The benefit of this method is to achieve up to 50% in average response time over current scheduling method such as First-Come First-Serve (FCFS) with the First-Fit backfill. Prabhat and Prasoon [13] designed and implemented a new process scheduling method that improves the performance of an MMOS (Multimedia Operating System) in batch jobs. Marcia et al. [14] present a scheduler module that is implemented with MPI-2. It demonstrates during the execution of tasks, on a processor that a newly spawned process runs with priority.

3 Research Methodology

After through study of a different research paper, a framework for processor scheduling using MPI is proposed for the said problem which is shown in Fig. 2.

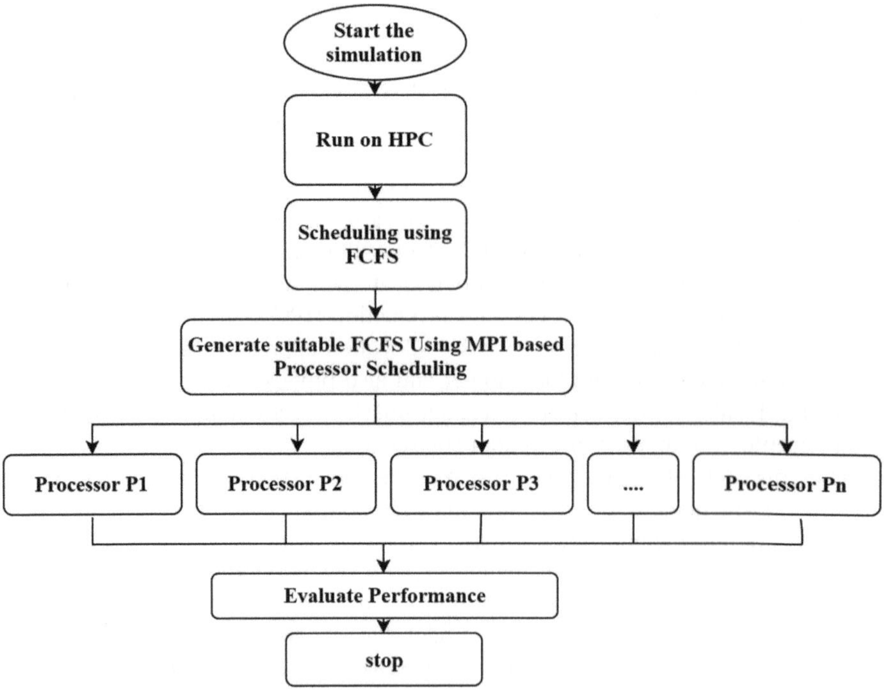

Fig. 2 Flow chart of FCFS using MPI scheduling on HPC

A step-by-step procedure for processor scheduling is given below.

Step 1: start the simulation on HPC.
Step 2: apply FCFS Algorithm using MPI

i. Enter the number of processors.
ii. Enter the burst time (bt).
iii. Apply MPI
iv. Assign waiting time (wt) for all the processors.
Waiting time for process 1 will be 0, i.e., wt [0] = 0.
The manager sends tp $p - 1$ number X_1, $X_2 \ldots X_{p-1}$ to the workers.
The worker receives the ith worker X_i in N.
The manager copies X_0 to N: $N = X_0$.
v. Every node process $N \rightarrow$ wt [N] = bt [$N - 1$] + wt [$N - 1$].

vi. Every worker sends f to the manager.
The manager receives X_i from the ith worker, $i = 1, 2 \ldots p - 1$.
Manager copies N to X_0: $X_0 = N$,
vii. Print the result on every processor.
viii. Calculate [turnaround time = wt + bt for all processes.
ix. Calculate average waiting time = total_waiting_time/no_of_processes.
x. Similarly, find average turnaround time = total_turn_around_time/no_of_processes.

Step 3: Evaluate the performance by comparing the result with the existing algorithm in result and discussion section.

Step 4: Stop the simulation and achieve the planning horizon.

4 Results and Discussion

Simulation work has been performed on HPC with Master Node-2 no of Intel Xeon E5-2630 v3 2.4 GHz, 8-core processors and 64 GB memory, 2*1 TB HDD, 500 GB Disk capacity FDR InfiniBand for a number of processors 5, FCFS scheduling algorithm using MPI. Figure 2 shows the comparisons of different scheduling approaches on the basis of average waiting time. Table 1 shows five processors along with their burst time and the respective average waiting time comparison for different scheduling techniques. The results of FCFS using MPI scheduling algorithm accomplish is better result than traditional FCFS. FCFS using MPI-based algorithms shows that it is more efficient than existing FCFS that requires future knowledge about processor bursts. FCFS using MPI technique the results may vary if we change the basic parameters of FCFS like a number of jobs, etc. Thus, some more experiments can be done by changing these parameters in order to solutions.

In Fig. 3, shows that the comparison of FCFS and FCFS using MPI where we have taken the execution time as a parameter for the evaluation. So it is proved that FCFS running in the single processor has taken more time then parallel. The FCFS using MPI provides the environment where the program executes in parallel. And the result shows less execution time taken by multiprocessor in compared to a single processor.

According to, Fig. 3 and Table 2, we can analyze that the results of FCFS using MPI are more optimal than FCFS, SJF, and GA.

Table 1 Summary of research development in the field of processor scheduling

Research	Methodology	Technique used	Advantages	Disadvantages
Xingwu et al. (2010) [8]	Simulated annealing used as optimize engine that supports resource organization of high-performance computing system	Scheduling is based on plan	(i) Reduce the job, wait time by 40% (ii) Reduce the job response time by 30%	(i) Repeatedly annealing, (ii) The scheduling is very slow, (iii) The cost function is expensive to compute
Byung and Chita (1997) [9]	Processor management schemes	Stack-based allocation algorithm	(i) Prevent starvation (ii) Reduced response time (iii) Minimizing the queuing delay	(i) Only work well with a larger number of jobs
Hsiu-Jy and Wei-Ming (2004) [10]	Self-adjust in real time in between two order according to the size of the job in the queue	Largest-job-first (LJF)	(i) Preserved while significantly diminishing the chance for the blocking situation to occur (ii) Largest job allocated first (iii) Minimize fragmentation problem	(i) It's impossible to predict how much processor time left for a job (ii) Long-running processor jobs can starve
Shun-LI et al. (2004) [11]	Grid resource management	First-Come First-Served	Simple	(i) The short job has to wait until long jobs have to finish so average wait time become high (ii) Lead to the poor overlap of I/O and CPU
Williams et al. (1999) [12]	Based on resource balancing	Generalized K-resource scheduling algorithms	Performance of average response time improve	Not good for online scheduling

(continued)

Table 1 (continued)

Research	Methodology	Technique used	Advantages	Disadvantages
Prabhat and Prasoon (1999) [13]	Designed various generic scheduling algorithms	Improve earliest deadline first (EDF)	Multimedia operating system performance improved in case of mixed task traffic	(i) The implementation of hardware is difficult (ii) When system is over loaded and processes miss deadlines is largely impulsive in nature
Marcia et al. (1999) [14]	During implementation MPI-2 standard provides primitives to spawn processes	Computed scheduling is based on hypotheses on a divide and conquer that follows MPI-2 program	(i) The balance of the load (ii) Distribution occurs transparently	(i) Process node is not dynamically programmed
Sandeep Singh and Lakshmi (2009) [6]	Grid Computing (GCs) technique	CPU scheduling algorithm	Reducing the computational time	(i) Resource sharing is complex (ii) Slow communication resources
Priya and S. Sahana (2017) [4]	Multiprocessor scheduling technique inspired by evolutionary techniques	Hybrid system based on ACO and GA with HEFT	Program's execution time is minimized	Only work with the homogeneous parallel multiprocessor system
Malhar et al. (2016) [5]	Allows the switching of processes between queues depending on their best time	Multilevel feedback Queue scheduling	Prevent the problem of starvation of huge burst time processes	Process around queue produce more CPU overhead

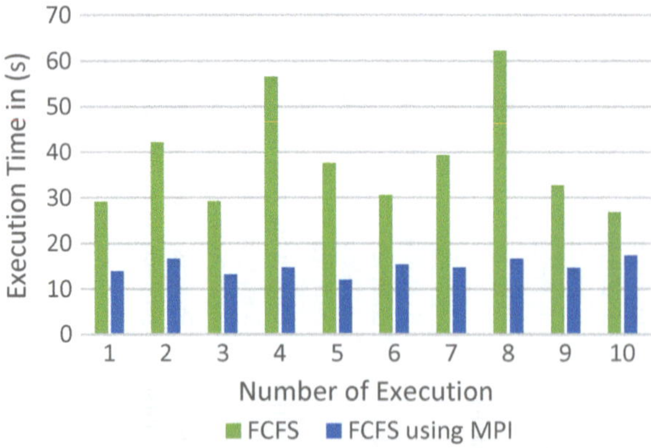

Fig. 3 Comparison of execution time for FCFS and FCFS using MPI

Table 2 Comparison of average waiting time

S. No	Processor burst time [P1, P2, P3, P4, P5]	FCFS	SJF	GA	FCFS using MPI
1	24, 20, 19, 15, 12	42	30.4	31.6	41
2	19, 30, 18, 16, 17	44	33	33.4	43
3	18, 24, 30, 12, 16	43	30.8	30.8	43
4	20, 15, 18, 12, 28	34	29.8	29.8	34
5	25, 23, 13, 13, 16	70	44	44	41
6	22, 27, 13, 18, 17	62	50.6	50.6	42
7	21, 27, 14, 18, 13	55	30.6	30.6	42
8	28, 23, 18, 15, 12	45	36	36.2	46
9	24, 28, 13, 17, 19	47	34.8	34.8	44
10	26, 23, 14, 19, 17	46	35.8	35.8	44

5 Conclusion

The novelty of this paper is the implementation of processor scheduling algorithm FCFS using MPI on high-performance computing system (HPC) to achieve high processor utilization, high throughput at a low response time. From the result section, it is evident that the proposed algorithm shows less execution time than FCFS which is an existing algorithm. Average waiting time of FCFS using MPI exhibits better results while matching with another existing algorithm. The implementation or other processor scheduling using MPI can be possible for large problems as per the proposed algorithm and also result can be analyzed to find out the scope of the processor scheduling for real-time scheduling area.

References

1. Toktam Ghafarian, Hossein Deldari, Mohammad-R "Multiprocessor Scheduling with Evolving Cellular Automata Based on Ant Colony Optimization" *IEEE Proceedings of the 14th International CSI Computer Conference (CSICC'09),* pp. 432–436, (2009).
2. T. Shih, M. C. Ruey, HYueh-Min, W. Chung-Lun "Multiprocessor system scheduling with precedence and resource constraints using an enhanced ant colony system" *Elsevier Ltd,* pp. 2071–2018, (2007).
3. L. Shih-Tang, C. Ruey-Maw, H. Yueh-Min, W. Chung-Lun "Multiprocessor system scheduling with precedence and resource constraints using an enhanced ant colony system" *Elsevier Ltd,* pp. 2071–2081, (2007).
4. P. Annu and S.K. Sudip "A Survey on Multiprocessor Scheduling using Evolutionary Technique" *3rd international conference (NCCS-2017) Springer,* 978-1-4673-9338-6, (2017).
5. T. Malhar, S. Rajiv, S. Priyam, C. Sheetal and R. Pooja "Efficient Implementation of Multilevel Feedback Queue Scheduling" *IEEE WiSPNETconference,* pp. 1950–1954, (2016).
6. R. Sandeep Singh and R. Lakshmi "Experiments with CPU Scheduling Algorithm on a Computational Grid" *IEEE International Advance Computing Conference IACC,* pp. 71–75, (2009).
7. Savas_ Balin "Non-identical parallel machine scheduling using genetic algorithm" *Elsevier Ltd,* pp. 6814–6821, (2010).
8. Z. Xingwu Zheng "Exploring plan-based scheduling for large-scale computing systems" *IEEE,* pp. 259–268, (2016).
9. Y. S. Byung and D. R. Chita "Good Processor Management = Fast Allocation + Efficient Scheduling" *IEEE,* pp. 280–287, (1997).
10. H. Hsiu-Jy and L. Wei-Ming "A Performance-Optimizing Scheduling Technique for Mesh-Connected Multicomputer Based on Real-Time Job Size Distribution" *IEEE Proceedings of the Tenth International Conference on Parallel and Distributed Systems (ICPADS'04),* 1521–9097, (2004).
11. D. Shun-LI, Y. Jing-BO, J. Jiu-Bin "An algorithm for Agent-based task scheduling in Grid Environments" *IEEE Proceedings of the Third International Conference on Machine Learning and Cybemetics, Shanghai,* 26–29 August, pp. 2809-2814, (2004).
12. L. William, K. George, and K. Vipin "Job Scheduling in the presence of Multiple Resource Requirements" *IEEE Proceedings of the ACM/IEEE SC99 Conference,* 1-58113-091-0, (1999).
13. S. k. Prabhat and G. Prasoon "Job Scheduling in the presence of Multiple Resource Requirements" *IEEE Proceedings of the ACM/IEEE Conference (SC'99),* pp. 513–517, (1999).
14. C. C. Marcia, P.P. Guilheme, P. L. Maurcio, M. B. Nicolas and N. O. A. Philippe. "Scheduling Dynamically Spawned Processes in MPI-2".
15. A. Hadi lotfii, A. Broumandnia, A. Shahriar "Task Graph Scheduling in Multiprocessor Systems Using a Coarse Grained Genetic Algorithm" *IEEE 2nd International Conference on Computer Technology and Development ICCTD*, pp. 259–268, (2010).

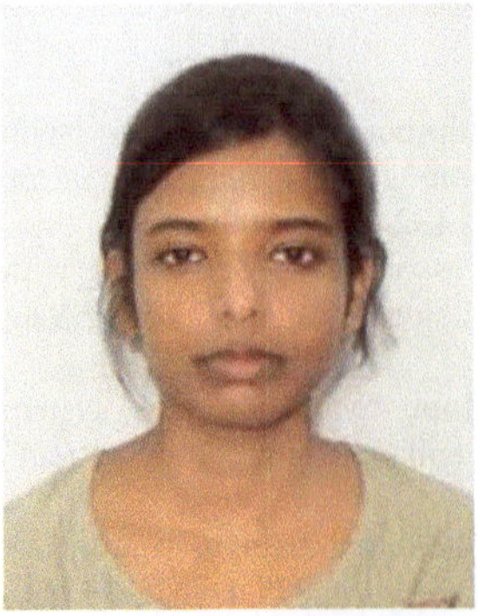

Annu Priya is a Ph.D. Student in Computer Science Engineering Department, Birla Institute of Technology, Jharkhand (Ranchi), India. She received her M.E. from BIT Mesra, India in 2016. Her research area of interest is Soft & Evolutionary Computing, and Image Processing.

Sudip Kumar Sahana is an Assistant Professor in Computer Science Engineering Department, Birla Institute of Technology, Jharkhand (Ranchi), India. He has completed his Ph.D. from Birla Institute of Technology, Mesra. His research area of interest is Artificial Intelligence, Soft & Evolutionary Computing, System Programming, and Grid Computing.

A Computer Vision-Based Gesture Recognition Using Hidden Markov Model

Keshav Sinha, Rashmi Kumari, Annu Priya and Partha Paul

1 Introduction

Human–computer interaction (HCI) plays an important role in increasing the performance of the tasks without using any external devices. In daily life, we solve many real-world problem by using the facial expression or by using hand signs which will express your feelings. To interact with any eletronic device by using human gesture, there are two types: (i) online gestures and (ii) offline gestures. Every day, we say that the gesture is a spatial-temporal pattern which was created in the form of static, dynamic, or both [1] when the user interacts with any objects. The gesture recognition will enable to ignore the use of conventional input devices such as mouse, keyboards. The gesture recognition provides the several features such as (i) accuracy, (ii) stability, and (iii) time saving. Gesture recognition has the variety of applications in different fields such as: (i) automobile sector, (ii) consumer electronics sector, (iii) gaming sector, (iv) smart phone production companies, and (v) transportation sector.

1.1 Gesture Taxonomy

There are several classifications for gesture recognition suggested by different researchers based on the psychological aspects and hand signs. Kendon [2] proposed the autonomous gesture technique which deals with the gesture which is independent of the speech. McNeill and Levy [3] have proposed the three different gesture recognition systems (i) iconic, (ii) metaphor, and (iii) beats. Quek [1 4]

K. Sinha (✉) · R. Kumari · A. Priya · P. Paul
Department of Computer Science Engineering,
Birla Institute of Technology, Mesra, Ranchi, India
e-mail: keshav.sinha@yandex.com

J. Chattopadhyay et al. (eds.), *Innovations in Soft Computing and Information Technology*, https://doi.org/10.1007/978-981-13-3185-5_6

proposed a taxonomy for the gesture recognition, whereas for this paper, a slightly different versions of the taxonomy are presented for HCI. Figure 1 shows that the movement of the arm or hand is classified into two types (i) gesture and (ii) unintentional movement. The unintentional movement is those movements which will not give any type of meaning, and it is often seen when someone is surprised or scared. The second one is the gesture which will give some meaning or signs and provide some information. The gesture is divided into two parts (i) communicative and (ii) manipulative. The movement of hand/arms is created such as rotation, moving the object is known as the manipulative gesture. The communicative gesture is used to express their feelings for communication purposes. For communication people, generally, they use the acts or symbols. In acts, people have given some patterns or expressions to perform like in the movies, whereas symbols are usually in the form of the model and allusion which will use to express their emotions. In the context of HCI, these gestures are the most common way to communicate with the computer. The taxonomy of the gesture largely influenced the recognition spaces.

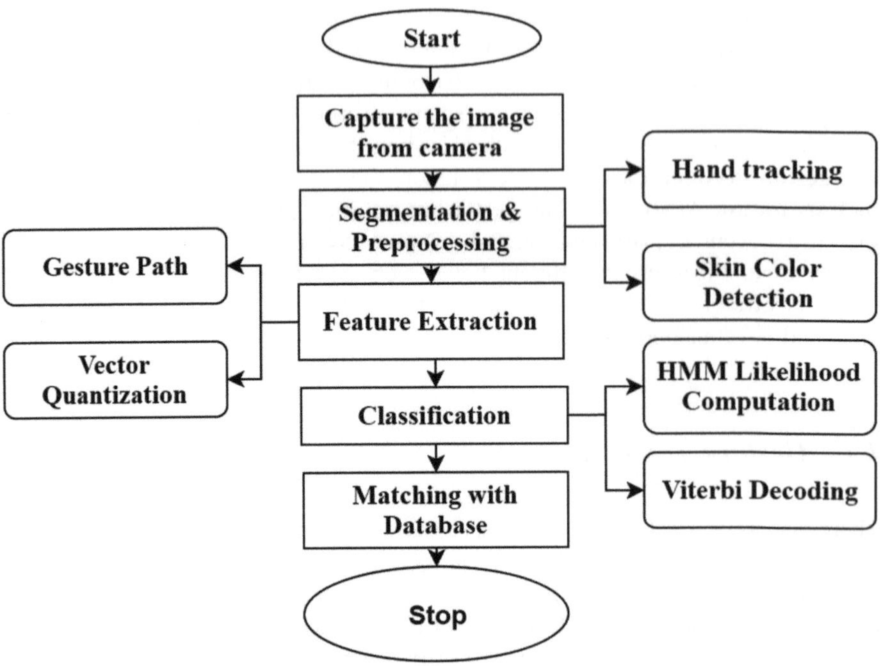

Fig. 1 Proposed framework for gesture recognition

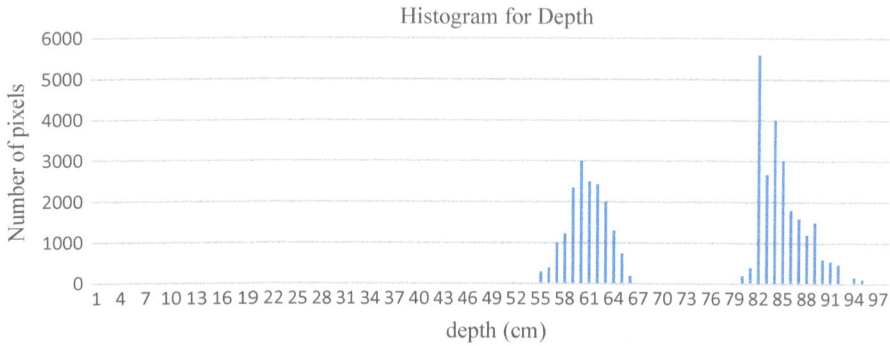

Fig. 2 Histogram for comparison of depth for face and gesture in binary image

1.2 Gesture Recognition Models

Based on the type of input data, the gesture recognition [5, 6] is done in different ways as shown in Fig. 2. The 3D coordinate system is used to track the motion of the gesture, accuracy, and quality of the input algorithm. Quek [1] proposed the several systems which will able to capture the whole space of the gesture. Many researchers have subdivided the gesture recognition into two types: (i) 3D model-based and (ii) appearance-based which are shown in Fig. 1. 3D model-based approach is usually used for the extraction of information about palm motion and joint angles, whereas the appearance-based approach is generally used for image and video for the direct interpretation.

1.2.1 3D Model-Based Algorithms

The 3D model approach is used for volumetric- and skeletal-based models for the gesture recognition. In volumetric approaches, it is most commonly used in the computer animation, mechanical industries where the model is developed in the form of 3D surfaces such as polygon, meshes, or even octahedral. The main drawback of this approach is it does not work for the real-time analysis because the system is not developed yet. In skeletal approach, the virtual skeleton of the human being is computed and the body parts are mapped in certain segments. For the analysis of the segments, it calculates the position and the orientation of the arms/hand. Using the skeletal approach, it will have several advantages such a—(i) the analysis will take place very quickly, (ii) the matching will be done with a database which already stores the image models, and (iii) it will directly focus on the key parts of the body.

1.2.2 Appearance-Based Models

These models directly take the parameters from the image and the video, so there is no need for taking anybody's representation. For processing the image in deformable 2D templates, it sets the point on the object so that it would use interpolation nodes for the outline approximation. Usually, it uses the linear function for the interpolation where it takes the average from the point sets, variability parameters, and external deformation. This approach is mostly used in hand tracking and classification of the gesture. Another approach is gesture detection where the sequence of the image is used as gesture template, and the parameters are taken from the image itself. Most of the time, it gets only monoscopic or stereoscopic view to use.

The rest of the paper is organized as follows: Sect. 1 deals with introduction followed by Sect. 2, having the literature survey of the existing methods. Sections 3 and 4 contain the research methodology and result and discussion about the proposed algorithm. Finally, Sect. 5 concludes the outline and the future scope of this work.

2 Literature Review

Gesture recognition is very demanding field in the era of Artificial Intelligence. The interaction of human gesture with the computer will reduce the workload and save lots of time for any personnel. Various researchers have proposed different techniques for gesture recognition such as Rehm et al. [7] introduced the Bayesian network model which was created using the GeNIe modeling environment. The working of this technique is to analyze the behaviors of the person and automatically understand the cultural background of that person. In [8], the author introduced the ergodic-based model using HMM where it categorizes the gesture from the background using the threshold of the image. The disadvantage of this method is it does not work on intricate background. Toshiaki et al. [9] develop the real-time system for recognition of the alphabet (A–Z) and the Arabic number (0–9) for LRB topology. Author uses the Gaussian mixture for the color detection, Baum–Welch (BW) for the training purposes, and the Viterbi path for the testing or matching. In [10], the author proposed the recognition of the figure posture using machine learning techniques. The figure posture is trained and tested with the linear artificial neural network. There are several researches which have been done with the gesture recognition by using various methods. Table 1 represents the review on the various methods on gesture recognition.

Table 1 Review of various methods for gesture recognition

Techniques	Methodology	Methods used	Advantages	Disadvantages
Elmezain et al. (2008) [11]	HMM	Forward algorithm for hand trajectory recognition	Get high accuracy rate	Still working on real-time recognition
Chen (2009) [12]	LLE	(i) Find lower-dimensional manifold (ii) In high-dimensional space	Better performance	Time taking
Pavlovic et al. (1997) [13]	HMM property for first orders	(i) Appearance-based computation (ii) Vision-based gesture recognition	Efficient recognition	Downgrade the effectiveness
Chen et al. (2003) [14]	Fourier Descriptor (FD), HMM	(i) Extraction for feature vector (ii) Characterize the temporal feature	Lowest error rate	Small search region
Kim et al. (2016) [15]	CNN	(i) Construct the classification model (ii) Classify the wave form for ASL	Influence from the environment	Its work only for ASL
Yu et al. (2016) [16]	HRI and ELm	(i) Recognition is done according to data glove and vision (ii) YoBu collects the data for the recognition	Better performance	Work only for the robotics
Pomboza-Junez et al. (2016) [17]	SVM	(i) Provides sEMG sensors around the forearm (ii) Connects the sensor with the Bluetooth (iii) Sends the real-time data	High effectiveness	Device controls the IR protocols which having the low distance

3 Research Methodology

After thorough study of different research papers, the framework for gesture recognition is proposed which is shown in Fig. 3.

The system is developed for recognizing the Arabic number, alphabet, and Roman number, with the help of tracking of the trajectory of the hand and finger sign in color image sequence. The whole process is divided into three main parts (i) segmentation, (ii) feature extraction, and (iii) classification. The step-by-step procedure is given below for the said problem.

Step 1: Segmentation and Preprocessing

Using the stereo color camera, the image of the hand is taken, whereas hand segmentation takes place using 3D depth map for the detection of hand area. For the colored hand skin, the chrominance is used for the color space YCbCr, whereas 'Y' channel represents the brightness of the image, and the (Cb, Cr) channels refer to chrominance. The 'Y' channel is only related to the brightness, and all the information is stored in the chrominance channel. For training of image, the Gaussian model is used on 30 different people and 64 different images. In recognition of skin color, the mathematical model is presented which will detect and evaluate the human body and the surrounding environment. For that, we transform the image color RGB to the YCbCr space. 'Y' takes the information of brightness, and Cb and Cr take the information about environment light. The mathematical model is shown as Eq. (1).

$$\begin{bmatrix} Y \\ \mathrm{Cb} \\ \mathrm{Cr} \end{bmatrix} = \begin{bmatrix} 16 \\ 128 \\ 128 \end{bmatrix} + \begin{bmatrix} 64.481 & 128.554 & 24.965 \\ -37.794 & -74.204 & 112.0 \\ 111.9 & -97.788 & -18.217 \end{bmatrix} \begin{bmatrix} R \\ G \\ B \end{bmatrix} \tag{1}$$

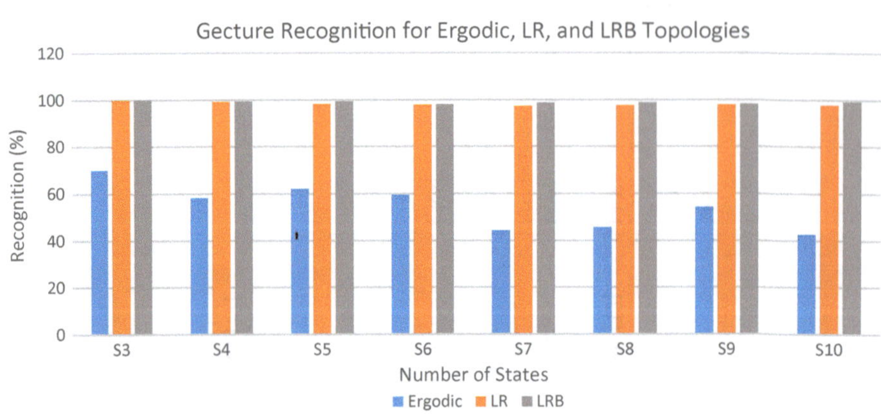

Fig. 3 Using HMM results for gesture recognition for LRB topology from the range of 3–10

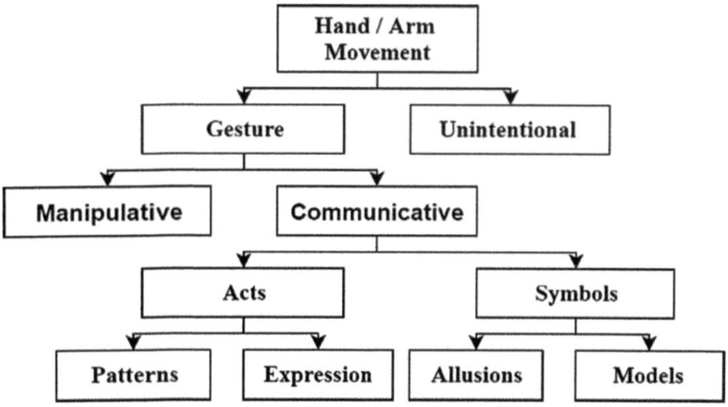

Fig. 4 Taxonomy for hand gesture recognition for HCI

So for the different color image, the ranges of Cr and Cb values are set as shown in (2).

$$\mathrm{Cr}\epsilon[132, 174] \cap \mathrm{Cb}\epsilon[76, 127] \tag{2}$$

Different range of Cr and Cb values cause mis-identification in color image. So for that, Gaussian model is used and it separates the binary image from the environment and its labeled black. For separation of binary image from the environment, the in-depth threshold is calculated. The pixel threshold of the face, body, and gesture is different in-depth value because they are not coplanar.

As shown in Fig. 4, we get two depth peaks; the pixel ranges from 54 to 67 cm which is generated by the hand and the 80–96 is generated by the face. The variation in the pixel range is due to the camera placing the face is much farther then hand. So the intensity of light reflection is higher than hand gesture. So using this, we can separate the face from the hand and we set the adaptive threshold which will capture only the gesture and rest will be discarded.

Step 2: Feature Extraction

The next is feature extraction where we recognize the path, location, and orientation of the hand. The feature of any extraction relies on the result of the accuracy. So to maintain the accuracy, we first find the spatiotemporal pattern which consists of centroid points. To the hand gesture, $(x_{\text{hand},}\ y_{\text{hand}})$ point is used in Eq. (3), which is given in the form of

$$\theta_t = \tan\left(\frac{y_{t+1} - y_t}{x_{t+1} - x_t}\right); \quad [t = 1, 2, \ldots, T - 1] \tag{3}$$

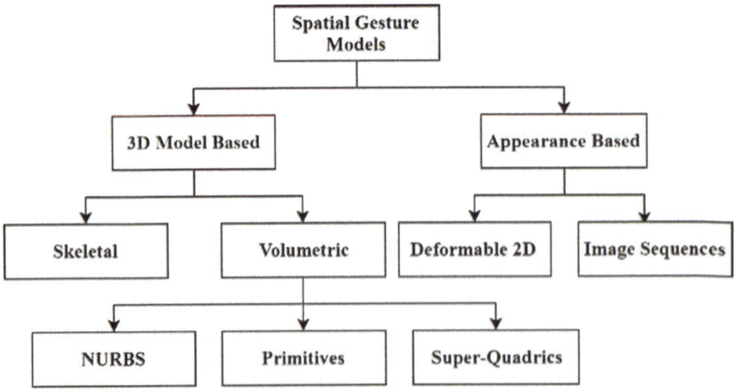

Fig. 5 Gesture recognition models

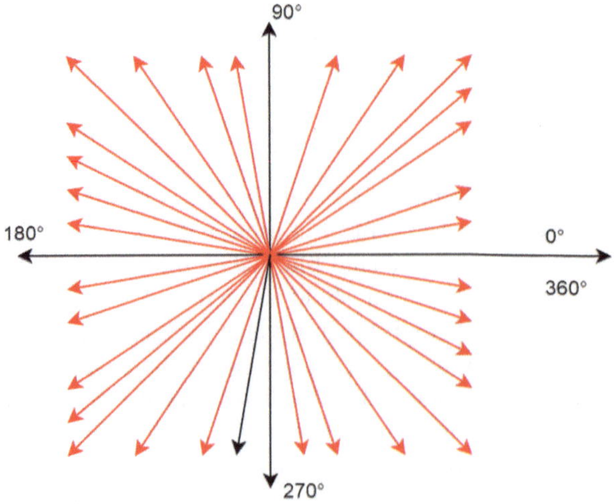

Fig. 6 Code word for the hand movement

where

T = Length of the gesture.
θ_t = It is quantized by dividing it by 10° for the code word from 1 to 36.

The discrete vector of the code word is used for the input of HMM. In Figs. 5 and 6, shows the orientation of the code word and who the small change in the angle is calculated mathematically.

Step 3: Hand Gesture Classification

For path classification, the hidden Markov model (HMM) is used which include the three parameters:

$$\lambda = \left(\prod, A, B\right)$$

where

$\prod$ = Initial vector.
A = Transition matrix.
B = Emission matrix.

The classification is done in two phases: (i) training and (ii) decoding.

In training phase, we applied the likelihood computation for the HMM forward algorithm. The computation likelihood is given as:

λ = (*A*, *B*) and the observation sequence is *O* then the likelihood *P* (*O*/λ).

Pseudocode for HMM Forward Algorithm

function Forward (observation T, graph_length N)
Probability matrix for HMM forward is [N + 2, T]
//Initialization steps
for 1 to N,
do for each state q,
forword [q, 1] $\leftarrow$ $a_{0,q}$ * b_q (o_1)
for each time step t from 2 to T
do
for 1 to N do, for each q state
forward [q, t] $\leftarrow \sum_{q'=1}^{N} forward\ [s', t-1] * a_{q',q} * b_q(o_t)$
// Termination step
Forward [s_F, T] $\leftarrow \sum_{q=1}^{N} forward[q, T] * a_{q,sF}$
return forward [s_F, T]
Return forward probability
Whereas the forward [q, t] represented as α_t (q).
Single element = α_t (q)
Previous value = α_{t-1}
Weighted transition probability = a
Observation probability = b_i (o_{t+1})
Transition probability = 0
While using this we compute the likelihood of the HMM forward algorithm. The next part is to decoding to find the sequence of the variables.
λ = (A, B) and the observation sequence is O then the likelihood P (O/ λ).

Pseudocode for Viterbi Algorithm

function Viterbi (observation T, graph_length N)
Return Best Path
Probability matrix for Viterbi is [N + 2, T]
//Initialization steps
for 1 to N,

do for each state q,
Viterbi [q, 1] ← $a_{0,q}$ * b_q (o_1)
backpointer [q, 1] ← 0
for each time step t from 2 to T
do
for 1 to N, for each q state

$$\text{Viterbi}[q,t] \leftarrow \max_{s'=1}^{N} \text{Viterbi}[s', t-1] * a_{q',q} * b_q(o_t)$$

$$\text{Backpointer}[q,t] = \underset{s=1}{\overset{N}{argmax}} \text{Viterbi}[s', t-1] * a_{q',q}$$

// Termination step

$$\text{Viterbi}[s_F, T] \leftarrow \max_{s=1}^{N} Viterbi[q, T] * a_{q,sF}$$

$$\text{Backpointer}[s_F, T] = \underset{s=1}{\overset{N}{argmax}} \text{Viterbi}[q, T] * a_{q,sF}$$

return the backtrack path by backpointer [s_F, T]

Step 4: For matching, there are several templates created for the comparison, and it is presented in the result and discussion section.
Step 5: Stop the simulation.

4 Result and Discussion

Simulation work has been performed using MATLAB toolbox version R2015a and C++ on Windows 8.1 having the hardware specification is Intel Core i5 CPU with frequency 1.70 GHz of clock with 4 GB RAM on ×64-bit operating system. The proposed algorithm is LRB and is compared with ergodic and LR topology. For finding the recognition ratio, the formula is given as Eq. (4).

$$\text{Recognition ration} = \frac{\text{Recognized Gesture}}{\text{Test Gesture}} \times 100 \tag{4}$$

The experimental results, for the numeric value of (0–9) the sample is created based on 40 video sequences, whereas 30 video samples for trained and approximately 15 video samples for use for the test. More than 150 video sequences are present for training, and 75 video sequences are used for the testing of gestures. The combination of HMM forward algorithm with Viterbi path decoding while increasing the recognition rate for real-time frame. The input images were taken by a stereo camera system that has 6 mm focal length, which takes the picture of

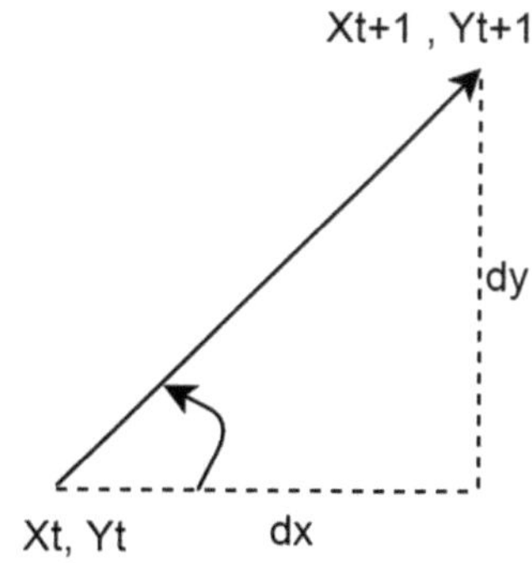

Fig. 7 Orientation and calculate slight change in the angle

240×320 pixels image resolution. Figure 7 shows the LRB topology from 3 to 10 states where the average recognition percentage is 99.05%, which shows the superiority of the algorithm for the said problem.

5 Conclusion

In this paper, we simulated the hand gesture recognition system for alphabet, Arabic, and Roman number using HMM. The recognition is done according to capturing of the hand trajectory which is very appropriate for the real-time application. The modified zero-codeword detection is used for tracking the velocity and continuous movement of the hand. The accuracy of the algorithm is tested on the LRB topology, and it shows 99.05% of average recognition rate. There are several future works remaining in the field of gesture recognition like the development of system which will help the abnormal people to come nearer to the world.

References

1. F.K.H. Quek: "Toward a Vision-Based Hand Gesture Interface," Virtual Reality Software and Technology Conf., pp. 17–31, Aug. 1994.
2. A. Kendon: "Current Issues in the Study of Gesture," The Biological Foundations of Gestures: Motor and Semiotic Aspects, J. -L. Nespoulous, P. Peron, and A. R. Lecours, Eds., pp. 23–47. Lawrence Erlbaum Assoc., 1986.
3. D. McNeill and E. Levy: "Conceptual Representations in Language Activity and Gesture," Speech, Place and Action: Studies in Deixis and Related Topics, J. Jarvella and W. Klein, Eds. Wiley, 1982.
4. F.K.H. Quek: "Eyes in the Interface," Image and Vision Computing, vol. 13, Aug. 1995.
5. L. Nianjun, C. L. Brian, J. K. Peter, and A. D. Richard: "Model Structure Selection & Training Algorithms for an HMM Gesture Recognition System," In International IWFHR, pp. 100–106, 2004.

6. S. L. Phung, A. Bouzerdoum, and D. Chai: "A Novel Skin Color Model in YCbCr Color Space and its Application to Human Face Detection," In IEEE International Conference on Image Processing (ICIP), pp. 289–292, 2002.
7. Matthias Rehm, Nikolaus Bee, Elisabeth André: "Wave Like an Egyptian — Accelerometer Based Gesture Recognition for Culture Specific Interactions," Published by the British Computer Society, 2007.
8. H. Lee and J. Kim: "An HMM-Based Threshold Model Approach for Gesture Recognition," IEEE Trans. TPAMI, Vol. 21 (10), pp. 961–973, 1999.
9. D. B. Nguyen, S. Enokida, and E. Toshiaki: "Real-Time Hand Tracking and Gesture Recognition System," IGVIP05 Conference, CICC, pp. 362–368, 2005.
10. Md Iqbal Quraishi, Krishna Gopal Dhal, J Paul Choudhury, Pulak Ghosh, Pranav Sai, Mallika De, "A Novel Human Hand, Finger Gesture Recognition using Machine Learning," 2nd IEEE International Conference on Parallel, Distributed and Grid Computing, 2012.
11. Mahmoud Elmezain, Ayoub Al-Hamadi, Jorg Appenrodt, Bernd Michaelis: "A Hidden Markov Model-Based Continuous Gesture Recognition System for Hand Motion Trajectory," IEEE, 2008.
12. Tianding Chen: "Novel Machine Learning in Hand Gesture Recognition Using Multiple View," IITA International Conference on Control, Automation and Systems Engineering, 2009.
13. Vladimir I. Pavlovic, Rajeev Sharma, Thomas S. Huang: "Visual Interpretation of Hand Gestures for Human-Computer Interaction: A Review," IEEE Transactions on Pattern Analysis and Machine Intelligence, Vol. 19, No. 7, July 1997.
14. Feng-Sheng Chen, Chih-Ming Fu, Chung-Lin Huang: "Hand gesture recognition using a real-time tracking method and hidden Markov models," Image and Vision Computing, Vol-21, pp- 745–758, 2003.
15. SEO Yul Kim, Hong Gul Han, Jin Woo Kim, Sanghoon Lee: "A Hand Gesture Recognition Sensor Using Reflected Impulses," IEEE Sensors Journal, 2016.
16. Danling Lu, Yuanlong Yu, and Huaping Liu: "Gesture Recognition Using Data Glove: An Extreme Learning Machine, Method," Proceedings of the 2016 IEEE, International Conference on Robotics and Biomimetics Qingdao, China, December 3–7, 2016.
17. Gonzalo Pomboza-Junez, Juan Holgado Terriza: "Hand Gesture Recognition based on sEMG signals using Support Vector Machines," International Conference on Consumer Electronics-Berlin, 2016.

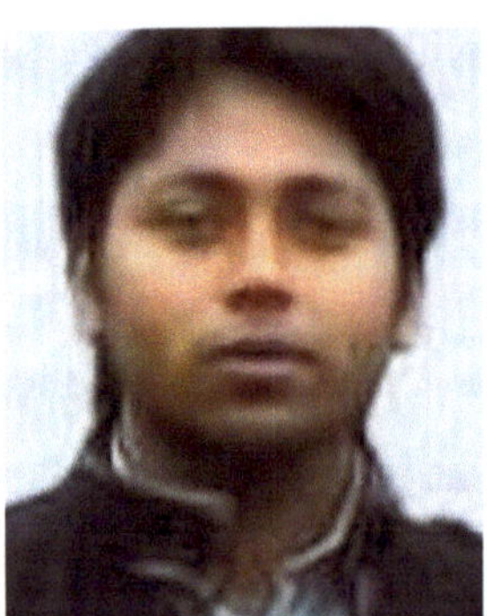

Keshav Sinha is a Ph.D. student in Computer Science Engineering Department, Birla Institute of Technology, Jharkhand (Ranchi), India. He received his M.E. From BIT Mesra, India in 2016 and B. E from Sri Chandrasekharendra Saraswathi Viswa Mahavidyalaya, Kanchipuram, India, in 2013. He also work as Project Assistant in CSIR-CIMFR in coal quality division. His research area of interest is Soft & Evolutionary Computing, IoT based Computer Vision, and Cryptography and Network Security which provides the flexibility in the computer science society.

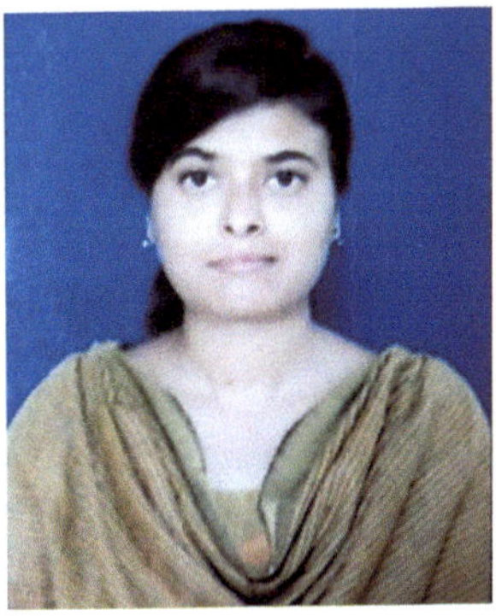

Rashmi Kumari is a PhD. student in Computer Science Engineering Department, Birla Institute of Technology, Jharkhand (Ranchi), India. She received her M.E. From BIT Mesra, India in 2016. Her research area of interest is machine learning, deep learning, and data mining.

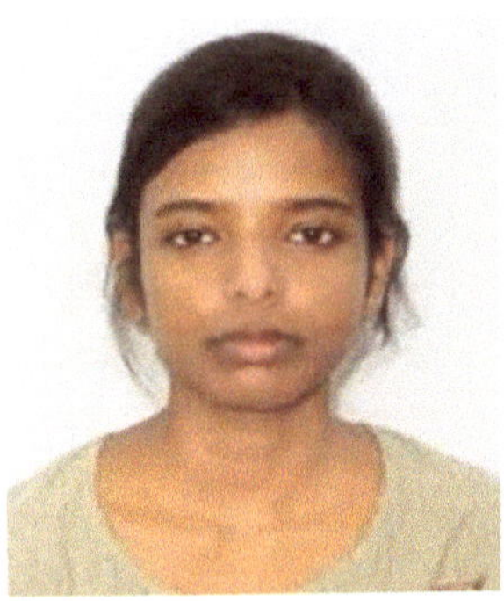

Annu Priya is a PhD. student in Computer Science Engineering Department, Birla Institute of Technology, Jharkhand (Ranchi), India. She received her M.E. From BIT Mesra, India in 2016. Her research area of interest is Soft & Evolutionary Computing, and Image processing.

Dr. Partha Paul received the B.E (CS) and M.E (CS) degree from Moscow State University, Russia in 1998 & 1999 respectively. He did his Ph.D. degree from Birla Institute of Technology, Mesra, and Ranchi, India in 2014. He is currently Assistant Professor in the Department of Computer Science & Engineering, Birla Institute of Technology, Mesra, Ranchi, India.

Schnorr Digital Signature to Improve Security Using Quantum Cryptography

Prity Kumari, Upendra Kumar and Shyam Krishna Singh

1 Introduction

For a secret communication, quantum cryptography provides amazing security assurance. Quantum physics breaks the inviolability of a law of classical physics. It has the potential to fundamentally change the face of cryptography. Quantum computers can break many existing classical cryptosystems [1], in particular most public key systems used today. Integer factorization problem and the discrete logarithm problem are two algorithms which are widely used in classical cryptography. Through Shor's quantum algorithms [2], these two mathematical problems are solved in polynomial time instead of exponential time.

The other significant algorithm by Lov Grover for quantum search invented in the year 1996 is Grover's algorithm [3]. Classically, for a single element search within n elements, one needs to look up $n/2$ elements on average before finding the nonzero output that requires $O(n\log_2 n)$. Grover's algorithm does the job by looking at an average of $n^{1/2}$ times. It is useful for search in unstructured database.

Classical cryptography is based on a combination of mathematics and logic. It is based on the problem of computational complexity which is not sufficient to fast-growing technique to compute the secret key. Depending only on the difficulty of computational complexity to create secret key is not enough secure communication because finding a fast calculation algorithm to calculate the secret key will compromise the key security of the communication systems.

P. Kumari (✉)
Magadh University, Bodh Gaya, India
e-mail: joiti.prity2@gmail.com

U. Kumar
Birla Institute of Technology, Patna Campus, Patna, India

S. K. Singh
A. N. College, Patna, India

J. Chattopadhyay et al. (eds.), *Innovations in Soft Computing and Information Technology*, https://doi.org/10.1007/978-981-13-3185-5_7

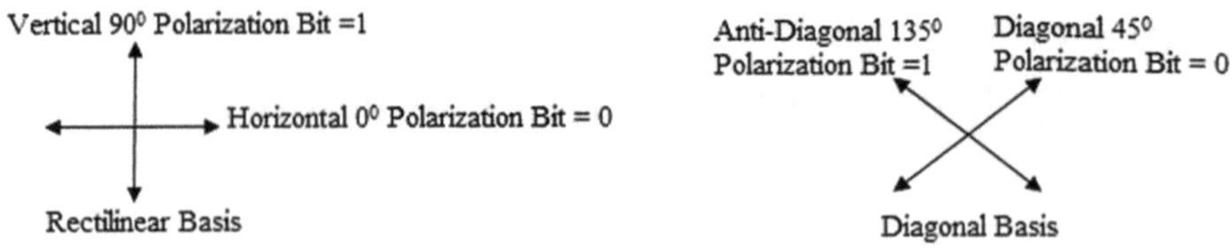

Fig. 1 Rectilinear and diagonal basis sets

Quantum computer uses a different technique for a different operation. It uses a quantum bit instead of bit. Quantum bit, i.e., qubit, poses quaternary nature. Quantum mechanics is fully different from the classical mechanics. Classical bit contains only in the states corresponding to value 0 and 1, but qubit contains the superposition of the two given eigenstates. So, choose two eigenstates $< 0|$, $< 1|$ $\varepsilon\ H$ from the Hilbert space. The possible states of the qubit are expressed as $\varphi = a{<}0| + b{<}1|$, where $a^2 + b^2 = 1$. This means that qubit is in state 0 with probability a^2 and in 1 with probability b^2 [4] and qubits are single polarized photon that exist into one of four polarization states |, −, /, and \ selected from one of the two conjugate basis sets. The rectilinear basis is of 0°/90° polarization photon (|,−), and the diagonal basis is composed of 45°/135° polarizations photon (/, \) as shown in Fig. 1.

Quantum cryptography uses the laws of quantum mechanics for unconditionally secure communication. Quantum mechanics also provides the ability to detect the presence of eavesdropper who is attempting to read the key as the quantum state on the transmitted data will collapse to a single state and therefore get disturbed. Quantum key distribution (QKD) provides a very secure way to distribute or exchange secret keys on quantum channel.

If a sender uses a filter in the 0° and 90° basis to given initial polarized photon which is either horizontal or vertical and a receiver uses the same polarized photon filter, then he/she can determine the information but if receiver uses a filter in the 45° and 135° basis to measure the polarized photon, he/she cannot determine the right information [5].

If an intruder uses a polarized photon same as sender's polarized photon, he/she can measure the original bits, and if he/she uses a misaligned polarized photon not as sender's polarized photon, then he/she will receive only wrong information, but also influenced the original photon so that he/she will be not able to measure the original information. In this condition, receiver either will receive wrong information or also will be able to detect the intruder's presence.

Quantum exchange over quantum channel															
Sender's randomly generated candidate bits	0	1	1	0	1	1	0	0	1	0	1	1	0	0	1
Sender's random basis	×	+	×	+	+	+	+	+	×	×	+	×	×	×	+
Send polarized photon by sender	/	|	\	—	|	|	—	—	\	/	|	\	/	/	|
Receiver's random basis	+	×	×	+	+	×	×	+	×	+	×	×	×	×	+
Receiver's measured received bits	0	0	1	0	1	0	1	0	1	1	1	1	0	0	1
Receiver's basis compared with sender's basis?	N	N	Y	Y	Y	N	N	Y	Y	N	N	Y	Y	Y	Y
Public discussion over classical channel															
Receiver reports bases of received bits	+	×	×	+	+	×	×	+	×	+	×	×	×	×	+
Sender checks which bases were correct	N	N	Y	Y	Y	N	N	Y	Y	N	N	Y	Y	Y	Y
Sifted key			1	0	1			0	1			1	0	0	1
Receiver reveals some key bits at random					1									0	
Sender confirms them					Y									Y	
Outcome															
Remaining shared secret bits			1	0				0	1			1	0		1

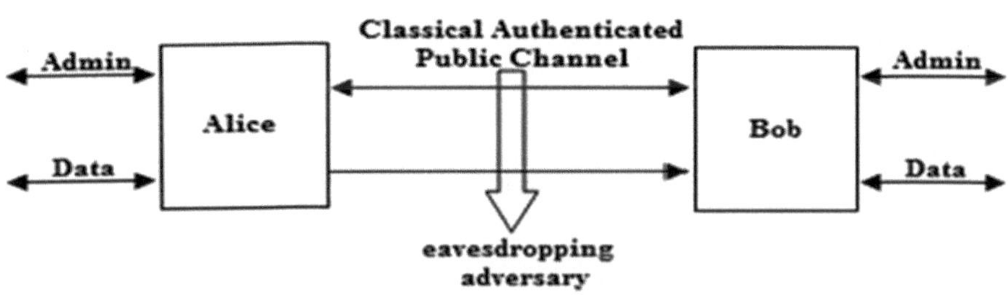

Fig. 2 A QKD block diagram

Sender and receiver are connected by a classical public channel and a quantum channel as shown in Fig. 2. The classical channel is a classically authenticated public communications channel and is used to conduct the key exchange process for error detection. The quantum channel is used to exchange the quantum key.

2 State of the Art

Quantum cryptography was introduced by Stephen Weisner in early 1970, this idea was issued in 1983 in SIGACT News, and at the same time, Bennet and Brassard agreed with him. In 1984, they delivered their own ideas. They issued the first quantum cryptography protocol called the "BB84". This protocol provides more security and depends on the quantum mechanics. It is a function over a 32-cm distance. After that, the technique is extended to kilometers from centimeters. In June 2004 communication is secured with quantum cryptography is up and running in Cambridge, Massachusetts. The leader of the quantum engineering team at BBN Technologies in Cambridge, Chip Elliott, transmitted the first packets of data across the quantum network. In June 2003, a team at the University of Vienna transferred entangled photons across the River Danube, through free space. The first money transfer occurred between two Austrian banks in April 2004 was encrypted by quantum keys. The two buildings were 500 meters away from each other, yet fiber optics were fed through 1.5 km of sewage system to link them together.

3 Proposed Technique

In this technique, use email ID or name and the encrypt/decrypt algorithm that is deployed to encrypt/decrypt the message, setup algorithm to generate the public parameter and master secret key, a hash function to ensure data integrity, a Schnorr digital signature for authentication, and QKD to exchange key between sender, receiver, and server.

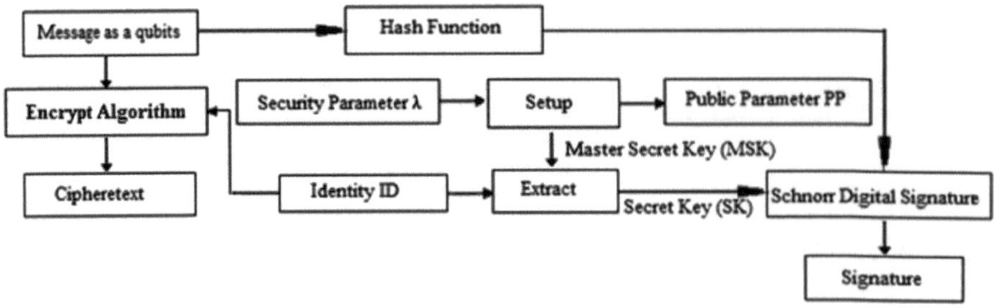

(1) Setup Algorithm—This algorithm is run by the private key generator (PKG) one time for creating the whole IBE environment. The master key is kept secret and used to derive users' private keys, while the system parameters are made public. It accepts a security parameter (i.e., binary length of key material) and outputs:

- A set P of system parameters, including the message space M and cipher C
- A master key K_m.

(2) Extract Algorithm—This algorithm is really the main difference between the identity-based encryption (IBE) and public key encryption (PKE), so this algorithm is run by the PKG when a user requests his private key. The verification of the authenticity of the requestor and the secure transport of d are problems with which IBE protocols do not try to deal. It takes as input P, K_m, and an identifier ID ε {0,1}* and returns the private key d for user ID.

(3) Encrypt Algorithm—The qubit message is encrypted with receiver identity ID ε {0,1}*, message m ε M and P, and given ciphertext c ε C. Here, ID works as a public key.

$$\text{Encrypt}(M, P.\text{ID}) = C$$

(4) Quantum Hash Function—The proposed algorithm consists of the following steps:

Initialization step—First of all, choose the parameters (n, (α, β, χ, δ)) and the *message* M, and after that generate a random sequence number S by using function pseudocode random number generator (PRNG) given below

$$S = \text{PRNG}(n, (\alpha, \beta, \chi, \delta), M).$$

(a) Each value of the sequence number $\boldsymbol{S}$ is a floating-point number that lies between 0 and 1.

Encryption procedure

(a) Original message M is divided into four parts and the same size. To determine the encryption order of four blocks M_1, M_2, M_3, and M_4 according to the size of the first four values of the sequence number S,

$$M_1 \rightarrow M_2 \rightarrow M_3 \rightarrow M_4$$

(b) The two-round one encryption process is

EM_1 = encrypt(M_1, M_4, S) → EM_2 = encrypt(M_2, EM_1, S)
→ EM_3 = encrypt(M_3, EM_2, S) → EM_4 = encrypt(M_4, EM_3, S)
→ EM'_1 = encrypt(EM_1, EM_4, S) → EM'_2 = encrypt(EM_2, EM'_1, S)
→ EM'_3 = encrypt(EM_3, EM_2, S) → EM'_4 = encrypt(EM_4, EM'_3, S).

Decryption procedure—In this process, determine the decryption order to use the secret key.

$EM_4 = EM'_3 \otimes EM'_4 \rightarrow EM_3 = EM'_2 \otimes EM'_3 \rightarrow EM_2 = EM'_1 \otimes EM'_2 \rightarrow EM_1 = EM_4 \otimes EM'_1 \rightarrow M_4 = EM_3 \otimes EM_4 \rightarrow M_3 = EM_2 \otimes EM_3 \rightarrow M_2 = EM_1 \otimes EM_2 \rightarrow M_1 = M_4 \otimes EM_1$ end

Here, ⊗ denotes the decryption operation and it is the reverse process of encryption [6].

(5) Schnorr Digital Signature—Its advantage is to increase the speed.

(a) Parameters: p, q, a, d, v, δ, x, y

where

- p is a prime number typically 1024 bits,
- q is a factor of $p - 1$ typically 160 bits,
- a is $a^q = 1 \bmod p$,
- d is $0 < d < q$, and
- v is $a^{-d} \bmod q$.

q is used to verify our signature. In this, (a, p, q) are global parameters for all, d is a secret key, and v is a public key.

(b) Signing:

- Choose random r, $0 < r < q$.
- Compute $x = a^r \bmod p$.
- Concatenate $e = H(M\|x)$ (message with x), and H is a hash function.
- Compute $y = (r + \text{de}) \bmod q$.

(c) Send:

- Send the message M and signature (e, y) so that receiver can verify that.

(d) Verification:

- Received (M, e, y),
- Known publically a, p, q, v,
- Compute $x' = a^y v^e \bmod p$, and this is not computed by receiver.

$x' = a^y v^e = a^y a^{-de} = a^{y-de} = a^r = x \bmod p$, where $y = r + \text{de}$. x is computed by the sender.

(e) Check:

- $e = H(M\|x')$, and this e value is same as e value sent by sender [7].

(6) Decrypt Algorithm
Through decrypt algorithm, decrypt encrypted message using secret key d, public parameter P, and ciphertext $c \in C$.
Decrypt(P,C,d) = M

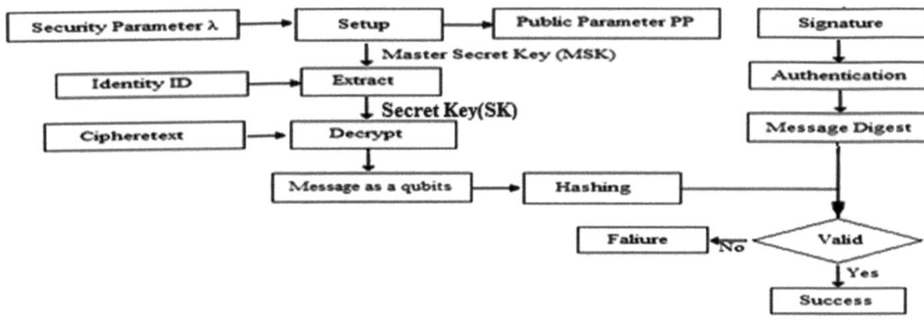

4 User Authentication and Quantum Basis Distribution

In this technique, the client and the server first communicate for the parameters of the secure key. These parameters consist of the master secret and public parameter. Then, the client and the server authenticate each other. The authentication is performed by using ID or name [8]. Public key certificates are installed in both the client and the server [9, 10]. When a client requests the private key from server, then server authenticates him through his ID or name. After that, server sends private key to client according to his ID or name.

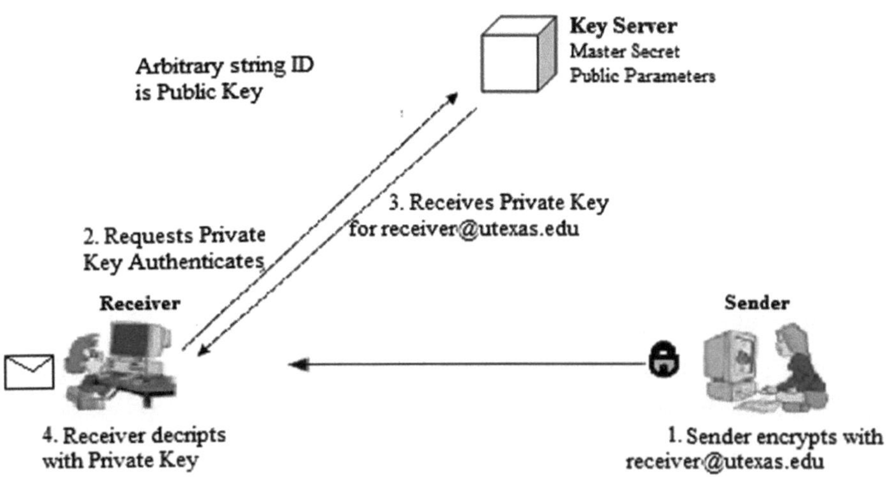

Idea with ID-based encryption is that any arbitrary string should now be our public key. Remember that in regular public key encryption, the key generation algorithm generates the public key and private key and we have no control over a public key we get. But through ID, we can choose to have our name and our email ID as our public key and the way that it would work. We have the probable sender who wants to send the message to receiver so he/she is just going to encrypt the message with his email ID and just it going to sent into him. Receiver has one-time setup cost, so receiver needs to contact a key server. The key server has some master secret, and once receiver authenticates himself to the key server and requests the private key corresponding to his email ID, the key server will return that private key to him, using which he/she can then decrypt message that sender sends to him. The following steps are involved in the algorithm:

1. Sender wants to connect with receiver, so he/she sends request to him.

$$\text{Sender} \rightarrow \text{QKD} : \text{Enc}_{\text{PR-S}}(\text{ID}_\text{S} \| \text{ID}_\text{R})$$

QKD will check the ID or name of sender for user authentication and record the connection request status in file. After that, QKD checks whether receiver's ID status is busy or free. If receiver is free, then QKD moves to step 2; otherwise, he waits.

2. In second step, QKD sends a connection request to receiver.

$$\text{QKD} \rightarrow \text{Receiver} : \text{Enc}_{\text{PU-R}}(\text{ID}_\text{S} \| \text{ID}_\text{R})$$

3. Receiver will send a confirmation message to QKD after he accepts the connection with sender.

$$\text{Receiver} \rightarrow \text{QKD} : \text{Enc}_{\text{PR-R}}(\text{ID}_\text{S} \| \text{ID}_\text{R})$$

Both are authenticated to send and receive message, and QKD decrypts the message and merges connection's status between them.

4. To encode the message, QKD begun to distribute quantum base (+, x) in a sequence to encrypted message using their public keys.

   - $\text{QKD} \rightarrow \text{Sender} : \text{E}_{\text{PU-S}}(\text{ID}_\text{S} \| \text{ID}_\text{R} \| \text{QB})$.
   - $\text{QKD} \rightarrow \text{Receiver} : \text{E}_{\text{PU-R}}(\text{ID}_\text{S} \| \text{ID}_\text{R} \| \text{QB})$.

Transfer Data over Quantum Channel

5. In this step, sender sends an encrypted data to receiver after they receive the quantum base.

$$\text{Sender} \rightarrow \text{Receiver} : E_{\text{PR-S}}(E_{\text{QB}}(M) \| E_{\text{PU-R}}(\text{ID}_\text{S}))$$

6. Then, by using private key of sender, both send a randomly selected part of the message to QKD.

$$\text{Receiver} \rightarrow \text{QKD} : E_{\text{PR-R}}(E_{\text{QB}}(M) \| E_{\text{PU-QKD}}(\text{ID}_{\text{R}}))$$

$$\text{Sender} \rightarrow \text{QKD} : E_{\text{PR-S}}(E_{\text{QB}}(M) \| E_{\text{PU-QKD}}(\text{ID}_{\text{S}}))$$

QKD decrypts the messages and compares between those messages; if there are any dissimilarities in bits, then QKD knows the presence of eavesdropper.

7. When detecting the presence of eavesdropper, QKD informs sender and receiver by sending the notification message.

   $\text{QKD} \rightarrow \text{Receiver} : E_{\text{PU-R}}(E_{\text{QB}}(\text{Notify}))$

   $\text{QKD} \rightarrow \text{Sender} : E_{\text{PU-S}}(E_{\text{QB}}(\text{Notify}))$ [11, 12].

5 Conclusion

This paper has implemented the security quantum algorithm that employs identity, ID, as a public key encryption algorithm to generate keys to improve security over quantum communication channel. The Schnorr digital signature algorithm has been used to enhance authentication and data privacy. A method of producing random like digits is called as pseudo random number generator is defined. The rectilinear and diagonal basis sets are discussed. The security aspect of quantum communication has been improved.

References

1. Mayers, D. (2001). "Unconditional security in quantum cryptography". Journal of the ACM (JACM), 48(3), 351–406. Retrieved from http://arxiv.org/pdf/quant-ph/9802025.
2. P. Shor. "Polynomial-time algorithms for prime factorization and discrete logarithms on a quantum computer." SIAM Journal on Computing, 26(5):1484{1509, 1997}.
3. L. Grover. "A fast quantum mechanical algorithm for database search." In Proceedings of the 28th ACM Symposium on the Theory of Computing, pages 212{219. ACM, New York, 1996.
4. Bennett, C.H., F. Bessette, G. Brassard, L. Salvail, and J. Smolin, "Experimental quantum cryptography", Journal of Cryptology 5(1), 328 (1992).
5. Brassard, G. (1993). "A Bibliography of Quantum Cryptography". ACM SIGACT News, pp. 16–20.
6. Yu-Guang Yang[1,2,3,4], Peng Xu[1], Rui Yang[1], Yi-Hua Zhou[1] & Wei-Min Shi[1.] "Quantum Hash function and its application to privacy amplification in quantum key distribution, pseudo-random number generation and image encryption." www.nature.com/scientificreports, Published: 29 January 2016.
7. The Golden Bough, Sir James George Frazer. "Cryptography and Network Security" Chapter 13 Fifth Edition by William Stallings Lecture slides by Lawrie Brown.

8. P. Eronen, et. al., “Pre-Shared Key Ciphersuites for Transport Layer Security (TLS)”, RFC4279, December 2005.
9. Tim Dierks, Eric Rescorla, “ The Transport Layer Security (TLS) Protocol, Version 1.2”, RFC5246, August 2008.
10. C. H. Bennett and G. Brassard, “Quantum cryptography: Public key distribution and coin tossing,” in Proc. IEEE Int. Conf. Computers, Systems and Signal Processing, New York, Bangalore, India, 1984, pp. 175–179.
11. Ammar Odeh, Khaled Elleithy, Muneer Alshowkan, Eman Abdelfattah. “Quantum Key Distribution by Using Public Key Algorithm(RSA)” (PDF Download Available). Available from:https://www.researchgate.net/publication/256455756_Quantum_Key_Distribution_by_Using_Public_Key_AlgorithmRSA [accessed Oct 21 2017].
12. S. Ranganathan, N. Ramasamy, S. K. K. Arumugam, B. Dhanasekaran, P. Ramalingam, V. Radhakrishnan, and R. Karpuppiah, “A Three Party Authentication for Key Distributed Protocol Using Classical and Quantum Cryptography,” International Journal of Computer Science Issues(IJCSI), vol. 7, 2010.

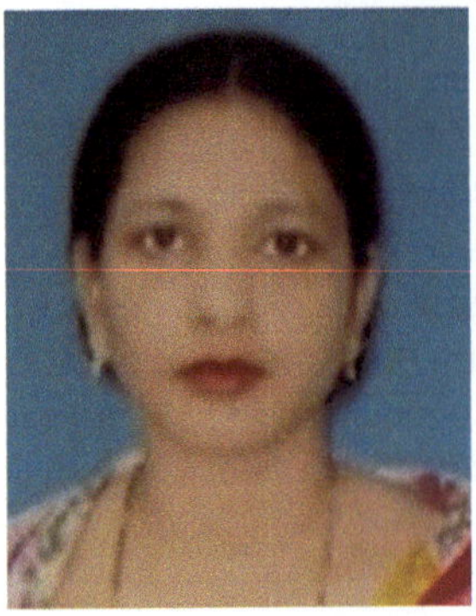

Prity Kumari Received her Bachlor degree in Mathematics and Master in Computer Science and Application from Magadh University Bodh Gaya in the Department of Mathematics and Computer Science. She is persuing Ph.D in Computer Science from same university. She is currently working as a Resource Person in the BCA Department from last 2 years. Her research field is related with Optimized Data Security in Computer Network. She has published 2 research papers, first is “N-tier Encryption/Decryption for Data Security” in International Journal of Advance Engineering and Research Development Journals and second is “Dual-Layer Video Encryption Using RSA and ECC Algorithm” in International Journal of Scientific and Research Publications and 2 Conference paper first is “Image Detection and Security Analysis” in National Seminar on “Digital India: Use of Technology for Transforming Security” organized by Gaya College, Gaya and second is “Full RSA Technique” in National Seminar on “Cloud Computing, Big Data Analysis and Mathematical Applications” organized by Department of Mathematics and Computer Science, Magadh University, BodhGaya.

Upendra Kumar Received his Master in Technology and Doctor in Philosophy from Birla Institute of Technology, Mesra Ranchi in the Department of Computer and Engineering. He is currently working as an Assistant Professor in the same Department from last 6 years. His research field is related with Computer Network Security and Data Mining. He has published more than 25 research papers in Journals of repute and Conferences. He has act as a reviewer of Journals and members of International Conferences of ACM, IEEE and Springer.

Shyam Krishna Singh Received his Master in Mathematics from Magadh university Bodh Gaya and Doctor in Philosophy from Mithla university Dharbhanga in the Department of Mathematics. He is currently working as an Associate Professor in the same Department from last 30 years. His research field is related with Gaussion Mesoures in Theory of Probability and Stochastic process. He has published more than 8 research papers in Journals of repute and Conferences.

Convergence of Common Solution of Variational Inequality and Fixed Point of a Pseudocontractive Mapping

Poonam Mishra

1 Introduction

Let E be a real Banach Space. Let J denote the normalized duality mapping from E into 2^{E^*} defined by

$$J(x) = \{f \in E^* : \langle x, f\rangle = \|x\|^2 = \|f\|^2\}$$

For all $x \in E$, where E^* denotes the dual space of E and $\langle .,. \rangle$ denotes the generalized duality pairing, we denote the single-valued duality mapping by j. In a Hilbert Space H, j is the identity mapping.

Let C be a nonempty closed convex subset of E. A mapping f: $C \to C$ is said to be a strict contraction if there exists a constant $\lambda \in (0, 1)$ satisfying

$$\|f(x) - f(y)\| \le \lambda\|x - y\| \quad \text{for all } x, y \in C$$

A mapping T: $C \to C$ is said to be an asymptotically nonexpansive if there exists a sequence $\{k_n\}$ with $k_n \to 1$ such that

$$\|T^n x - T^n y\| \le k_n\|x - y\|, \quad \forall x, y \in C$$

An asymptotically nonexpansive mapping contains strict contraction, nonexpansive mapping as a special case.

A mapping T: $C \to C$ is said to be an asymptotically pseudocontractive mapping in Banach Spaces if there exists a sequence $\{k_n\}$ with $k_n \to 1$ and $j(x - y) \in J(x - y)$ for which the following inequality holds

P. Mishra (✉)
Department of Applied Mathematics, Amity School of Engineering and Technology, Amity University, Raipur, Chhattisgarh, India
e-mail: ppoonam22@gmail.com

J. Chattopadhyay et al. (eds.), *Innovations in Soft Computing and Information Technology*, https://doi.org/10.1007/978-981-13-3185-5_8

$$\langle T^n x - T^n y, j(x-y)\rangle \le k_n \|x-y\|^2,$$
$$\forall x, y \in C, n \ge 1$$

A mapping T: $C \to C$ is said to be uniform L-Lipschitzian if there exists some $L > 0$ such that

$$\|T^n x - T^n y\| \le L\|x-y\|, \quad \forall x, y \in C, \forall n \ge 1$$

It is easy to see that if T is an asymptotically nonexpansive mapping, then it is both asymptotically pseudocontractive and uniformly L-Lipschitzian but the converse is not true in general.

Many authors investigated the viscosity iterative algorithms in order to find the common element of the set of fixed point of nonexpansive mappings and the set of solution of variational inequality problem see [1–4] and the references therein. In 2000, Moudafi [5] introduced the viscosity iterative algorithm for strong convergence of nonexpansive mappings in real Hilbert Space. Later, in 2004, Xu [1] extended results of Moudafi [5] and introduced the following viscosity technique for nonexpansive mapping in a uniformly smooth Banach Space:

$$x_{n+1} = \alpha_n f(x_n) + (1-\alpha_n)Tx_n, \quad n \ge 0$$

where f is a contraction and $\{\alpha_n\}$ is a sequence in $\{0,1\}$.

The implicit midpoint rule is a powerful method for solving ordinary differential equations; see [6, 7] and the references therein. Recently in 2015, Xu et al. [3] extended Alghamadi et. al. [8] and applied the viscosity technique to the implicit midpoint rule for a nonexpansive mapping. They introduced the following viscosity implicit midpoint rule:

$$x_{n+1} = \alpha_n f(x_n) + (1-\alpha_n)T\left(\frac{x_n + x_{n+1}}{2}\right), \quad n \ge 0 \tag{1}$$

They proved that the sequence generated by Eq. (1) converges strongly to a fixed point of T, which also solves the following variational inequality in Hilbert Space:

$$\langle (I-f)q, x-q\rangle \ge 0, \quad x \in F(T) \tag{2}$$

In 2017, Luo et al. [9] proved strong convergence for strict pseudocontractive mapping with some appropriate conditions on parameters by using above Eq. (1) implicit midpoint rule of nonexpansive mappings in uniformly smooth Banach Spaces which also solves the following variational inequality problem:

$$\langle (I-f)q, x-q\rangle \ge 0, \quad x \in F(T)$$

Recently, Yan et al. [10] extended the result of Luo et al. [9] from nonexpansive mapping to asymptotically nonexpansive mapping and introduced the generalized viscosity implicit rule for asymptotically nonexpansive mapping in Hilbert Space and proved the following result:

Theorem 1.1 (Yan et al. [10]) *Let H be a Hilbert Space, C a nonempty closed convex subset of H. Let T: C → C be an asymptotically nonexpansive mapping with a sequence* $\{\theta_n\}$ *such that* $F(T) \neq \phi$ *and f: C → C a strict contraction with coefficient* $\alpha \in [0,1]$. *Pick any* $x_0 \in C$. *Let* $\{x_n\}$ *be a sequence generated by*

$$x_{n+1} = \alpha_n f(x_n) + (1 - \alpha_n)T^n\left(\frac{x_n + x_{n+1}}{2}\right), \quad n \geq 0$$

where $\{\alpha_n\}$ *is a real sequence in [0,1] satisfying the conditions:*

(i) $\lim_{n\to\infty} \alpha_n = 0; \quad \sum_{n=0}^{\infty} \alpha_n = \infty;$

(ii) $\lim_{n\to\infty} \frac{\theta_n}{\alpha_n} = 0;$

(iii) $\sum_{n=1}^{\infty} \left|\alpha^{n+1} - \alpha^n\right| < \infty;$

(iv) $\sum_{n=0}^{\infty} \sup_{x\in C'} \left\|T^{n+1}x - T^n x\right\| < \infty,$

where C′ is a closed convex subset of C that contains sequence $\{x_n\}$.

Then $\{x_n\}$ converges strongly to a fixed point q of the asymptotically nonexpansive mapping T, which is also the solution of the variational inequality

$$\langle (I - f)q, y - q\rangle \geq 0, \quad \text{for all } y \in F(T)$$

In this paper, we propose an iterative algorithm and prove strong convergence of an asymptotically pseudocontractive mapping in a reflexive smooth Banach Space E which also solves a certain variational inequality. Our algorithm improves some conditions of above iterative algorithm and extends the result of Yan et al. [10] from asymptotically nonexpansive mapping to asymptotically pseudocontractive mapping and also from Hilbert Space to reflexive smooth Banach Space. Also, our result improves and extends many recent results.

2 Preliminaries

We shall make use of the following Lemmas:

Lemma 2.1 (Xu [1]) *Assume* $\{a_n\}$ *is a sequence of nonnegative real numbers such that*

$$a_{n+1} \leq (1-\alpha_n)a_n + \delta_n, \quad n \geq 0$$

where $\{\alpha_n\}$ is a sequence in (0,1) and δ_n is a sequence in R such that

(i) $\sum_{n=0}^{\infty} \alpha_n = \infty;$

(ii) $\limsup_{n\to\infty} \frac{\delta_n}{\alpha_n} \leq 0$ or $\sum_{n=1}^{\infty} |\delta_n| < \infty$

Then, $\lim_{n\to\infty} a_n = 0$.

Lemma 2.2 (Wang et al. [11]) *Let* **E** *be a reflexive smooth Banach Space with a weakly sequential continuous duality mapping J. Let C be a nonempty bounded and closed convex subset of* **E** *and T: C $\to$ C be a uniformly L-Lipschitzian and asymptotical pseudocontraction. Then I-T is demiclosed at zero, where I is the identical mapping, i.e., if $x_n \to x$ weakly and $x_n - Tx_n \to 0$ strongly, then $x \in F(T)$.*

Lemma 2.3 (Yao et al. [12]) *Let $\{x_n\}$ and $\{y_n\}$ be bounded sequence in a Banach Space E and $\{\beta_n\}$ be a sequence in [0,1] with $0 < \liminf_{n\to\infty} \beta_n \leq \limsup_{n\to\infty} \beta_n$. Suppose that $x_{n+1} = (1 - \beta_n) + \beta_n y_n$ for all $n \geq 0$ and* $\limsup_{n\to\infty}(\|y_{n+1} - y_n\| - \|x_{n+1} - x_n\|) \leq 0$. *Then* $\lim_{n\to\infty} \| y_n - x_n \| = 0$.

3 Main Result

Theorem 3.1 *Let E be a reflexive smooth Banach Space with weak sequential continuous duality mapping J. Let C be a nonempty bounded and closed convex subset of E and T: C $\to$ C be a uniformly L-Lipschitzian and asymptotically pseudocontractive mapping with a sequence $\{k_n\}$ such that $F(T) \neq \phi$ and f: C $\to$ C be a contraction with coefficient $\alpha \in (0,1)$. Pick any $x_0 \in C$. Let $\{x_n\}$ be a sequence generated by*

$$x_{n+1} = \alpha_n f(x_n) + (1-\alpha_n)T^n\left(\frac{x_n + x_{n+1}}{2}\right) \tag{3}$$

where $\{\alpha_n\}$ is a real sequence in [0,1] satisfying the following conditions:

(i) $\lim_{n\to\infty} \alpha_n = 0, \quad \sum_{n=0}^{\infty} \alpha_n = \infty;$

(ii) $\lim_{n\to\infty} \frac{k_n}{\alpha_n} = 0;$

(iii) $\sum_{n=0}^{\infty} |\alpha_{n+1} - \alpha_n| < \infty;$

(iv) $\sum_{n=0}^{\infty} \sup_{x\in C'} |T^{n+1}x - T^n x| < \infty,$

where C' is a closed convex subset of C that contains sequence $\{x_n\}$.

Then $\{x_n\}$ converges strongly to a fixed point q of the asymptotically pseudo-contractive mapping T, which is also the solution of the variational inequality

$$\langle (I-f)q, y-q\rangle \geq 0, \quad \text{for all}\, y \in F(T)$$

Proof First, we show that $\{x_n\}$ is bounded. Indeed, take $p \in F(T)$ arbitrarily, since $\lim_{n\to\infty}\frac{k_n}{\alpha_n} = 0$, then there exists $N \in \mathbf{N}$ such that for all $n \geq N$, $\frac{k_n}{\alpha_n} \leq \left(\frac{1-\alpha}{2}\right)$. Choose a constant $M_1 > 0$ sufficiently large such that

$$\|x_N - p\| \leq M_1, \quad \|f(p) - p\| \leq \frac{1-\alpha}{2}M_1$$

We proceed by induction to show that $\|x_n - p\| \leq M_1, \forall n \geq 1$. Assume $\|x_n - p\| \leq M_1$, for some $n \geq N$. We show that $\|x_{n+1} - p\| \leq M_1$. We observe

$$\begin{aligned}
\|x_{n+1} - p\| &= \left\|\alpha_n f(x_n) + (1-\alpha_n)T^n\left(\frac{x_n + x_{n+1}}{2}\right) - p\right\| \\
&= \left\|\alpha_n[f(x_n) - p] + (1-\alpha_n)\left[T^n\left(\frac{x_n + x_{n+1}}{2}\right) - p\right]\right\| \\
&\leq \alpha_n\|f(x_n) - f(p)\| + \alpha_n\|f(p) - p\| + (1-\alpha_n)\left\|T^n\left(\frac{x_n + x_{n+1}}{2}\right) - p\right\| \\
&\leq \alpha_n\alpha\|x_n - p\| + \alpha_n\|f(p) - p\| + (1-\alpha_n)L\left\|\left(\frac{x_n + x_{n+1}}{2}\right) - p\right\| \\
&\leq \alpha_n\alpha\|x_n - p\| + \alpha_n\|f(p) - p\| + \frac{L(1-\alpha_n)}{2}\|x_n - p\| \\
&\quad + \frac{L(1-\alpha_n)}{2}\|x_{n+1} - p\|
\end{aligned}$$

It follows that

$$\begin{aligned}
\left[1 - \frac{L(1-\alpha_n)}{2}\right]\|x_{n+1} - p\| &\leq \left[\alpha_n\alpha + \frac{L(1-\alpha_n)}{2}\right]\|x_n - p\| + \alpha_n\|f(p) - p\| \\
\|x_{n+1} - p\| &\leq \frac{2\alpha_n\alpha + L(1-\alpha_n)}{2 - L(1-\alpha_n)}\|x_n - p\| + \frac{2\alpha_n}{2 - L(1-\alpha_n)}\|f(p) - p\| \\
&= \left[1 - \frac{2\alpha_n(1-\alpha) + 2L(1-\alpha_n) - L\alpha_n}{2 - L(1-\alpha_n)}\right]\|x_n - p\| + \frac{2\alpha_n}{2 - L(1-\alpha_n)}\|f(p) - p\| \\
&\leq \left[1 - \frac{\alpha_n(1-\alpha)}{2 - L(1-\alpha_n)}\right]\|x_n - p\| + \frac{\alpha_n(1-\alpha)}{2 - L(1-\alpha_n)}\frac{\|f(p) - p\|}{1-\alpha} \\
&\leq \text{Max}\left\{\|x_n - p\|, \frac{\|f(p) - p\|}{1-\alpha}\right\} \\
&\leq M_1
\end{aligned} \tag{4}$$

This implies that $\{x_n\}$ is bounded.

It turns out that $f(x_n)$, $T^n\left(\frac{x_n + x_{n+1}}{2}\right)$ are also bounded.

Next, we prove that $\lim_{n\to\infty}\|x_{n+1}-x_n\| = 0$. It follows from Eq. (3) that

$$\begin{aligned}
\|x_{n+1}-x_n\| &= \Big\|\alpha_n f(x_n) + (1-\alpha_n)T^n\Big(\frac{x_n+x_{n+1}}{2}\Big) - \alpha_{n-1} f(x_{n-1}) \\
&\quad + (1-\alpha_{n-1})T^{n-1}\Big(\frac{x_{n-1}+x_n}{2}\Big)\Big\| \\
&= \left\| \begin{array}{l} \alpha_n(f(x_n)-f(x_{n-1})) + (\alpha_n-\alpha_{n-1})f(x_{n-1}) \\ +(1-\alpha_n)\left[T^n\left(\frac{x_n+x_{n+1}}{2}\right) - T^n\left(\frac{x_{n-1}+x_n}{2}\right)\right] \\ +(1-\alpha_n)T^n\left(\frac{x_{n-1}+x_n}{2}\right) \\ -(1-\alpha_{n-1})T^n\left(\frac{x_{n-1}+x_n}{2}\right) \\ +(1-\alpha_{n-1})\left[T^n\left(\frac{x_{n-1}+x_n}{2}\right) - T^{n-1}\left(\frac{x_{n-1}+x_n}{2}\right)\right] \end{array} \right\| \\
&= \left\| \begin{array}{l} \alpha_n(f(x_n)-f(x_{n-1})) + (1-\alpha_n)\left[\begin{array}{c} T^n\left(\frac{x_n+x_{n+1}}{2}\right) \\ -T^n\left(\frac{x_{n-1}+x_n}{2}\right)\end{array}\right] \\ +(\alpha_n-\alpha_{n-1})\left[f(x_{n-1}) - T^n\left(\frac{x_{n-1}+x_n}{2}\right)\right] \end{array}\right. \\
&\quad \left. + (1-\alpha_{n-1})\Big[T^n\Big(\frac{x_{n-1}+x_n}{2}\Big) - T^{n-1}\Big(\frac{x_{n-1}+x_n}{2}\Big)\Big]\right\| \\
&= \alpha\alpha_n\|x_n-x_{n-1}\| + (1-\alpha_n)L\Big\|\frac{x_{n+1}+x_n}{2} - \frac{x_{n-1}+x_n}{2}\Big\| \\
&\quad + |\alpha_n-\alpha_{n-1}|\Big\|f(x_{n-1}) - T^n\Big(\frac{x_{n-1}+x_n}{2}\Big)\Big\| \\
&\quad + \sup_{x\in c'}\|T^n x - T^{n-1}x\| \\
&\le \frac{2\alpha\alpha_n + (1-\alpha_n)L}{2}\|x_n-x_{n-1}\| \\
&\quad + \frac{(1-\alpha_n)L}{2}\|x_{n+1}-x_n\| + |\alpha_n-\alpha_{n-1}|M_2 \\
&\quad + \sup_{x\in c'}\|T^n x - T^{n-1}x\|
\end{aligned}$$

where M_2 is a constant such that

$$M_2 = M_2 = \sup_{n\ge 0}\Big\|f(x_{n-1}) - T^n\Big(\frac{x_{n-1}+x_n}{2}\Big)\Big\|$$

It follows that

$$\begin{aligned}
\frac{2-(1-\alpha_n)L}{2}\|x_{n+1}-x_n\| &\le \frac{2\alpha\alpha_n + (1-\alpha_n)L}{2}\|x_n-x_{n-1}\| + |\alpha_n-\alpha_{n-1}|M_2 \\
&\quad + \sup_{x\in c'}\|T^n x - T^{n-1}x\|
\end{aligned}$$

This implies

$$\begin{aligned}\|x_{n+1}-x_n\| &\leq \frac{2\alpha\alpha_n+(1-\alpha_n)L}{2-(1-\alpha_n)L}\|x_n-x_{n-1}\| \\ &\quad+\frac{2M_2}{2-(1-\alpha_n)L}|\alpha_n-\alpha_{n-1}|+\frac{2}{2-(1-\alpha_n)L}\sup_{x\in C'}\left\|T^nx-T^{n-1}x\right\| \\ &\leq\left(1-\frac{2[1-\alpha\alpha_n-(1-\alpha_n)L]}{2-(1-\alpha_n)L}\right)\|x_n-x_{n-1}\| \\ &\quad+\frac{2M_2}{2-(1-\alpha_n)L}|\alpha_n-\alpha_{n-1}| \\ &\quad+\frac{2}{2-(1-\alpha_n)L}\sup_{x\in C'}\left\|T^nx-T^{n-1}x\right\|\end{aligned} \tag{5}$$

Let $\gamma_n=2\frac{[1-\alpha\alpha_n-(1-\alpha_n)L]}{2-(1-\alpha_n)L}$. We note

$$\begin{aligned}\gamma_n&=\frac{2[(1-\alpha)\alpha_n+(1-\alpha_n)(1-L)]}{2-(1-\alpha_n)L}\\ &\geq\frac{2[(1-\alpha)\alpha_n+(1-\alpha_n)(1-L)]}{2-L}\\ &=[(1-\alpha)\alpha_n+\alpha_nL]\\ &\geq\alpha_n(1-\alpha)+L\geq\frac{1-\alpha}{2}\alpha_n\end{aligned}$$

By condition (1), we have $\sum_{n=0}^{\infty}\gamma_n=\infty$. Apply Lemma 2.1 to Eq. (5), we get

$$\lim_{n\to\infty}\|x_{n+1}-x_n\|=0 \tag{6}$$

Next, we prove that $\lim_{n\to\infty}\|x_n-Tx_n\|=0$. In fact, we have

$$\left\|x_{n+1}-T^n\left(\frac{x_n+x_{n+1}}{2}\right)\right\|=\alpha_n\left\|f(x_n)-T^n\left(\frac{x_n+x_{n+1}}{2}\right)\right\|\to 0\quad\text{as } n\to\infty \tag{7}$$

Moreover, we get

$$\begin{aligned}\|x_n - T^n x_n\| &= \left\|x_n - x_{n+1} + x_{n+1} - T^n\left(\frac{x_n + x_{n+1}}{2}\right) + T^n\left(\frac{x_n + x_{n+1}}{2}\right) - T^n x_n\right\| \\ &\leq \|x_{n+1} - x_n\| + \left\|x_{n+1} - T^n\left(\frac{x_n + x_{n+1}}{2}\right)\right\| + \left\|T^n\left(\frac{x_n + x_{n+1}}{2}\right) - T^n x_n\right\| \\ &\leq \|x_{n+1} - x_n\| + \left\|x_{n+1} - T^n\left(\frac{x_n + x_{n+1}}{2}\right)\right\| + L\|x_{n+1} - x_n\| \\ &= (L+1))\|x_{n+1} - x_n\| + \left\|x_{n+1} - T^n\left(\frac{x_n + x_{n+1}}{2}\right)\right\|\end{aligned}$$

Combining Eqs. (6) and (7), we can obtain

$$\lim_{n\to\infty} \|x_n - T^n x_n\| = 0 \tag{8}$$

We notice

$$\begin{aligned}\|x_n - Tx_n\| &= \left\|x_n - T^n x_n + T^n x_n - T^{n+1}x_n + T^{n+1}x_n - Tx_n\right\| \\ &\leq \|x_n - T^n x_n\| + \left\|T^n x_n - T^{n+1}x_n\right\| + L\|T^n x_n - x_n\| \\ &\leq \|x_n - T^n x_n\| + \left\|T^n x_n - T^{n+1}x_n\right\| + L\|T^n x_n - x_n\| \\ &\leq \|x_n - T^n x_n\| + \sup_{x\in C'}\left\|T^n x_n - T^{n+1}x_n\right\| + L\|T^n x_n - x_n\|\end{aligned}$$

By condition (iv) and Eq. (8), we have

$$\lim_{n\to\infty} \|x_n - Tx_n\| = 0 \tag{9}$$

Next, we claim that

$$\limsup_{n\to\infty} \langle q - f(q), j(q - x_n)\rangle \leq 0 \tag{10}$$

where $q = P_{F(T)} f(q)$. Since $\{x_n\}$ is bounded and C is a reflexive Banach Space, there exists a subsequence of $\{x_n\}$ which converges weakly to p. Without loss of generality, we may assume that $x_{ni} \to p$ weakly. Indeed, there exists a subsequence $\{x_{ni}\}$ of $\{x_n\}$ such that

$$\begin{aligned}&\limsup_{n\to\infty} \langle q - f(q), j(q - x_n)\rangle \\ &\quad = \lim_{i\to\infty} \langle q - f(q), j(q - x_{ni})\rangle\end{aligned}$$

Since C is a smooth Banach Space, it follows from Eq. (9) and Lemma 2.2, we have $p \in F(T)$. This together with the property of weakly sequential continuous duality mapping implies that

$$\limsup_{n\to\infty}\langle q-f(q),j(q-x_n)\rangle$$
$$=\lim_{i\to\infty}\langle q-f(q),j(q-x_{ni})\rangle=\langle q-f(q),j(q-p)\rangle\le 0$$

Then Eq. (10) holds. Finally, we show that $x_n \to q$ as $n \to \infty$. In fact, we have

$$\begin{aligned}
\|x_{n+1}-q\|^2 &= \left\langle \alpha_n f(x_n)+(1-\alpha_n)T^n\left(\frac{x_n+x_{n+1}}{2}\right)-q,j(x_{n+1}-q)\right\rangle\\
&=\left\langle \alpha_n(f(x_n)-q)+(1-\alpha_n)(T^n\left(\frac{x_n+x_{n+1}}{2}\right)-q),j(x_{n+1}-q)\right\rangle\\
&=\alpha_n\langle f(x_n)-q,j(x_{n+1}-q)\rangle+\alpha_n\langle f(q)-q,j(x_{n+1}-q)\rangle\\
&\quad+(1-\alpha_n)\left\langle T^n\left(\frac{x_n+x_{n+1}}{2}\right)-q,j(x_{n+1}-q)\right\rangle\\
&\le \alpha\alpha_n\|x_n-q\|.\|x_{n+1}-q\|+(1-\alpha_n)L\left\|\frac{x_n-q}{2}+\frac{x_{n+1}-q}{2}\right\|\\
&\quad\cdot\|x_{n|+1}-q\|+\alpha_n\langle f(q)-q,j(x_{n+1}-q)\rangle\\
&\le\frac{\alpha\alpha_n}{2}\|x_n-q\|^2+\frac{\alpha\alpha_n}{2}.\|x_{n+1}-q\|^2\\
&\quad+\frac{(1-\alpha_n)L}{4}\|x_n-q\|^2+\frac{(1-\alpha_n)L}{4}\|x_{n+1}-q\|^2\\
&\quad+\frac{(1-\alpha_n)L}{2}\|x_{n+1}-q\|^2+\alpha_n\langle f(q)-q,j(x_{n+1}-q)\rangle
\end{aligned}$$

which implies

$$\begin{aligned}
&\frac{4-2\alpha\alpha_n-3(1-\alpha_n)L}{4}\|x_{n+1}-q\|^2\\
&\le\frac{2\alpha\alpha_n+(1-\alpha_n)L}{4}\|x_n-q\|^2+\alpha_n\langle f(q)-q,x_{n+1}-q\rangle
\end{aligned}$$

That is,

$$\begin{aligned}
\|x_{n+1}-q\|^2&\le\frac{2\alpha\alpha_n+(1-\alpha_n)L}{4-2\alpha\alpha_n-3(1-\alpha_n)L}\|x_n-q\|^2\\
&\quad+\frac{4\alpha_n}{4-2\alpha\alpha_n-3(1-\alpha_n)L}\langle f(q)-q,j(x_{n+1}-q)\rangle\\
&=1-\frac{4(\alpha_nL+1-\alpha\alpha_n-L)}{4-2\alpha\alpha_n-3(1-\alpha_n)L}\|x_n-q\|^2\\
&\quad+\frac{4\alpha_n}{4-2\alpha\alpha_n-3(1-\alpha_n)L}\langle f(q)-q,j(x_{n+1}-q)\rangle
\end{aligned}\tag{11}$$

Put

$$\gamma_n = \frac{4(\alpha_n L + 1 - \alpha\alpha_n - L)}{4 - 2\alpha\alpha_n - 3(1 - \alpha_n)L}$$

We have

$$\begin{aligned}\gamma_n &= \frac{4[(\alpha_n - 1)(1 - L) + \alpha_n(1 - \alpha)]}{1 - 2\alpha\alpha_n + 3(\alpha_n - 1)L + 3\alpha_n} \\ &\geq \frac{4[(\alpha_n - 1)(1 - L) + \alpha_n(1 - \alpha)]}{1 + 3\alpha_n} \\ &\geq (1 - L)(\alpha_n - 1) + \alpha_n(1 - \alpha) \\ &\geq \alpha_n(1 - \alpha) - L \geq \frac{1 - \alpha}{2}\alpha_n\end{aligned}$$

Apply Lemma 2.1 to Eq. (11), we obtain $x_n \to q$ as $n \to \infty$.
This completes the proof.

Since nonexpansive mapping is asymptotically nonexpansive and also asymptotically nonexpansive mapping is asymptotically pseudocontractive, we obtain the results of Xu et al. [3] and Yan et al. [10].

References

1. Xu, H.K., Viscosity approximation methods for nonexpansive mappings, J. Math. Anal. Appl. 2004;298: 279–29.
2. Nakajo, K., Takahashi, W., Strong convergence theorems for nonexpansive mappings and nonexpansive semigroups, J. Math. Anal. Appl. 2003;279: 372–379.
3. Xu, H.K., Alghamdi, M.Ali, Shahzad, N., The viscosity technique for the implicit midpoint rule of nonexpansive mappings in Hilbert spaces, Fixed Point Theory and Applications 2015:41.
4. Lou, J., Zhang, L., He, Z., Viscosity approximation methods for asymptotically nonexpansive mappings, Appl. Math. Comput. 2008, 203:171–177.
5. Moudafi, A. Viscosity approximation methods for fixed points problems, J. Math. Anal. 241 2000; 241: 46–55.
6. Bader, G., deuflhard, P: A semi-implicit mid-point rule for stiff systems of ordinary differential equations, Numer. Math. 1983;41: 373–398.
7. Deuflhard, P: Recent progress in extrapolation methods for ordinary differential equations., SIAM Rev. 1985; 27(4):505–535.
8. Alghamadi, M.A., Alghamdi, M. Ali., Shahzad, N., Xu, H.K. The implicit midpoint rule for nonexpansive mappings, Fixed Point Theory and Applications 2014:96.
9. Luo, Ping., Cai, Gang, Shehu, Yekini. The viscosity iterative algorithms for the implicit midpoint rule of nonexpansive mappings in uniformly smooth Banach spaces, Journal of Inequalities and Applications (2017) 2017:154.

10. Yan, Q., Hu, Shaotao: Strong convergence theorems for the generalized viscosity implicit rules of asymptotically nonexpansive mappings in Hilbert spaces. J. Computational Analysis and Applications, 2018; 24(3):486–496.
11. Wang, Y.H., Xia, Y.H., Strong Convergence for asymptotically pseudocontractions with the demiclosedness principle in Banach spaces, Fixed Point Theory Appl., 2012, 8 pages, 1.2.
12. Yao, Y.H, Shahzad, N., Liou, Y.,C., Modified semi-implicit midpoint rule for nonexpansive mappings, Fixed Point Theory Appl., 2015, 15 pages,1,1.3,2.5 theory and Applications 2014:96.

Poonam Mishra is working as Assistant Professor in the Department of Applied Mathematics, Amity University Chhattisgarh. She is a M.Sc, M.Phil in Mathematics and has done her MBA in Operations Management. Currently she is pursuing her Ph.D in Applied Mathematics from Chhattisgarh Swami Vivekananda Technical University, Bhilai(C.G). The thrust area of her research is Variational inequalities, its applications and its equivalence with fixed point theory. She has been teaching Applied Mathematics and Statistics for more than 13 years to various courses of Engineering, IT, Commerce, Management and Biotech. She has authored/coauthored 5 publications in peer reviewed national and International Journals and has presented her work in several conferences/ seminars. Her current research proposes an iterative algorithm which gives a common element of fixed point of asymptotically pseudocontractive mapping and a certain variational inequality.

A Literature Review on Dynamic Modeling and Epidemic Dynamics

Biswarup Samanta

1 Introduction

Epidemiology is nothing but the learning of occurrence and spreading of diseases in altered inhabitants. An endemic denotes to an ailment that occurs in an inhabitant over comparatively lengthier period of time during which there is a regeneration of vulnerable inhabitants because of fresh births or expiry of brief immunity which boons the ailment from getting totally enriched from the population.

In our study, an effort has been made to analyze the works of epidemic modeling for malware spread to find the scope of applying the philosophies of epidemiological modeling to study and comprehend the contaminating nature, transmission mode, and destructive behavior of worms in computer networks.

2 Literature Survey

To discourse the unidentified admittance validation issue for CPS Wireless Mesh Network (CPS-WMN), an innovative unspecified entrée validation system grounded on proxy ring signature has been projected by **Tianhan Gao et al**. Authors offered a recognized safety immune of the projected procedure with SVO logic. The recreation and performance study establish that the projected structure possesses advanced efficacy and adaptability than the typical one [1].

The goal of this paper, inscribed by **Lianwen Wang**, was to outspread the occurrence proportion of an SEIR endemic model through decline and changing whole inhabitants scope to a overall nonlinear arrangement, which did not only

B. Samanta (✉)
Department of Computer Science & Information Technology, Amity Institute of Information Technology, Amity University, Ranchi, Jharkhand, India
e-mail: biswarup.k@gmail.com; bsamanta@rnc.amity.edu

J. Chattopadhyay et al. (eds.), *Innovations in Soft Computing and Information Technology*, https://doi.org/10.1007/978-981-13-3185-5_9

comprise a extensive series of monotonic and dipped occurrence rates but similarly receipts on certain neither monotonic nor concave cases, which might be used to replicate mass media teaching or psychosomatic outcome. By submission of the renewed geometric approach built on the third additive compound matrix, authors engrossed on launching the global stability of the SEIR model. Their inferences are functional to two distinct incidence functions reflecting mass media effect [2].

Jinliang Wang et al. examined the global asymptotic stability of multi-group SIR and SEIR age-structured models in their paper. These representations permit the infectiousness and the demise rate of vulnerable individuals to differ and depend on the vulnerability, through which they considered the heterogeneity of population. They recognized global dynamics and prove that the heterogeneity does not alter the dynamical structure of the basic SIR and SEIR with age-dependent susceptibility [3].

Vyacheslav Kharchenko and Oleg Illiashenko have stated in their paper that the industrial safety-critical instrumentation and control systems (I & Cs) are fronting extra with information (in all-purpose and cyber, in specific) security coercions and bouts. The submission of programmable logic, first of all, field-programmable gate array (FPGA) in critical systems causes precise safety discrepancies. The goal of their paper was in portrayal of the method and tools for situation-based security valuation of MV FPGA-based I & Cs [4].

Cyber combat impose a main latent threat to critical infrastructures (CIs). Decision makers who want to develop robust CIs must deliberate both strategic and operational characteristics of CIs as well as nonlinear dynamics characterizing such cyber-physical systems. The paper, written by **Elisa Canzani and Stefan Pickl**, cartels System Dynamics (SD) with a game-theoretic method to comprehend cyber epidemics dynamics of CI procedures generated by the invader and defender strategic relations. This research effort was grounded on their earlier work by encompassing a new block building demonstrating framework for commotion influence study in networked CIs [5].

The key feature of the investigation effort done by **Kuldeep Kaur et al**. was that the data is innocuous in banking organization for extended period and exposed any account after an extended time. The imminent scope of the study of security is used to diminish threats. Security is used in the extended run results in the decrease of quantity of branches, saying rentals of linked and properties [6].

As cyber-physical schemes incline to become further intricate, abstraction and modification methods become more tempting. In 2009, Oehlerking and Theel offered a (de-)composition technique for amalgam systems. This method is graph-based and builds a Lyapunov function for hybrid schemes having an intricate disconnected state space. The method comprises of (1) decomposing the fundamental graph of the hybrid scheme into subgraphs, (2) figuring numerous local Lyapunov functions for the subgraphs, and lastly, (3) composing the local Lyapunov functions into a piecewise Lyapunov function. A Lyapunov function can attend numerous purposes, e.g., it endorses the stability or expiry of a system or permits building invariant groups, which in turn may be used to endorse safety and security. In their article, **Eike Mohlmann and Oliver Theel** suggested an upgrade to the decomposing method, which reduces the graph structure before smearing the decomposition method [7].

In their paper, **Liping Feng et al.** have offered an enhanced SIRS model for investigating dynamics of worm proliferation in WSNs. This model can describe the procedure of worm spread through the energy consumption and unlike dispersed density of nodes. Grounded on this model, a control parameter $R0 < 1$ that entirely governs the global dynamics of worm proliferation has been acquired by the explicit mathematical investigations. It has been found that worm will be controlled in WSNs when $R0 < 1$, and they will be widespread otherwise. Lastly, grounded on $R0$, author conferred the threshold of worm spread about communication circle and distributed densities of nodes in WSNs. Mathematical simulations confirm the accuracy of theoretical investigation. Research outcomes demonstrate that declining the worth of communication radius or dipping disseminated thickness of nodes is an effective process to stop worms' spread in WSNs. Exploration of this paper delivers the theoretical foundation for envisaging and monitoring worm propagation in WSNs [8].

Conferring to the writer, **Jinhua Ma et al.**, they are the first to investigate the intricate communications between benign worms and malicious worms in dissimilar M2M network. They evidenced that the global dynamics are determined by the threshold value $R0$. In the nonappearance of birth, death, and the treatment effect of users, they acquire the final size formula in diverse vaccination schemes. The outcomes demonstrate that the nodes with higher node degrees are more susceptible to infection than those with lesser node degrees. Additionally, the effects of numerous vaccination schemes are studied. Mathematical simulations authenticate the outcomes. Their paper delivers a robust theoretical foundation to take effective actions to regulate the large-scale spread of malicious worms in mixed M2M network [9].

The dynamics of catastrophic circumstances is the main research area of **Canzani et al.** Their study on the existing body of literature in epidemiology exemplifies the diversity of methodologies used for displaying dynamics. The intricacy of epidemics modeling augmented from compartmental SIR models that contemplate only organic pathogens to the numerous social network models. Appreciating relations and modeling social architecture is vital for understanding the dynamics of epidemics. Traditional models and revolutionary outcomes, such as the Kermack–McKendrick threshold condition, have been enormously successful in informing public health policy. In particular, this review represents a preliminary work to explore high dynamics of crisis situations such as the spread of computer viruses in the cyber world and of the information diffusion on the IT infrastructure of a firm [10].

In their paper, **Meng Wang et al.** projected a fixing mechanism for benign worm spread based on the mobile network. In the discovery and mending mechanism, after gathering the difficulties of the entire mobile network, they place the effective benign worms into the mobile network atmosphere to progress the mending productivity when malicious worms outburst. For benign worm spread mechanism, they first used the lively mode of benign worm to hurriedly handle malevolent worms and later after the malicious worms are under control; they switched to the passive mode and release mobile network resources further. Thus, they confirm the safety of mobile networks and also optimize the network correspondingly [11].

The paper written by **Elisa Canzani et al**. reports publicly accessible information on IT security, its accessibility, and excellence and offers an investigation of publicly available IT security information for mobile devices in Germany. They established that there are barely any patterns in terms of timeliness, content, and quality [12].

In their paper, an SEIQV epidemic model with saturated occurrence rate has been well-thought-out by **Xingbo Liu and Lijuan Yang**. The elementary reproduction number $R0$ is found. If $R0 \leq 1$, the disease-free equilibrium is globally asymptotically stable; if $R0 > 1$, endemic equilibrium is globally asymptotically stable and the disease is persistent [13].

In their paper, an original model has been wished for by **Yajuan Zhang et al**. to assist e-commerce specialists in the valuation of e-commerce security. The projected model is based on analytical hierarchy process (AHP) and Dempster–Shafer (DS) theory of evidence. First, conferring to the features of e-commerce, an ordered structure of e-commerce security is recognized to compute the weights of related matters using AHP. Then Dempster–Shafer theory of evidence is applied to combine all the matters, regarded as evidences, in order to originate a unanimity conclusion for the degree of e-commerce security [14].

Ryan E Hohimer and Frank L Greitzer termed a predictive modeling framework that mixes a varied set of data sources from the cyber-sphere, as well as incidental psychological/motivational influences that may motivate malicious insider adventures. This comprehensive risk valuation method proposes automated care for the finding of high-risk behavioral "triggers" to help emphasize the analyst's attention and notify the study [15].

A compartmental e-epidemic model SIRS for the spread of worms in computer network is considered by **B. K. Mishra and Samir Kumar Pandey**. They investigated the three cases of epidemic control approaches as—when the volume of contamination will be low, worms will not be in the network; for the high quantity of contamination, worm will attack; and for the moderate amount of contamination, worm may or may not attack the computer network [16].

In their article, centered on SIR and SEIR epidemic models with an overall nonlinear occurrence rate, **Huang and Gang** assimilated time delays into the normal differential equation models. By building suitable Lyapunov functional and consuming the Lyapunov–LaSalle invariance principle, they showed the global stability of the endemic equilibrium and the virus-free equilibrium for time delays of any span in each model. Their outcomes display that the global properties of equilibria also only depend on the basic reproductive number and that the latent period in a vector does not affect the stability, but the latent period in an infested host plays an optimistic role to control ailment development [17].

Two unlike models for the dynamics of computer viruses spread were equated by **Jose R. C. Piqueira et al.**: autoregressive and Fourier analysis giving alike results. In spite of not being entirely acceptable, these models present the likelihood of forecasting, growing, and declining tendencies in the spread of a certain type of virus by using the gathered knowledge with another one [18].

Encouraged on infection spread models, a compartmental model was established by **Jose R. C. Piqueira et al**. for the spread of viruses in computer networks. The SAIC model appears to be adherent to the actual data in an acceptable way and, besides, is healthy under parameter differences. As the SAIC model has modest execution, its use might help in approximating the dynamic behavior of viruses in real systems [19].

Fangwei Wang et al. wished for a delayed SEIRS model with online rate, off-line rate, and death rate by the use of the epidemic dynamics. The reproduction outcomes demonstrate the spread of passive worms being primarily ruled by the number of hops, the size and the quantity of portions of sharing files, and the temporary resistance period [20].

In their article, a SEIV epidemic model with a nonlinear occurrence rate is explored by **Li-Ming Cai and Xue-Zhi Li**. The model displays two equilibria, viz, the ailment-free equilibrium and the pervasive equilibrium. It is shown that if the basic reproduction number $R0 < 1$, the ailment-free equilibrium is globally asymptotically stable and in such a case, the endemic equilibrium does not occur. Furthermore, they exhibited that if the basic reproduction number $R0 > 1$, the ailment is homogeneously persistent and the unique endemic equilibrium of the system with saturation occurrence is globally asymptotically steady under certain conditions [21].

In their paper, an SEIR epidemic disease model with time delay and nonlinear incidence rate is calculated, and the dynamical behavior of the model under pulse immunization is examined by **Zhong Zhao et al**. Using the separate dynamical system determined by the stroboscopic map, they showed that there exists a contamination-free periodic result. Further, they exhibited that the infection-free periodic result is globally striking when the period of impulsive consequence is less than certain serious value. Using a novel modeling process, they acquired an adequate condition for the permanence of the epidemic model with pulse immunization. The central feature of this paper is to acquaint time delay and impulse into the SEIR epidemic model and to give pulse vaccination strategies [22].

The prototypes measured by **Richard J. Boys and Philip R. Giles** in their paper signify generalizations of the multi-type SEIR compartmental model to comprise a set of models for which the elimination rate is time inhomogeneous [23].

An endeavor has been made by **B. K. Mishra et al**. to develop mathematical models on computer viruses contaminating the system beneath diverse situations. The mathematical models established in their paper will benefit in discovery the likelihood of a system being infested by some computer virus or a collection of computer viruses at some time precisely dealing with the speed of upbringing of the viruses. These models will benefit in carrying out the sensitivity investigation and can be confirmed by simulation [24].

A variable population SEIRS epidemic malevolent body spread model in computer network with constant latent and immune phases has been articulated by **B. K. Mishra et al**. In their model when a node is detached from the disease-ridden class, it recuperates provisionally, obtaining provisional protection with probability p $(0 <= p <= 1)$ and dies [25].

B. K. Mishra et al. have established a SIRS epidemic model with a static period of temporary protection, subsequent provisional rescue from the contamination of malevolent substances in place of an exponentially dispersed period of provisional protection. Provisional resistance is perceived in the computer network when anti-malicious software is run after a node becomes affected by a malevolent body. Mathematically, it has been confirmed that the endemic equilibrium is not asymptotically stable for all factor values [26].

An effort has been made by **B. K. Mishra et al.** to scientifically articulate a compartmental differential susceptibility-infectiousness (DSI), susceptible-infectious-removed-susceptible (SIRS) epidemic transmission model of malevolent agents in a computer network, with death rate other than attack of malevolent agent and birth rate being measured as constant [27].

An effort has been made by **B. K. Mishra** to articulate the final size formulation for infected nodes in a computer network due to the outbreak of diverse malevolent representatives like viruses, Trojan horse, and worms. The writer presumed that the inhabitant of the nodes in a computer network is consistent and there does not occur any diverse mixing. The stability of the result is specified in the terms of reproductive number R0. The system is stable if $R0 < 1$, and unstable $R0 > 1$ [28].

In her article, **Hisashi Inaba** has established a novel method to deal with asymptotic behavior of the age-structured similar epidemic schemes and converse its submission to the MSEIR epidemic model. For the consistent scheme, there is no appealing nontrivial equilibrium; as an alternative, the author had to scrutinize the presence and steadiness of persistent results [29].

An independent SIRS epidemic model with time delay is considered by **Tailei Zhang and Zhidong Teng**. The rudimentary reproductive number R_0 is found which controls whether the infection is destroyed or not. When the basic reproductive number is larger than 1, it is demonstrated that the disease is everlasting in the population, and explicit formula is found by which the ultimate lower bound of the fraction of infectious entities can be calculated [30].

An age-structured model that describes the rotavirus spread dynamics of infections is announced by **E. Shim et al.** in their article. Circumstances that promise the local and global stability investigation of the infection-free stable state spreading as well as the presence of an endemic stable state dissemination are recognized. A convergent mathematical model is presented but not employed [31].

Arunabha Mukhopadhyay et al. have established a framework, centered on copula-aided Bayesian Belief Network (BBN) model, to measure the threat connected with online commercial transactions, arising out of a safety rupture, and thereby benefit in designing e-insurance products. They have replicated borderline data for each BBN nodes [32].

Junling Ma et al. have revised earlier effort and launch extra all-purpose circumstances under which Kermack and McKendrick's formula is effective. They presented that the final size method is unaffected if there is a latent phase, any number of different transmittable phases and/or a phase through which infectious nodes are quarantined [33].

M. E. Alexander et al. has witnessed that despite the usefulness of vaccines in melodramatically declining the quantity of novel transferable cases and severity of infections, defective vaccines may not totally stop contagion. The study of the model offers the threshold state for ailment regulator in relations of selected main parameters: coverage of the principal vaccine; efficacy of the vaccine; fading rate; and the rate of booster supervision [34].

J. Lopez Gondar and R. Cipolatti familiarized a simplified theoretical model to define a cybernetic virus transmission method in a set of interrelating processors. The transmission mechanisms considered here are those connected to the reception of communications through the Internet as well as the ones regarding the modest conversation of files by means of recording devices as compact disks or the usually used floppy disks. In spite of its intrinsic effortlessness, this model offers a decent knowledge of the contamination procedure and inclinations [35].

A self-adaptive disseminated agent-based protection resistant scheme established on organic plans has been established within a graded layered architecture by **Paul K. Harmer et al.** A sample interactive structure is premeditated and employed in Java, and the same is verified also. The outcomes authenticate the use of a dispersed-agent organic-system method toward the computer security difficulties of virus removal [36].

Wang Wendia and Ma Zhiena have discoursed the asymptotic behavior of the SIS model with time delay. They have revealed that $R0 > 1$ suggests that the infection will be consistently tenacious. Their outcomes specify that a huge contamination span, together with a reproduction number near to 1, will lead to the global stability of the endemic equilibrium [37].

Thresholds, equilibria, and their stability are found for SIQS and SIQR epidemiology models with three methods of the occurrence by **Herbert Hethcote et al.** For most of these models, the endemic equilibrium is asymptotically stable, but for the SIQR model with the isolation-adjusted occurrence, the endemic equilibrium is an unbalanced spiral for some parameter values and periodic results rise by Hopf bifurcation [38].

3 Conclusion

In this paper, we have considered the philosophies of epidemiological modeling as offered by diverse writers to comprehend the contaminating nature, transmission mode, and destructive conduct of malware through diverse modeling, e.g., SEIR epidemic model, SIR model, SIRS model, SEIQV epidemic model, SEIV model, SIQS model, and SIQR model in different network, viz, mesh network, M2M network, and WSN. We have come to know that, if, basic reproduction number, $R0 < 1$, the disease-free equilibrium is globally asymptotically stable and if,

$R0 > 1$, endemic equilibrium is globally asymptotically stable and the infection is insistent. This study shows that the principle of epidemiological modeling for malware propagation can easily be applicable in e-commerce network also.

References

1. Tianhan Gao, Quanqi Wang, Xiaojie Wang, and Xiaoxue Gong, "An Anonymous Access Authentication Scheme Based on Proxy Ring Signature for CPS-WMNs", Mobile Information Systems, Volume 2017 (2017), Article ID 4078521, 11 pages, 2017.
2. Lianwen Wang, Xingan Zhang, Zhijun Liu, "An SEIR Epidemic Model with Relapse and General Nonlinear Incidence Rate with Application to Media Impact", Qualitative Theory of Dynamical Systems, 2017.
3. Jinliang Wang, Xianning Liu, Toshikazu Kuniya, "Global stability for multi-group SIR and SEIR epidemic models with age-dependent susceptibility", Discrete and Continuous Dynamical Systems—Series B (DCDS-B), American Institute of Mathematical Science, Pages: 2795–2812, Volume 22, Issue 7, September 2017, https://doi.org/10.3934/dcdsb.2017151, 2017.
4. Vyacheslav Kharchenko, and Oleg Illiashenko, "Diversity for security: case assessment for FPGA-based safety-critical systems", 20th International Conference on Circuits, Systems, Communications and Computers (CSCC 2016), Volume 76, 2016, MATEC Web Conf., 02051, 2016.
5. Elisa Canzani, Stefan Pickl, Cyber Epidemics: Modelling Attacker-Defender Dynamics in Critical Infrastructure Systems", Advances in Human Factors in Cyber security pp. 377–389, 2016.
6. Kuldeep Kaur, Dr. Ashutosh Pathak, Parminder Kaur, Karamjeet Kaur, "E-Commerce Privacy and Security System", Int. Journal of Engineering Research and Applications, ISSN: 2248–9622, Vol. 5, Issue 5, (Part-6) May 2015, pp. 63–73, 2015.
7. Eike Möhlmann, Oliver Theel, "Breaking Dense Structures: Proving Stability of Densely Structured Hybrid Systems", Electronic Proceedings in Theoretical Computer Science, 2015; 184(Proc. ESSS 2015):49–63 https://doi.org/10.4204/eptcs.184.4, 2015.
8. Liping Feng et el.; "Modeling and Stability Analysis of Worm Propagation in Wireless Sensor Network", Mathematical Problems in Engineering (Hindawi Publishing Corporation), Volume 2015 (2015), Article ID 129598, 8 pages 2015; 2015 https://doi.org/10.1155/2015/129598 1024-123X (Print); 1563-5147 (Online), 2015.
9. Jinhua Ma et el., "Analysis of Two-Worm Interaction Model in Heterogeneous", M2M Network", "information", www.mdpi.com/journal/information, ISSN 2078-2489 2015, 6, 613-632; https://doi.org/10.3390/info6040613, 2015.
10. Canzani et al., "Insights from Modeling Epidemics of Infectious Diseases – A Literature Review", Proceedings of the ISCRAM 2015 Conference—Kristiansand, 2015.
11. Meng Wang et al., "Spread and Control of Mobile Benign Worm Based on Two-Stage Repairing Mechanism", Journal of Applied Mathematics (Hindawi Publishing Corporation) Volume 2014 (2014), Article ID 746803, 14 pages, http://dx.doi.org/10.1155/2014/746803 2014;2014, https://doi.org/10.1155/2014/746803 1110-757X (Print); 1687-0042 (Online)", 2014.
12. Elisa Canzani, Hans-Christian Heldt, Stephan Meyer and Ulrike Lechner, "Towards an Understanding of the IT Security Information Ecosystem", Autonomous Systems 2014, Proceedings of the 7th GI Conference. VDI Reihe, 2014.
13. XingboLiu, Lijuan Yang, "Stability analysis of an SEIQV epidemic model with saturated incidence rate", Nonlinear Analysis: Real World Applications, 2671–2679, 2012.

14. Yajuan Zhang, Xinyang Deng, Daijun Wei, Yong Deng, "Assessment of E-Commerce security using AHP and evidential reasoning", Expert Systems with Applications 39 (2012) 3611–3623 Elsevier, 2012.
15. "Ryan E Hohimer and Frank L Greitzer, "Modeling Human Behavior to Anticipate Insider Attacks", Journal of Strategic Security, ISSN: 1944-0464 (Print); 1944-0472 (Online) 2011;4 (2):25–48, 2011.
16. Bimal Kumar Mishra, Samir Kumar Pandey, "Fuzzy epidemic model for the transmission of worms in computer network Nonlinear Analysis: Real World Applications 11, 5, 4335–4341, Oct, 2010.
17. Huang, Gang, "Global stability for delay SIR and SEIR epidemic models with nonlinear incidence rate", Bulletin of Mathematical Biology. 72(5), p. 1192–1207, 01-07-2010.
18. Jose R. C., Piqueira and Felipe Barbosa Cesar, "Dynamical Models for Computer Viruses Propagation", Mathematical Problems in Engineering, Volume 2008, Article ID 940526, 11 pages, https://doi.org/10.1155/2008/940526, 2008.
19. Jose R.C. Piqueira, Adolfo A. de Vasconcelos, Carlos E.C.J. Gabriel, Vanessa O. Araujo, "Dynamic models for computer viruses", computers & security, 27 (2008) 355–359, 2008.
20. Fangwei Wang, Yunkai Zhang, Jianfeng Ma, "Modeling and Analyzing Passive worms over Unstructured Peer-to-Peer Networks", Transactions of Tianjin University, 14(1):66–72, 2008.
21. Li-Ming Cai, Xue-Zhi Li, "Analysis of a SEIV Epidemic Model with a Nonlinear Incidence Rate Applied Mathematical Modelling", 7-Jan-2008.
22. Zhong Zhao, Lansun Chen, Xinyu Song, "Impulsive vaccination of SEIR epidemic model with time delay and nonlinear incidence rate", Mathematics and Computers in Simulation, Elsevier, Volume 79, Issue 3, 1 December 2008, Pages 500–510, 2008.
23. Richard J. Boys and Philip R. Giles, "Bayesian inference for stochastic epidemic models with time-inhomogeneous removal rates", Mathematical Biology, J. Math. Biol. (2007) 55:223–247, 15-March-2007.
24. Bimal Kumar Mishra, Dinesh Saini, "Mathematical models on computer viruses", Applied Mathematics and Computation, Elsevier, 187, 2, 929–936, 2007, 15-April-2007.
25. Bimal Kumar Mishra, Dinesh Kumar Saini, "SEIRS epidemic model with delay for transmission of malicious objects in computer network", Applied Mathematics and Computation, Elsevier, 188, 1476–1482, 2007.
26. Bimal Kumar Mishra, Navnit Jha, "Fixed period of temporary immunity after run of anti-malicious software on computer nodes Applied Mathematics and Computation, Elsevier, 190, 1207–1212, 2007, 15-July-2007.
27. B.K. Mishra et al., "Differential susceptibility-infectiousness epidemic model of propagation of malicious agents with self-replication in a computer network", Applied Mathematics and Computation, xxx (2007) xxx–xxx, 2007.
28. Bimal Kumar Mishra, "Generality of the final size formula for infected nodes due to the attack of malicious agents in a computer network, "Applied Mathematics and Computation, xxx (2007) xxx–xxx, 2007.
29. Hisashi Inaba, "Age-structured homogeneous epidemic systems with application to the MSEIR epidemic model Mathematical Biology J. Math. Biol. (2007) 54:101–146, 2007.
30. Tailei Zhang, Zhidong Teng, "Global behavior and permanence of SIRS epidemic model with time delay", Nonlinear Analysis: Real World Applications, ELSEVIER, 14 March 2007.
31. E. Shim, Z. Feng, M. Martcheva, C. Castillo-Chavez, " An age-structured epidemic model of rotavirus with vaccination", Mathematical Biology, J. Math. Biol. (2006) 53:719–746, 2006.
32. Arunabha Mukhopadhyay, Samir Chatterjee, Debashis Saha, Ambuj Mahanti, Samir K Sadhukhan, "e-Risk Management with Insurance: A framework using Copula aided Bayesian Belief Networks ", Proceedings of the 39th Hawaii International Conference on System Sciences—2006.
33. Junling Ma, David J.D. Earn, "Generality of the Final Size Formula for an Epidemic of a Newly Invading Infectious Disease, "Bulletin of Mathematical Biology (2006), 68: 679–702, 2006.

34. M.E. Alexander, S.M. Moghadas, P. Rohani, A.R. Summers, " Modelling the effect of a booster vaccination on disease epidemiology", Mathematical Biology, J. Math. Biol. 52, 290–306 (2006), 10-November-2005.
35. J. Lopez Gondar and R. Cipolatti, "A mathematical model for virus infection in a system of interacting computers", Computational and Applied Mathematics, Vol. 22, N. 2, pp. 209–231, 2003.
36. Paul K. Harmer, Paul D. Williams, Gregg H. Gunsch, and Gary B. Lamont, An Artificial Immune System Architecture for Computer Security Applications", IEEE TRANSACTIONS ON EVOLUTIONARY COMPUTATION, VOL. 6, NO. 3, JUNE 2002.
37. Wang Wendia, Ma Zhiena, "Global dynamics of an epidemic model with time delay", Nonlinear Analysis: Real World Applications, 3(2002)365–373, 2002.
38. Herbert Hethcote, Ma Zhien b, Liao Shengbing, "Effects of quarantine in six endemic models for infectious diseases", Mathematical Biosciences 180 (2002) 141–160, 26-March-2002.

Mr. Biswarup Samanta has more than **15 years** of academic (both **Teaching & Academic Administration**) experience in MCA, B.Tech, M.Sc-IT, B.Sc-IT, BCA, MBA-IT, BBA-IT and Diploma programs of various national level Colleges, Universities & B-Schools, since Aug'2003. Presently he is working as an **Asst. Professor** and **Program Leader (CSIT)** in The Department of Computer Science and IT, with Amity University Jharkhand, Ranchi, since Aug'2016. He has more than 5 years of **research** experience since 2013. Mr. Samanta has **already been submitted the Thesis** of his **Ph.D.** in **Computer Science** (*as per UGC 2009 norms*). His research area is "Developing and Analyzing Dynamic Models on the Attack of Malware in E-Commerce Network". He had presented his research papers in various International & National Conferences/Seminar and published his research papers in various International & National Journals with good impact factor. Mr. Samanta is a Life Member of ***Computer Society of India (CSI),*** Computer Science Teachers Association **(CSTA)** & International Associations of Engineers **(IAENG)**.

Mr. Samanta, a Chemistry Graduate, holds **double** Post Graduate degrees in "Computer Applications" (MCA) and "Business Administration" (MBA in Systems & Mkt.), respectively. Apart from attending **one month long FDP in IT** at ICFAI Head Quarter, **Hyderabad** and 2 weeks **FDP** in **Entrepreneurship** at **Usha Martin**, Ranchi, sponsored by NSTEDB, Department of Science & Technology, Govt of India, he has also conducted and attended many seminars, workshops, FDPs & MDPs in IT and Management, respectively, throughout India. He was also appointed two times as a resource person for IGNOU **Interactive Live Radio** Counseling Program at **Gyan Vani FM (105.4 MHz)** for MCA program in Kolkata.

Prior to joining **Amity University Jharkhand**, Mr. Samanta had worked with various organizations, like, **Jharkhand Rai University**, Ranchi as an Asst. Professor in The Department of Computer Science for 1.5 years, **Usha Martin Education & Solutions Ltd.** as an Asst. Professor in The Department of Computer Science for 5 years, **ICFAI University** for 5 years and 2 years for **Sikkim Manipal University**.

Q-MRAS-Based Speed Sensorless Permanent Magnet Synchronous Motor Drive with Adaptive Neural Network for Performance Enhancement at Low Speeds

Badini Sai Shiva, Vimlesh Verma and Yawer Abbas Khan

1 Introduction

Permanent magnet synchronous motor (PMSM) has collected an increasing demand in the passing years, due to their high efficiency, large torque to volume ratio and reliability. Compared to synchronous motor, PMSM is more advantageous in considerations of elimination of rotor copper losses, slip rings, and brushes in field winding. Since losses are concentrated in the stator, there will be no problem of cooling.

The vector-controlled drive is more popular in industrial applications due to its superior dynamic performance [1, 2]. From the viewpoint of cost, reliability, compatibility, signal transmission, hazardous, and environment issues, the PMSM is operated in closed loop without speed sensor. So, we require some speed estimation techniques to estimate the speed in order to have successful speed control of PMSM drive [3]. Several speed estimation techniques have been recommended in recent years in the literature to accomplish speed sensorless operation of vector-controlled PMSM drive.

The back emf-based method [4], signal injection-based method [5], model reference adaptive system-based techniques [6], sliding model observer [7], state observer-based method [8] are some ways of speed estimation techniques available in the literature. Each one of them has their own merits and demerits [9]. Some of them give the good performance at low speeds and some for high speeds of

B. Sai Shiva (✉) · V. Verma · Y. A. Khan
Department of Electrical Engineering, National Institute of Technology, Ashok Rajpath, Patna 800005, Bihar, India
e-mail: badinisai.eepg16@nitp.ac.in

V. Verma
e-mail: vimlesh.verma@nitp.ac.in

Y. A. Khan
e-mail: yawer.abbaskhan@gmail.com

J. Chattopadhyay et al. (eds.), *Innovations in Soft Computing and Information Technology*, https://doi.org/10.1007/978-981-13-3185-5_10

operation. In model reference adaptive system (MRAS) family [10] of estimators, reactive power-based MRAS-based speed estimation technique is designed from this paper [9, 11]. The Q-MRAS-based speed estimator gives better results for both (motoring and regenerative) modes of operations [9]. Also the Q-MRAS gives good operating performance for high speeds to zero speed with high and medium torques. The existing Q-MRAS-based speed estimator has very poor performance at low speeds with no-load or lightly loaded conditions. The performance of the existing drive with MRAS-based speed estimator is slightly improved using artificial neural networks (ANNs) [12, 13]. This paper makes use of ANN in the Q-MRAS-based speed estimator [14, 15]. Thus in this paper, we are working with ANN-based Q-MRAS, which is trained by Bayesian regularization method. Modeling of PMSM and Q-MRAS-based speed estimator are presented in Sect. 2. The proposed ANN-based Q-MRAS speed estimator is discussed in Sect. 3, and in Sect. 4, all simulation results carried out in MATLAB/Simulink platform are shown and discussed. Finally, Sect. 5 concludes the work.

2 Modeling of PMSM and Q-MRAS-Based Speed Estimator

2.1 Modeling of PMSM

The stator voltages of PMSM are expressed in *d*- and *q*-axes with rotor reference frame [2].

$$v_{\mathrm{ds}} = R_{\mathrm{s}} i_{\mathrm{ds}} + L_d \frac{\mathrm{d}i_{\mathrm{ds}}}{\mathrm{d}t} - \omega_{\mathrm{s}} L_q i_{\mathrm{qs}} \tag{1}$$

$$v_{\mathrm{qs}} = R_{\mathrm{s}} i_{\mathrm{qs}} + L_q \frac{\mathrm{d}i_{\mathrm{qs}}}{\mathrm{d}t} + \omega_{\mathrm{s}} L_d i_{\mathrm{ds}} + \omega_{\mathrm{s}} \lambda_{\mathrm{af}} \tag{2}$$

$$T_{\mathrm{e}} = \left(\frac{3}{2}\right) p \left\{ i_{\mathrm{qs}} \lambda_{\mathrm{af}} + \left(L_d - L_q\right) i_{\mathrm{ds}} i_{\mathrm{qs}} \right\} \tag{3}$$

$$T_{\mathrm{e}} - T_{\mathrm{L}} = J \frac{\mathrm{d}\omega_{\mathrm{r}}}{\mathrm{d}t} + B\omega_{\mathrm{r}} \tag{4}$$

where 'p' is number of 'pole pair', 'λ_{af}' is the 'mutual flux' linkage between rotor and stator due to permanent magnet, 'L_q' and 'L_d' are the 'q-axis' stator inductance and 'd-axis' stator inductance, respectively, 'R_{S}' is the 'stator resistance', and '$\omega_{\mathrm{s}} = p\omega_{\mathrm{r}}$'; here the PMSM is non-salient type with a sinusoidal back emf waveform. The developed electromagnetic torque is expressed in Eq. (3). The governing electromechanical equation is expressed in Eq. (4). Where 'T_{e}' and 'T_{L}' are electric torque and load torque.

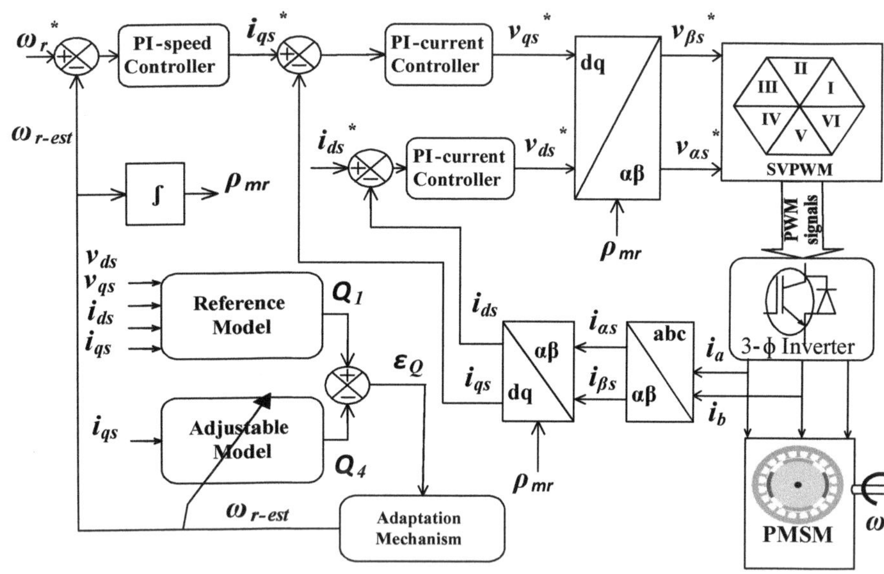

Fig. 1 Block diagram of the complete speed sensorless vector-controlled PMSM drive with ANN-based MRAS-based rotor speed estimator

Figure 1 represents the block diagram of the complete vector-controlled sensorless PMSM drive with Q-MRAS-based speed estimator. The reference model computes instantaneous values which are independent on machine parameters, and the adjustable model computes steady-state values which are dependent on the required unknown parameter.

Error signal is extracted from the difference of two differently expressed same quantities and passed through an adaptation mechanism (basically PI) to estimate required parameter. The estimated parameter is used to tune the adjustable model until the error becomes zero.

2.2 Q-MRAS-Based Speed Estimator

The instantaneous reactive power (Q) can be expressed as [9].

$$Q_1 = v_{qs} i_{ds} - v_{ds} i_{qs} \tag{5}$$

Substituting Eqs. (1) and (2) in Eq. (5)

$$Q_2 = \left(L_q i_{ds} \frac{di_{qs}}{dt} - L_d i_{qs} \frac{di_{ds}}{dt} \right) + \omega_s \left(L_d i_{ds}^2 + L_q i_{qs}^2 \right) + \omega_s i_{ds} \lambda_{af} \tag{6}$$

Under steady state, $\frac{\mathrm{d}}{\mathrm{d}t}$ terms are zero, and the new expression for Q becomes:

$$Q_3 = \omega_s\left(L_d i_{ds}^2 + L_q i_{qs}^2\right) + \omega_s i_{ds}\lambda_{af} \tag{7}$$

Now the condition for 'vector-controlled' PMSM drive ($i_{ds} = 0$) is imposed in Eq. (7). Therefore, the more simplified expression for Q is:

$$Q_4 = \omega_s L_q i_{qs}^2 = p\omega_r L_q i_{qs}^2 \tag{8}$$

Among all the expression, Q_2 and Q_3 are dependent on L_q, λ_{af} and L_d, and the presence of derivative will increase the noise. And when we come across Q_4, it is dependent only on L_q.

Q_1 and Q_4 equations are used to build reference model and adjustable model of speed estimation MRAS technique. Here Q_1 is independent on speed and machine parameters, and Q_4 is dependent on rotor speed.

Error is evaluated from Eq. (9) difference between Q_1 and Q_4 and passed through an adaption mechanism which is normally PI controller. The output of the adaptation mechanism is the estimated quantity ($\omega_{r,est}$), which is used for adjusting the adjustable modeling such a way to reduce the error to zero.

$$\varepsilon = (v_{qs} i_{ds} - v_{ds} i_{qs}) - \omega_s(\varepsilon, t) L_q i_{qs}^2 \tag{9}$$

3 The Proposed ANN-Based Q-MRAS Speed Estimator

The ANN-based Q-MRAS speed estimator was developed by using the MATLAB/ Simulink tools that was shown in Fig. 3. Once ANN was developed, it will be trained for the data acquired from speed sensor vector-controlled PMSM drive for various speeds and loads.

Figure 2 shows the existing Q-MRAS-based speed estimator which is available in the literature. The structure of the proposed ANN-based Q-MRAS speed estimator is shown in Fig. 3. In the proposed estimator, the adjustable part is modeled by ANN. Here in this paper, ANN is developed with single hidden layer of 24 neurons and trained by Bayesian regularization method.

Reference model was developed by instantaneous reactive power (Q_1 or Q_{ref}) of PMSM. Adjustable model computes steady-state reactive power (Q_4 or Q_{est}) which was developed by ANN. Error was evaluated by comparing the two reactive powers. In this paper, rotor speed was evaluated by passing error signal through adaption mechanism. Adaption mechanism consists of PI controller. Estimated speed is feedback to ANN-based adjustable model to tune it until error signal was zero. The proposed system is well trained, so that for lightly loaded conditions, very low values of i_{qs} (with its nonlinear nature in steady-state reactive power (i.e.))

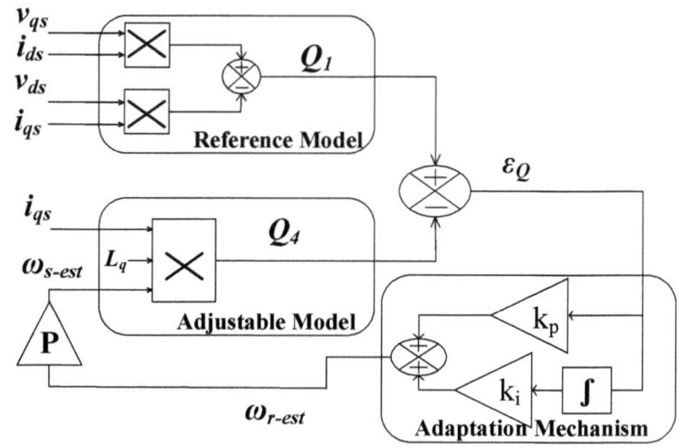

Fig. 2 Existing Q-MRAS-based speed estimator for vector-controlled PMSM drive

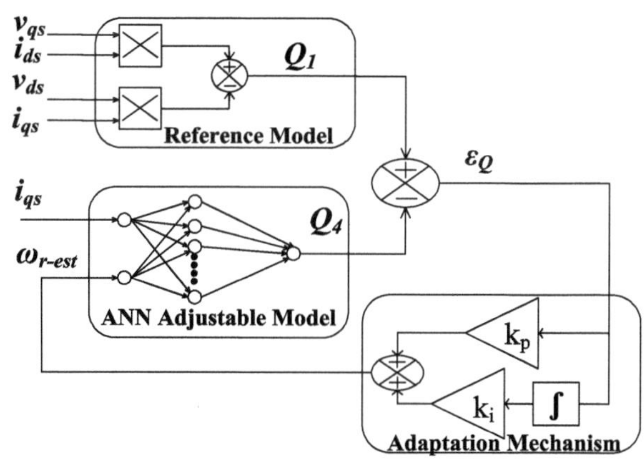

Fig. 3 Proposed ANN-based MRAS-based rotor speed estimator

$$\omega_s L_q i_{qs}^2 \tag{110}$$

will cause the error in existing Q-MRAS that can be overcome by ANN-based Q-MRAS.

Figure 4 shows the error histogram with 20 bins for the two steps, training and testing in ANN model. The zero error is denoted by a yellow line, each vertical bar represents the samples of data present in that particular bin. Figure 5 shows the training performance of ANN by least mean square error. It shows the validation performance and mean square error of the network, which starts with a large error and is trained to a small error.

Figure 6 shows comparison between linear regression and ANN model results. The three plots in Fig. 6 represent the training, validation, and testing. That plot shows how data was fitted to the target by ANN. They indicate an exact linear relationship between outputs and targets.

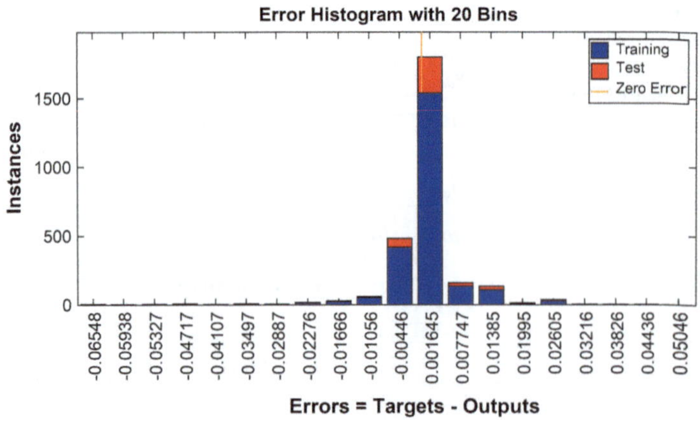

Fig. 4 Error histogram for the trained ANN

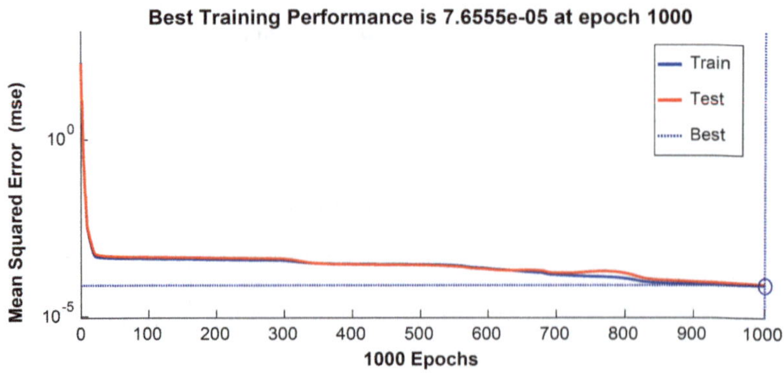

Fig. 5 Least mean square error (LMSE) of ANN

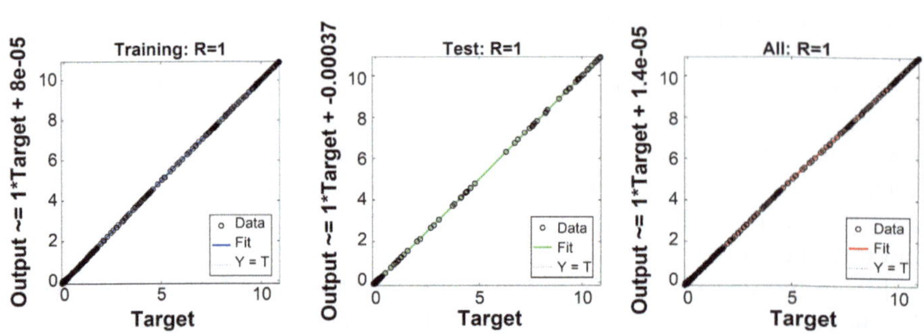

Fig. 6 Graph represents the network validation for ANN

4 Simulation Results

An exhaustive simulation has been done in MATLAB/SIMULINK to verify the performance of the proposed controller. The parameters of machine are available in Table 1 [9].

Figure 7 shows the simulation results of speed for existing Q-MRAS-based speed estimation, and Fig. 8 shows the corresponding stator current of Fig. 7. At time t = 50 s, motor decelerates to zero speed and then continues to accelerate in opposite direction of rotation up to time t = 100 s, and at time t = 160 s, motor again decelerates and reaches zero speed and starts rotating in opposite direction.

Figure 9 shows the simulation results of speed for the proposed ANN-based Q-MRAS speed estimated and shaft speed. Figure 10 shows the d- and q-axes stator currents of PMSM of Fig. 9. In Fig. 10, current i_{qs} represents torque component of current which has similar profile as that of torque developed.

ANN-based Q-MRAS speed estimator was closed to give speed sensorless vector-controlled drive, and simulation results of it are shown in Fig. 11; this figure shows actual shaft speed, estimated speed of existing Q-MRAS-based speed estimator, and estimated speed of the proposed ANN-based Q-MRAS speed estimator.

Table 1 PMSM machine parameters

Pole pair (P)	3
Nominal speed (ω_n)	1430 rpm
d-axis inductance (L_d)	6.66 mH
q-axis inductance (L_q)	5.80 mH
Mutual flux linkage between rotor and stator due to permanent magnet (λ_{af})	0.1546 Wb turn
Stator phase winding resistance (R_S)	1.4 Ω
Machine inertia (J)	0.00176 Kg m^2
Viscous coefficient (B)	0.00038818

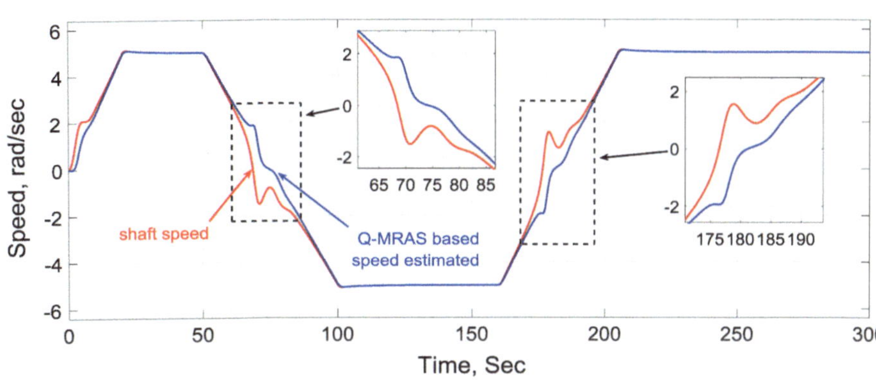

Fig. 7 Simulation results for speed with existing Q-MRAS in forward and reverse motoring modes

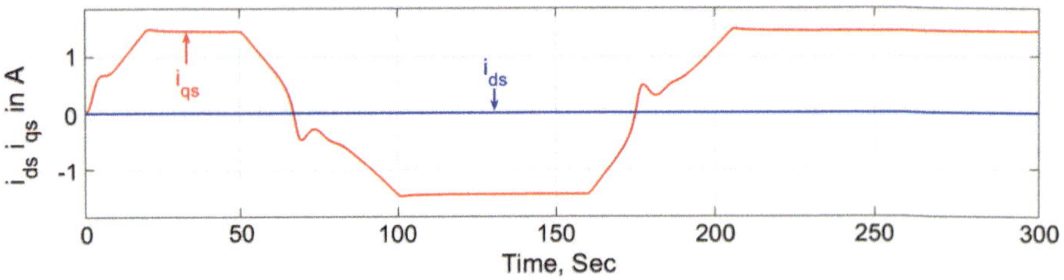

Fig. 8 Simulation results for *d*- and *q*-axes stator currents for Fig. 7

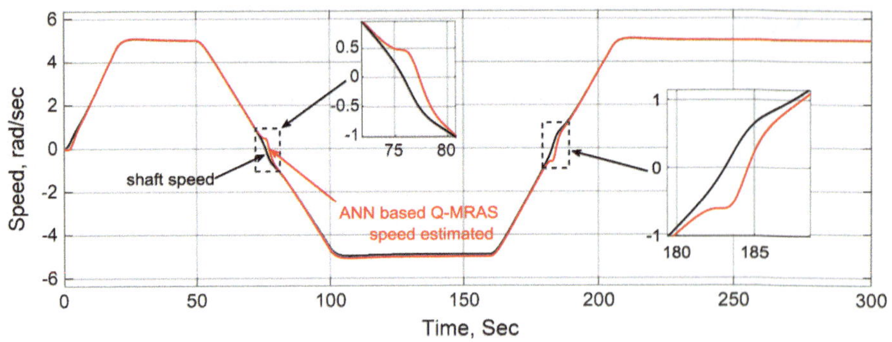

Fig. 9 Simulation results for speed with ANN-based Q-MRAS speed estimator for forward and reverse motoring modes

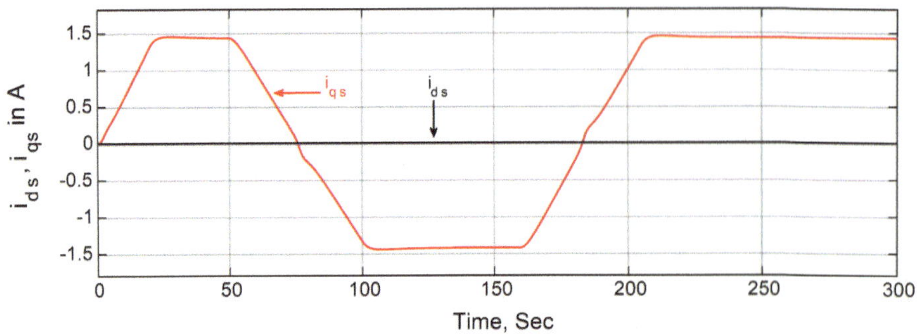

Fig. 10 Simulation results for *d*- and *q*-axes stator currents for Fig. 9

It clearly shows that ANN-based speed estimator estimates speed more accurately than existing Q-MRAS from Fig. 11. As simulation results show the improved performance of shaft speed at zero crossing from Figs. 7, 9 and 11. Here we can clearly notify that performance was improved at zero and very low speeds with the proposed ANN-based Q-MRAS.

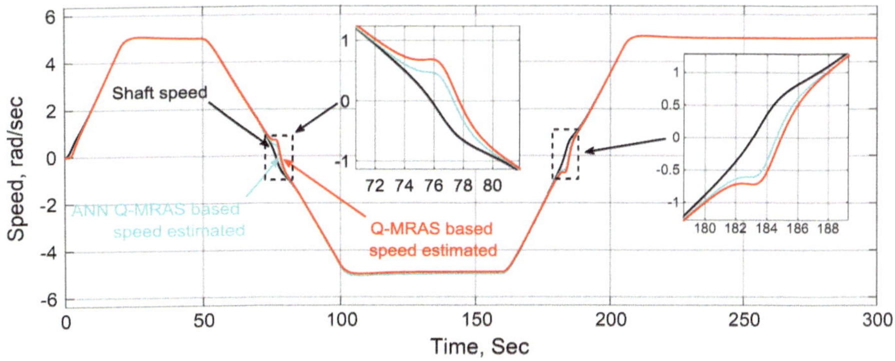

Fig. 11 Simulation results for speed with ANN-based Q-MRAS and existing Q-MRAS in forward and reverse motoring modes when vector-controlled loop closed with ANN-based Q-MRAS

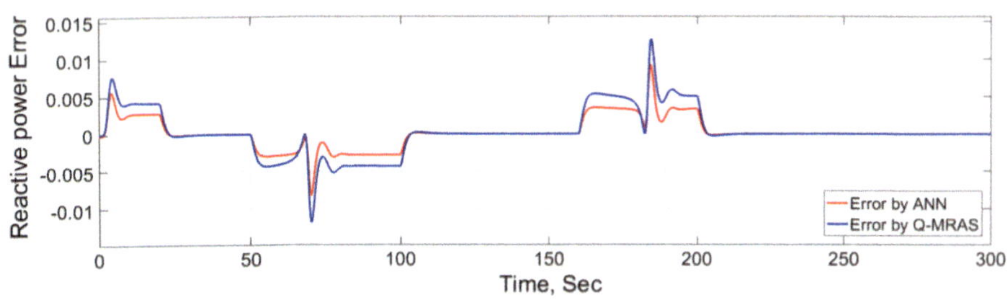

Fig. 12 Reactive power error for the proposed ANN-based Q-MRAS and existing Q-MRAS

Figure 12 shows the reactive power error between Q_1 and Q_4 for existing Q-MRAS and the proposed ANN-based Q-MRAS. As it shows that existing Q-MRAS will give more error than ANN-based Q-MRAS at lightly loaded conditions.

Figures 13 and 15 are the detailed view of Fig. 12 for 50–110 and 160–210 s. Figures 14 and 16 shows the corresponding speed errors for the same duration of Figs. 13 and 15. From the Fig. 11 we can observe that the actual shaft speed follows the reference speed over the time period 30–55, 100–160, and 210–300 s. The reactive power error and the speed error for zero crossing and low speeds with low torques are shown in Figs. 12, 13, 14, 15 and 16, that shows the improved performance of ANN-based Q-MRAS over existing Q-MRAS.

Figure 17 shows the simulation results for forward motoring and regenerating mode of operation with ANN-based Q-MRAS speed estimator, and Fig. 18 show the *d*- and *q*-axes stator currents corresponding to Fig. 17. Motoring operation can be seen clearly from 0 to 70 s and 180 to 300 s in Fig. 17, as the speed and the corresponding i_{qs} current are positive. Regenerative mode of operation can be seen

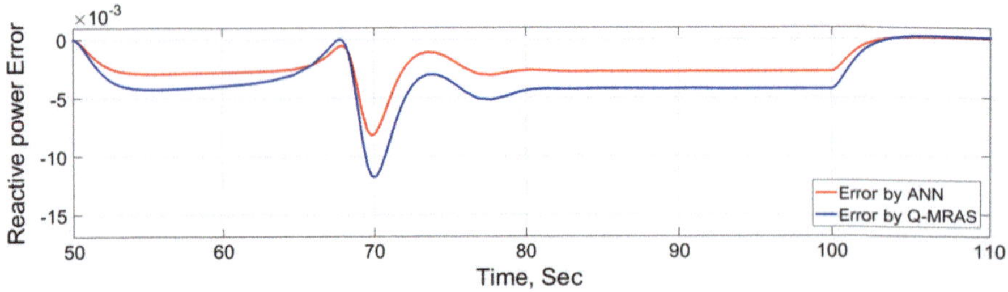

Fig. 13 Zoom around 50–110 s of Fig. 12

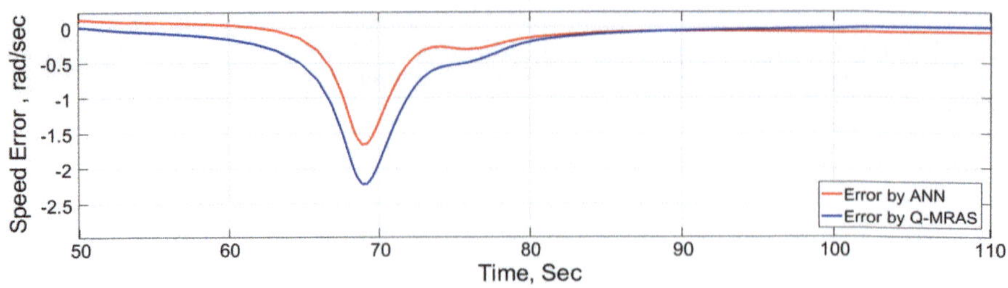

Fig. 14 Speed error for the time period of 50–110 s

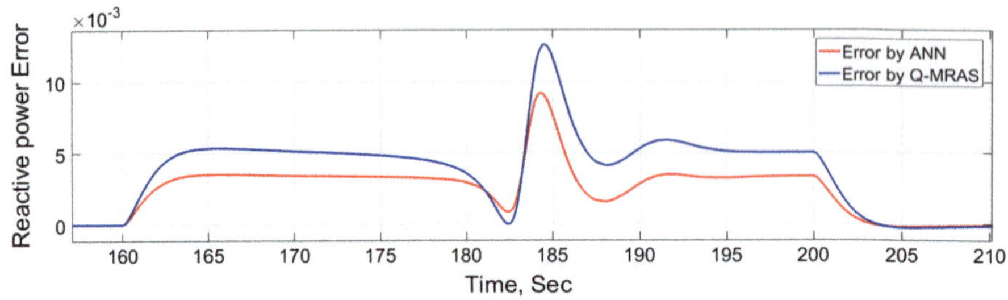

Fig. 15 Zoom around 155–210 s of Fig. 12

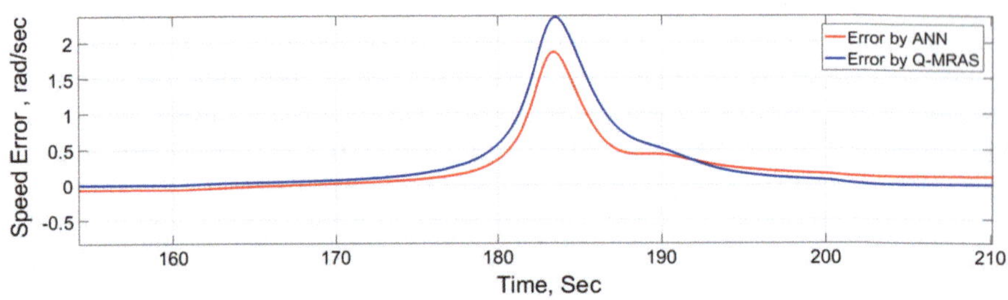

Fig. 16 Speed error for the time period of 155–210 s

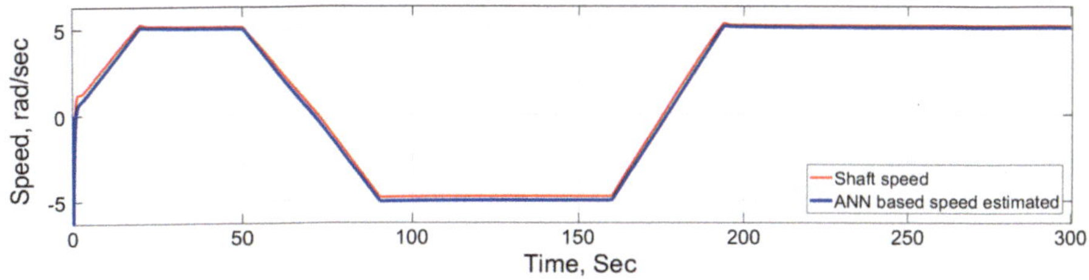

Fig. 17 Simulation results for forward motoring and regenerative mode of operation with ANN-based Q-MRAS speed estimation. Graph shows actual shaft speed and estimate speed by ANN

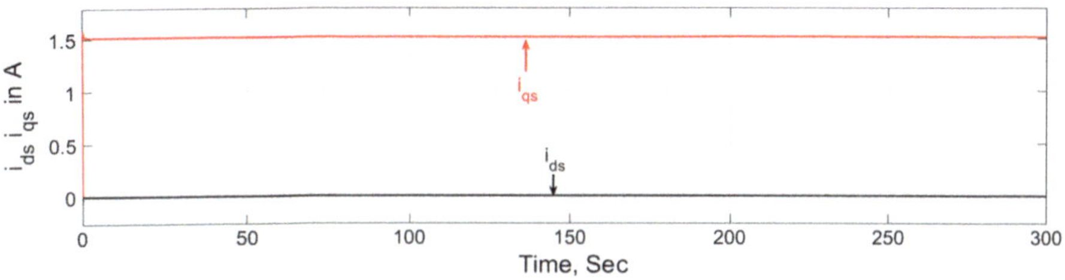

Fig. 18 *d*- and *q*-axes stator currents for Fig. 17

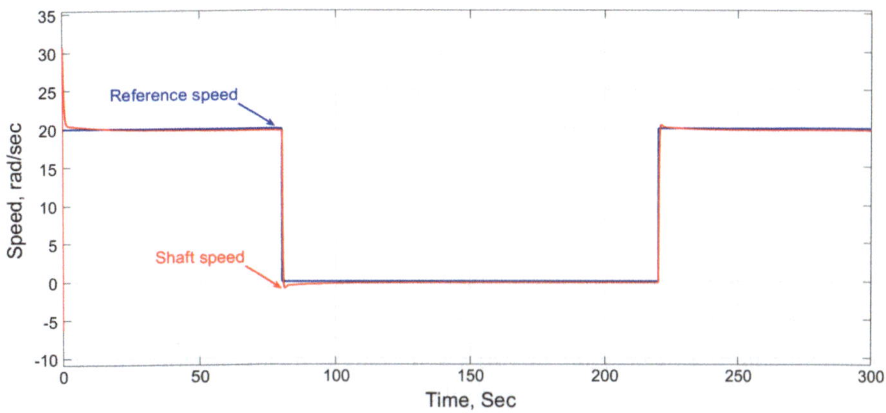

Fig. 19 Simulation results for zero speed operation with ANN-based Q-MRAS represents reference and actual rotor speed

clearly from 80 to 170 s in Fig. 17 as the speed reversed and the corresponding i_{qs} current are positive.

Figure 19 shows the speed reference command and actual shaft speed of PMSM. The command speed is changed to zero from 20 rad/s at 80 s and is again stepped

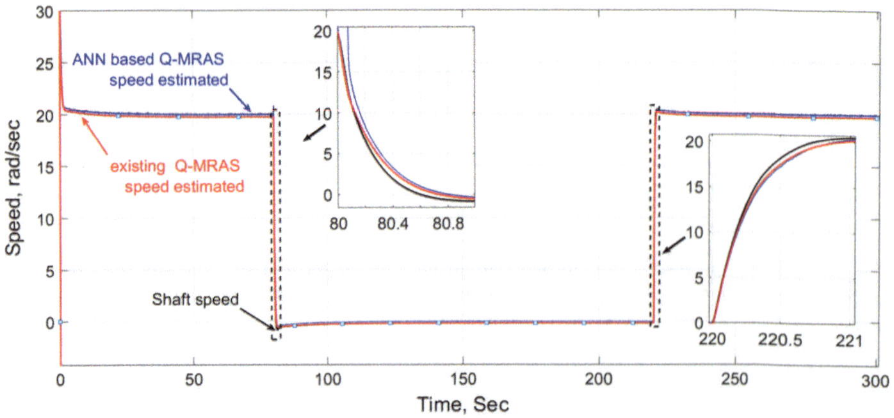

Fig. 20 Actual rotor speed, ANN-based Q-MRAS speed estimated, and Q-MRAS speed estimated

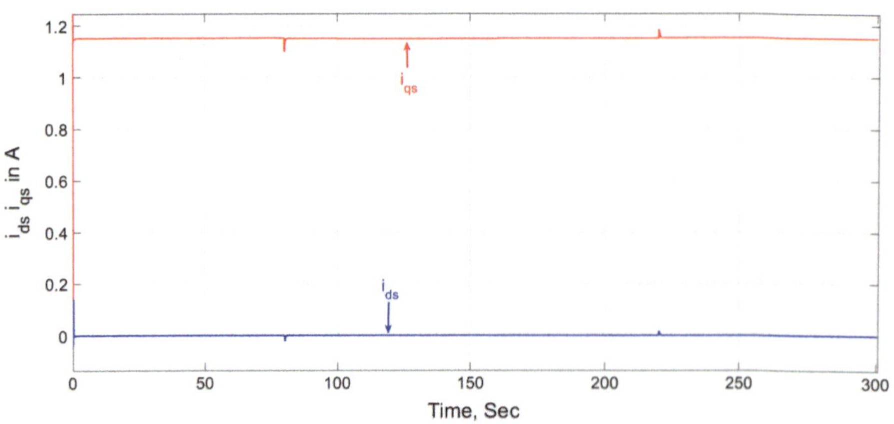

Fig. 21 *d*- and *q*-axes stator currents of PMSM for Fig. 19

up to 20 rad/s from zero rad/s at 220 s and continued. Figure 20 shows the performance of PMSM at zero speed with the proposed ANN-based Q-MRAS speed estimated and existing Q-MRAS speed estimated. Figure 21 shows the stator currents for Fig. 19 when controlled by the proposed model.

5 Conclusion

The performance of existing Q-MRAS-based speed estimator is improved specifically at low speeds and low torques by replacing the adjustable model by ANN. The ANN was developed by using MATLAB tools. The system is first trained with extensive data those are generated from simulations for a normal vector-controlled drive with speed sensor. Once training is complete, the weights are adjusted and the speed sensorless drive is tested for different operating conditions. This paper has used ANN to improve the performance for low speed with lightly loaded conditions in all four quadrants of operation.

Acknowledgements "This work was supported by the Science and Engineering Research Board (FILE NO. ECR/2016/000900), under Early Career Research Award".

References

1. Bose B.K, "Power Electronics and Variable Frequency Drives Technology and Applications", IEEE press, New York, 1996.
2. Krishnan R, "Permanent magnet synchronous and brushless DC motor drives", CRC press, Taylor & Francis, 2010.
3. Pillay P, Krishnan R, "Modeling simulation and analysis of permanent magnet synchronous motor drives", Part-I: The permanent magnet synchronous motor drive, IEEE Trans. Ind. Applicat. (1989) 265–273.
4. Mobarakeh B.N, Tabar F.M, Sargos F.M, "Back-EMF estimation based sensorless control of PMSM: robustness with respect to measurement errors and inverter irregularities", in: Conf. Rec. of 39th IAS Annual Meeting, vol. 3, 2004, pp. 1858–1865.
5. Corley M.J, Lorenz R.D, "Rotor position and velocity estimation for a salient pole permanent magnet synchronous machine at standstill and high speed, IEEE Trans. Ind. Applicat. 34 (1998) 784–789 (July/August).
6. Landau Y.P, Adaptive Control: "The Model Reference Approach", Marcel Dekker, New York, 1979.
7. Utkin V, Guldner J and Shi J, "Slide mode control in electromechanical systems," Taylor & Francis Press, 1999.
8. Bologani S, Oboe R, Zigliotto M, "Sensorless full digital PMSM drive with EKF estimation of speed and rotor position", IEEE Trans. Ind. Electron. 46 (1) (1999) 184–191 (February).
9. Maiti S, Chakraborty C, Sengupta S, "Simulation studies on model reference adaptive controller based speed estimation technique for the vector controlled permanent magnet synchronous motor drive", Elsevier, Simulation Modeling Practice and Theory 17 (2009) 585–596.
10. V. Verma and C. Chakraborty, "New series of MRAS for speed estimation of vector controlled induction motor drive," in IECON 2014 - 40th Annual Conference of the IEEE Industrial Electronics Society, 2014, pp. 755–761.
11. Kim Y.S, Choi Y.K, Lee J.H, "Speed sensorless vector control for PMSM based on instantaneous reactive power in the wide speed range", IEE Proc. Power Electr. Power Appl. 152 (2005) (September).
12. Bose B.K, "Neural network applications in power electronics and motor drives-An introduction and perspectives," IEEE Trans. Ind. Electron., vol. 54, no. 1, pp. 14–33, Jan. 2007.

13. Elbulk M.E, Tong L, Husain I, "Neural network based model reference adaptive systems for high performance motor drives and motion controls", IEEE Trans. Ind. Appticat. 38 (2002) (May/June).
14. Maiti S, Verma V, Chakraborty C and Hori Y, "An Adaptive Speed Sensorless Induction Motor Drive with Artificial Neural Network for Stability Enhancement". IEEE Trans. Ind. Info, vol. 8, no. 4, November 2012.
15. Batzel T.D, Lee K.Y, "An approach to sensorless operation of the permanent-magnet synchronous motor using diagonally recurrent neural networks", IEEE Trans. Energy Convers. 18 (2003) 100–106.

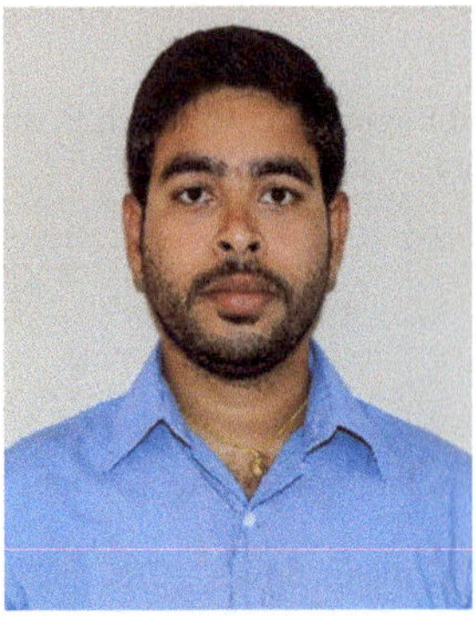

Badini Sai Shiva received the B.Tech. degree in electrical and electronics engineering from JNTU Hyderabad, Telangana, India, in 2012 and M. Tech degrees in Energy Systems from JNTUH College of Engineering Hyderabad, Telangana, India and in Control Systems from NIT Patna, India in 2015 and 2018, respectively. He is currently associated with the Department of Electrical Engineering, National Institute of Technology Patna, India, where he is working towards the Ph.D. degree. His research interest includes motor drives, control, estimation and renewable energy.

Vimlesh Verma was born in Mumbai, India. He received the B.Tech. degree in electrical and electronics engineering from Andhra University, Andhra Pradesh, India, in 2002, M.Tech. degree in power apparatus and systems from Nirma University, Gujarat, India, in 2005 and Ph.D. degree in Electrical Engineering from Indian Institute of Technology Kharagpur, Kharagpur, India in 2015. He is currently an Assistant Professor in the Department of Electrical Engineering, NIT Patna. His research interests include sensorless control of ac drives, fault diagnosis of induction-motor drives, and renewable energy. Mr. Verma was awarded the Queensland-India Friendship Scholarship to pursue a part of his research at the University of Queensland, Brisbane, Australia.

Yawer Abbas Khan received the B.Tech. degree in electrical engineering from NIT Srinagar, India, in 2012 and M. Tech degree in Power Systems from NIT Patna, in 2017. He is currently associated with the Department of Electrical Engineering, National Institute of Technology Patna, India, where he is working towards the Ph.D. degree. His research interest includes motor drives, control, estimation and Power quality issues.

Advance Anti-collision Device for Vehicles Using GPS and Zigbee

Satyajeet Patnayak, Anisha Swain and Manaswini Das

1 Introduction

Increase in population leads to subsequent increase in the number of automobiles plying regularly on the roads. As a consequence, the number of accidents in the world is on a hike. According to a report by WHO, around 1.25 million people perish each year owing to road accidents. The safety of mankind depends on us. Collision avoidance systems have proved their worth since the recent past nevertheless they are inadequate in their attempts. The major objective of this project was to introduce an advance collision avoidance system that can reduce the chance of accidents up to a significant extent. The key feature of this system is to access the location of all the vehicles within a demarcated area and spontaneously apply brakes if there exists a chance of imminent collision. This device is not just a conventional collision detection system but a smart collision avoidance system that can detect a chance of collision and take appropriate action even when a user is unaware of an obstacle in front of them. Hence, we hope this system brings forth a decline in the accident rates to a certain extent. Though the other vehicles do not have the required device, the user's controller can still sense and avoid head-on

S. Patnayak (✉)
Department of Mechanical Engineering, College of Engineering and Technology, Bhubaneswar, India
e-mail: satyajeetpatnayak@gmail.com

A. Swain
Department of Instrumentation and Electronics, College of Engineering and Technology, Bhubaneswar, India
e-mail: anishaswain2@gmail.com

M. Das
Department of Computer Science and Engineering, College of Engineering and Technology, Bhubaneswar, India
e-mail: dasmanaswini10@gmail.com

J. Chattopadhyay et al. (eds.), *Innovations in Soft Computing and Information Technology*, https://doi.org/10.1007/978-981-13-3185-5_11

collision by means of mmWave sensor. The best use of this system will be when the other vehicles contain a similar system for which the vehicles can even avoid side collisions, collisions in t-junctions, or collision in dividers.

2 Literature Survey

Till now innumerable devices have been invented, but the outcome is not up to the extent of avoiding accidents. Research has also been conducted in the area of adaptive cruise control to develop it further [1]. The front-end collision systems are being used in vehicles for a long time, and further researches are been conducted with the help of IR sensors to prevent front-end collisions [2]. Yet the current front-end collision warning system is not capable enough to establish their importance. Research has been extended further on the rear-end collision warning systems and systems where drivers can get a warning if there is a chance of collision from behind [3]. Engineers have found out that there may be a possibility that collisions can be avoided if there exists a peer-to-peer network in between the vehicles [4]. Recent works are more focused on the inter-vehicle communication by various means such as Tatchikou et al. [5]. In his paper, he has used dedicated short-range communication links for inter-communication between vehicles. Further researches have been conducted by Biswas et al. [5], for enhancing highway traffic safety by using dedicated short-range communication. Still some effective way has not been deciphered to achieve higher safety figures. Anurag et al. [6] in his paper has put some efforts to warn the drivers before collision. The system uses IEEE 802.15.4 for the communication. It operates on the free industrial scientific medal (ISM) of 2.4 GHz band. However, it had a few disadvantages as the search was basically for dumpers in mines and was unable to use the device in the real scenario. Apart from that the device was performing poor for unicast systems. Also using 802.15.4, the connectivity between two vehicles was highly depending on nature of path and most beneficial in LOS communication et al. [7]. Another paper by Uvaraja et al. [8] revealed a system using RFID and FLIR cameras, meant to be used in railways and not on roads.

3 Observation

In the recent past, there are less types of vehicle that use the conventional collision avoidance system with the help of cameras, radars, and infrared sensors. But the devices fail to work in places where there is no LOS contact of front vehicles and on road intersections. The other types of collision avoidance system which will be used in the near future are the interconnection of vehicles which keeps track of the speed and location of one vehicle with the other through a wireless channel. These types of systems are commonly known as CWS (CCWS). The demerits of the CWS systems are they cannot avoid an obstacle directly; however, by the help of local wireless networks, the vehicle can detect other vehicles without any visible line of contact between them. The system that was made by D. Anurag (GPS-based vehicular collision warning system using IEEE 802.15.4 MAC/PHY standard) has not yet been exercised on the standard vehicles using ECU.

In this paper, we have made use of progressive technology different from the conventional methods of anti-collision devices by two strategies. First, after detecting the collision, our system will automatically transfer the signal to the ABS of the vehicle for applying brakes, and secondly, machine learning has been applied to understand the response of the driver and sense the right time for the application of brakes, which is the time when it will be impossible for the driver to avoid the collision. To the best of our knowledge, this is the first work in developing a vehicle anti-collision device and integrating it with vehicle's ABS system. In addition to this, this system also makes use of machine learning in the ECU for better performance.

4 Proposed Model

From the many types of GPS and microcontrollers available, we zeroed in on NEO-6M GPS module and Arduino UNO as it was economic and could easily be implemented on microcontrollers.

4.1 *Finding the Safety Zone of Vehicles*

A safety zone or an imaginary rectangle surrounding the vehicle will be constructed and will be responsible for saving the coordinates of the safety zone that will be found by taking the size of vehicle in the device, i.e., its length and breadth as shown in Fig. 1.

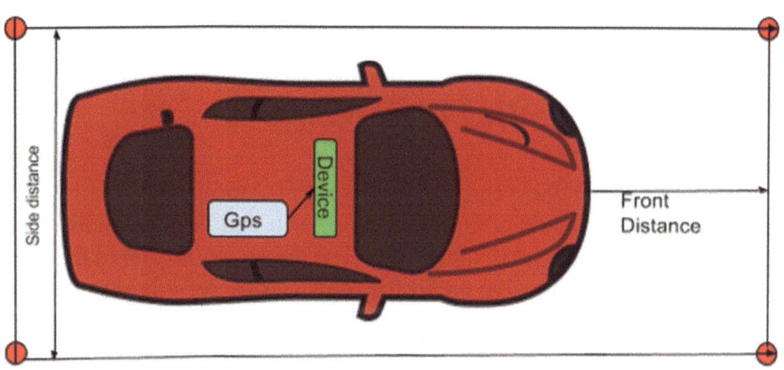

Fig. 1 Front and side distances of a vehicle

4.2 *Variation in the Safety Zones by Various Factors*

The safety zone or imaginary coordinates that the device will always detect will have varying coordinates, viz the size of the imaginary safety region will increase or decrease continuously taking into account certain conditions. For instance, when the speed of vehicle will alter the safety coordinates, the size will increase or decrease accordingly. Also the safety zone will decrease by taking into account the driving conditions; i.e., whether the vehicle is located in a crowded city or expressway.

4.2.1 The Variation of the Safety Zone with Speed is

If

v Velocity of vehicle,
μ Coefficient of friction (0.60 assumed),
g Acceleration due to gravity (9.8 m/s assumed),
d Stopping distance.

So,

$$d = \frac{v^2}{2\mu g} \tag{1}$$

Therefore, $d = X$ (the coordinate will be changed, and the distance will be made close to the stopping distance).

So, by calculating the stopping distance of the vehicle, we will change the coordinates of the rectangle accordingly. Apart from this, when the vehicle is travelling in crowded places, the imaginary coordinates will be chosen closer to vehicle in order to avoid any unnecessary warning. In this way, the device will adapt as per the surrounding and the behavior of the driving conditions of the driver.

4.3 *Automatically Braking of the Vehicle*

We altered the ECU for the vehicle comprising of Arduino by making a parallel connection with the vehicle's preinstalled ECU such that the preinstalled ECU will function smoothly without any hindrance. The ECU sends a signal to the ABS System where the motors receive the instructions and exercise brakes automatically by opening and closing the solenoid valves in order to prevent collision.

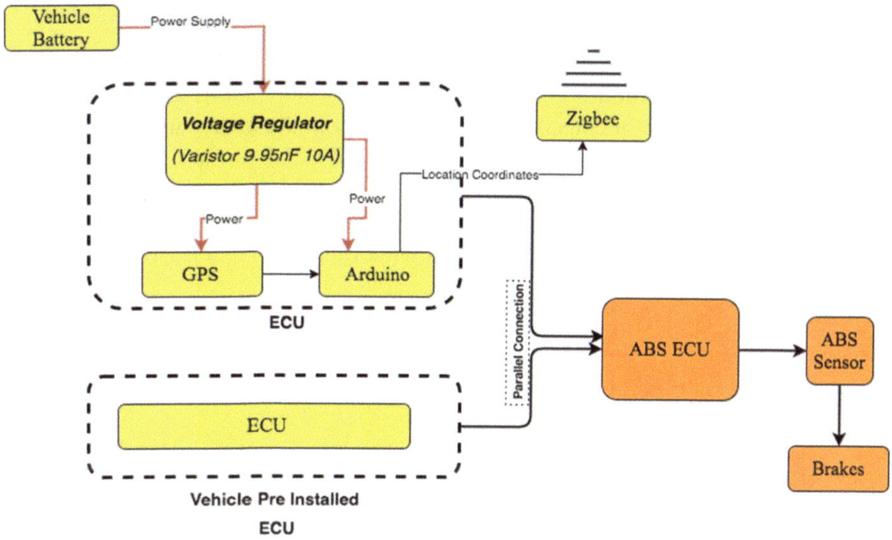

Fig. 2 Circuit Diagram

5 Circuit Diagram of Our Model

The circuit diagram depicted in Fig. 2 was used in our model. The system was made available to vehicles so that they can be connected to each other.

6 Results

We successfully demonstrated the vehicle collision avoidance system by using inter-vehicular communications. Our system comprised of four important steps:

1. Introduce an imaginary boundary in the vicinity of the car's outer surface.
2. Connect to other vehicles which are close to it.
3. Apply brakes automatically so as to avoid any chance of collision.
4. Analyze the behavior of the driver so that our device can understand driver's reaction time to determine when it has to automatically apply the brakes.

Our principal technical findings were as follows:

1. Connect the vehicles using standard Zigbee while using vehicle's power sources.
2. To interconnect the vehicle ECU with our own ECU comprised of Arduino UNO, Zigbee, GSM Module within a parallel connection such that we will not hinder the safety and electronics of the vehicle.

7 Conclusion

In this paper, we presented a collision avoidance system using Zigbee and GPS. The system was capable of detecting a vehicle absent in the user's immediate line of sight and automatically applying brakes using this ECU from a safe distance. We believe that our system will develop the safety of the vehicles and further lessen the chance of collisions up to a significant extent. We tried to make our system as robust as possible so that it can be used in a wide range of vehicles with significant accuracy.

8 Future Work

We intend to extend our research to the vehicles using non-ABS brakes as they make a majority of the automobiles. The vehicles which do not use the ABS systems use conventional drum brakes which are controlled directly through mechanisms. An advanced approach would be spontaneous application of brakes if there is a possibility of an imminent collision taking into account the driver's response and frequency of usage.

References

1. A. Vahidi and A. Eskandarian, (Sept. 2003) "Research advances in intelligent collision avoidance and adaptive cruise control," in IEEE Transactions on Intelligent Transportation Systems, vol. 4, no. 3, pp. 143–153.
2. Vignesh, R. (May 2017) "Collision Avoidance System In Vehicles" International Journal of Advanced Research Methodology in Engineering & Technology, ISSN 2456 6446 Volume 1, Issue 3.
3. Araki, H., Yamada, K., Hiroshima, Y., & Ito, T. (1996, September). Development of rear-end collision avoidance system. In Intelligent Vehicles Symposium, 1996, Proceedings of the 1996 IEEE (pp. 224–229).
4. Miller, R., & Huang, Q. (2002). An adaptive peer-to-peer collision warning system. In Vehicular technology conference, 2002. VTC Spring 2002. IEEE 55th (Vol. 1, pp. 317–321). IEEE.
5. Tatchikou, R., Biswas, S., & Dion, F. (2005, December). Cooperative vehicle collision avoidance using inter-vehicle packet forwarding. In Global Telecommunications Conference, 2005. GLOBECOM'05. IEEE (Vol. 5, 5 pp).
6. Anurag, D., Ghosh, S., & Bandyopadhyay, S. (2008, October). GPS based vehicular collision warning system using IEEE 802.15. 4 MAC/PHY standard. 8th International Conference on ITS Telecommunications, 2008: 154–159.
7. Lou F., What's the Difference Between IEEE 802.15.4 and Zigbee Wireless? Electronic Design: 2013.

8. Uvaraja, S. (2012). Advanced Pre-Warning System (Railways). International Journal of Engineering and Technology, 4(2): 213.
9. Tan, H. S., & Huang, J. (2006). DGPS-based vehicle-to-vehicle cooperative collision warning: Engineering feasibility viewpoints. IEEE Transactions on Intelligent Transportation Systems, 7 (4): 415–428.

Satyajeet Patnayak Satyajeet Patnayak is a final year undergraduate, pursuing Bachelor's in Mechanical Engineering from College of Engineering and Technology, Bhubaneswar, India. He is a former Summer Research Fellow under Defence Research and Development Organisation, Gov. of India. He is a automobile enthusiast and loves to devote his time in the field of research in automobile and its subdomain. His fields of interest include Vehicle Safety and dynamics, Machine Design, and Avionics.

Anisha Swain Anisha Swain is a final year undergraduate, pursuing Bachelor's in Instrumentation and Electronics Engineering from College of Engineering and Technology, Bhubaneswar, India. She is a former Summer Research Fellow under Indian Academy of Sciences and a newbie in the field of research. She is a technology enthusiast aiming to develop cutting edge technologies for future. Her fields of interest include image processing, machine learning, and data science.

Manaswini Das Manaswini Das is a final year undergraduate, pursuing Bachelor's in computer science from College of Engineering and Technology, Bhubaneswar, India. She is a former Outreachy intern at Open Humans and being new to the research field, contributes to open source software and is ambitious of developing futuristic technologies. Her fields of interest include open source, machine learning, and artificial intelligence.

Doubling Circuit-Based Hybrid Multilevel Inverter for Reduced Components

S. Majumdar, B. Mahato and K. C. Jana

1 Introduction

Since the last decades, the emergence of multilevel inverter [1] has been increased in various power electronic applications like renewable energy sources interconnected to both grid and off-grid network, medium as well as high power industries, drive applications like electric vehicle (EV), hybrid-electrical vehicle (HEV), high-voltage DC network, FACTS devices, and many more. Various multilevel inverter (MLIs) topologies have been reported since 1975. The cascaded H-bridge (CHB), diode-clamped MLI, and flying capacitor (FC) are considered to be the classical MLIs. Conventional MLIs have been accepted as the benchmark for high-voltage and high-power capability applications both in the industry and domestic applications compared to two-level inverters for several years [2]. In comparison to two-level inverters like half-bridge inverter, conventional MLI's have shown their voltage operation beyond classical semiconductor limits, small dv/dt's, smaller common-mode voltage, voltage output is near sinusoidal waveform are the main characteristics which justified the popularity of multilevel converters both in industry and research.

A hybrid MLI configuration is one of the promising inverter topologies for achieving a large number of inverter output voltage levels for generating near-sinusoidal output voltages that may not require any filter circuit. However, to restrict the number of components like switching devices, voltage sources, the combination of power electronic switches in the hybrid inverter should have reduced number of components [3–8]. Thus, the inverter cost and complexity can be minimized and reliability can also be improved. In this chapter, a hybrid MLI combining a symmetrical inverter for level generating part, a voltage doubling

S. Majumdar (✉) · B. Mahato · K. C. Jana
Electrical Engineering Department, Indian Institute of Technology (Indian School of Mines), Dhanbad, India
e-mail: majumdarsaikat2014@gmail.com

J. Chattopadhyay et al. (eds.), *Innovations in Soft Computing and Information Technology*, https://doi.org/10.1007/978-981-13-3185-5_12

circuit for doubling the voltage levels and polarity generating circuit for generating inverter voltage of either polarity is analyzed and presented. To achieve better quality of inverter output voltage waveform, carrier-based level-shifted PWM technique (LS-PWM) [9, 10] is adopted where triangular carriers compare with sinusoidal reference waveform for producing gate signals for the conducting switches. The proposed topology necessitates fewer component counts to boost up the output voltage levels, and also the % total harmonic distortion (THD) is lowered when compared to conventional multilevel inverters. Both simulation and experimental verifications are done to verify the working of the novel MLI configuration. The comparisons of the proposed inverter with the classic MLI are also presented.

2 Proposed Single-Phase MLI with Reduced Count

In this section, a novel single-phase 11-level hybrid MLI is proposed has two main parts. The 'first part' converter comprises a level generating part and the voltage doubling part that combines together to generate multiple voltage levels having only one polarity and requires high-frequency semiconductor devices to develop multiple voltage levels. The 'second part' is the polarity generation part that uses only four low-frequency switches for conversion of unipolar voltage waveform to bipolar voltage waveform at line frequency.

The working of the 'first part' can be described by separating the whole circuit into two sections like stage-1 and stage-2 converter and are vividly explained. The stage-1 converter structure consists of two insulated-gate bipolar transistors (IGBTs) and one bidirectional switch. For two identical DC sources in the stage-1 converter, only one bidirectional switch is required along with two IGBTs to generate three voltage levels across stage-1 converter having their magnitudes 0, V_{DC}, $2V_{DC}$. Stage-1 converter is called level generating part. Addition of one DC voltage source and one bidirectional switch causes increment of one voltage level. For the generalized structure, $(n + 1)$ number of voltage levels can be obtained by adding n numbers of DC voltage sources. The stage-2 converter is named as voltage doubling part, which is made up of one half-bridge comprising of two IGBTs and one DC voltage source having value equal to half of the magnitude of the identical DC voltage sources used in level generating circuit. Using desired switching signals, the voltage levels of the stage-1 converter having magnitudes of 0, $1V_{DC}$, $2V_{DC}$ get doubled to 0, $0.5V_{DC}$, $1V_{DC}$, $1.5V_{DC}$, $2V_{DC}$, $2.5V_{DC}$. The 'second part' named as polarity generation part is used to develop bipolar voltage levels using bridge converter. The output voltage levels across the single-phase 11-level inverter now becomes $-2.5V_{DC}$, $-2V_{DC}$, $-1.5V_{DC}$, $-1V_{DC}$, $-0.5V_{DC}$, 0, $0.5V_{DC}$, $1V_{DC}$, $1.5V_{DC}$, $2V_{DC}$, $2.5V_{DC}$. The inverter can boost up the voltage to two higher levels by the addition of one bidirectional switch.

The proposed MLI is capable of generating 11 voltage levels; the inverter is highlighted in Fig. 1. Table 1 shows the relationship between the voltage levels of the proposed inverter and their switching conditions under all possible modes of

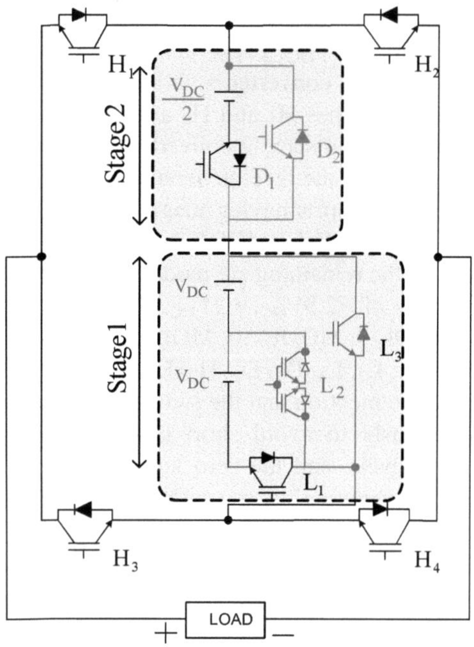

Fig. 1 Proposed single-phase 11-level MLI with reduced switching count

Table 1 Switching position of the proposed 11-level MLI and their corresponding output voltage levels obtained under various modes of operation

Modes	L_1	L_2	L_3	D_1	D_2	H_1	H_2	H_3	H_4	Stage-1	Stage-2	Output voltage
1	1	0	0	1	0	1	0	0	1	$2V_{DC}$	$0.5V_{DC}$	$2.5V_{DC}$
2	1	0	0	0	1	1	0	0	1	$2V_{DC}$	0	$2.0V_{DC}$
3	0	1	0	1	0	1	0	0	1	$1V_{DC}$	$0.5V_{DC}$	$1.5V_{DC}$
4	0	1	0	0	1	1	0	0	1	$1V_{DC}$	0	$1.0V_{DC}$
5	0	0	1	1	0	1	0	0	1	0	$0.5V_{DC}$	$0.5V_{DC}$
6	0	0	1	0	1	1	0	0	1	0	0	0
7	0	0	1	0	1	0	1	1	0	0	0	0
8	0	0	1	1	0	0	1	1	0	0	$0.5V_{DC}$	$-0.5V_{DC}$
9	0	1	0	0	1	0	1	1	0	$1V_{DC}$	0	$-1.0V_{DC}$
10	0	1	0	1	0	0	1	1	0	$1V_{DC}$	$0.5V_{DC}$	$-1.5V_{DC}$
11	1	0	0	0	1	0	1	1	0	$2V_{DC}$	0	$-2.0V_{DC}$
12	1	0	0	1	0	0	1	1	0	$2V_{DC}$	$0.5V_{DC}$	$-2.5V_{DC}$

operation. The 'first part' of the proposed inverter creates six staircases like voltage levels namely $2.5V_{DC}$, $2V_{DC}$, $1.5V_{DC}$, $1V_{DC}$, $0.5V_{DC}$, 0. Detailed operation of 'first part' converter and 'second part' converter is vividly described in Fig. 2. In the first six modes, low-frequency switches H_1 and H_4 are always ON and the switches of stage-1 converter (L_1, L_2, L_3) and stage-2 converter (D_1, D_2) are operated in such a fashion such that voltage across the level generating converter always develops six staircase pattern voltage waveforms having magnitudes of $2.5V_{DC}$, $2V_{DC}$, $1.5V_{DC}$, $1V_{DC}$, $0.5V_{DC}$, 0 under MODE-1, MODE-2, MODE-3, MODE-4, MODE-5, MODE-6, respectively. In the remaining six modes, switches H_2 and H_3 are ON and generate the voltage levels of $-2.5V_{DC}$, $-2V_{DC}$, $-1.5V_{DC}$, $1V_{DC}$, $0.5V_{DC}$, 0 under MODE-7, MODE-8, MODE-9, MODE-10, MODE-11, MODE-12, respectively, by controlling the switches L_1, L_2, L_3, D_1, D_2, H_2, H_3 in similar way as under previous six modes. It is important to mention that the switches H_1, H_2 and H_3, H_4 should not be conducted simultaneously to avoid short circuiting of the inverter. All the positive levels, negative levels, and the zero level are obtained by the periodical switching of the 'level generating' part switches namely H_1, H_2, H_3, H_4. The switches H_1, H_4 conduct for the positive half cycle of the fundamental voltage waveform, and the switches H_2, H_3 conduct unstoppably for the negative half cycle of the fundamental voltage waveform.

3 Modulation Techniques

The common modulation techniques used for developing switching states are categorized as low and high switching frequency modulation techniques [11–17] are shown in Fig. 3. The gate pulses are supplied to the MLI switches using level-shifted pulse-width modulation technique (LS-PWM) where a series of ten triangular waveforms named as carriers are stacked vertically due to the requirement of 11-level output voltage waveform. These ten carriers are now compared with the reference sinusoidal voltage to produce the desired gate pulses for the switches of the inverter. Further, the gate pulses are decoded for triggering the appropriate semiconductor device necessitates for the generation of any particular voltage level. Observing all the possible modes of operation of the proposed inverter, 11 voltage levels can be generated and for producing individual voltage level four switches must be conducted at a time.

4 Comparison Study

The comparison table shows that the proposed inverter requires lesser number of DC source when compared to cascaded H-bridge inverter. Whereas the diode clamped and the flying capacitor require lesser DC voltage sources than the proposed inverter, additional requirement of DC-bus capacitors and their complex

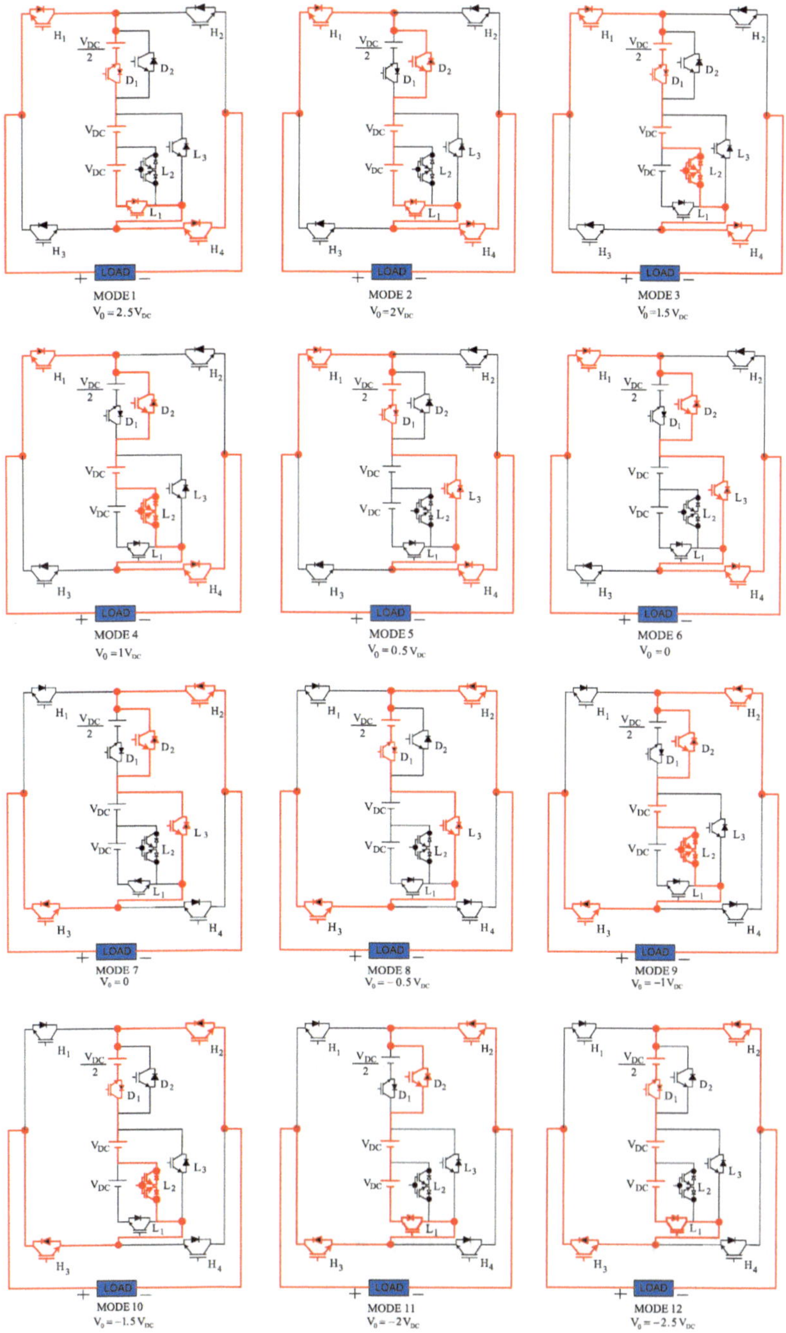

Fig. 2 Modes of operation of the single-phase 11-level MLI

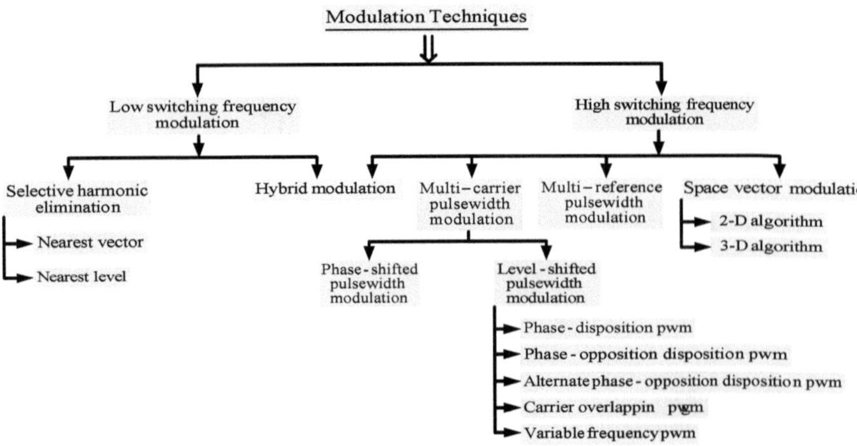

Fig. 3 Tabular presentation of the categorization of the low switching frequency modulation and high switching frequency modulation techniques

Table 2 Comparison shows the requirement of components required for both the traditional inverter and the proposed inverter

MLI type	No. of sources	No. of capacitors	No. of switches	Clamping diodes
Diode clamped	1	10	20	90
FC-type	1	10	20	–
CHB	5	–	20	–
Proposed	3	–	9	–

design of controllers required for voltage balancing of capacitors makes the classical inverters costly. The above table also depicts that the proposed inverter requires lesser number of switches and driver circuits than conventional MLI. Overall the proposed inverter has the capability to double the voltage levels by addition of only one switch and one DC voltage source and also the % total harmonic distortion (THD) is greatly reduced in comparison to conventional MLI which contributes greatly to both commercial and industrial purpose (Table 2).

5 Simulation and Experimental Results

Simulation and modelling of the proposed 11-level MLI are carried out with the help of MATLAB simulation software to verify the working of the designed circuit. The simulation result of the 11-level inverter is shown in Fig. 4. The DC voltage source magnitude of stage-1 converter is considered 50 V and the DC voltage

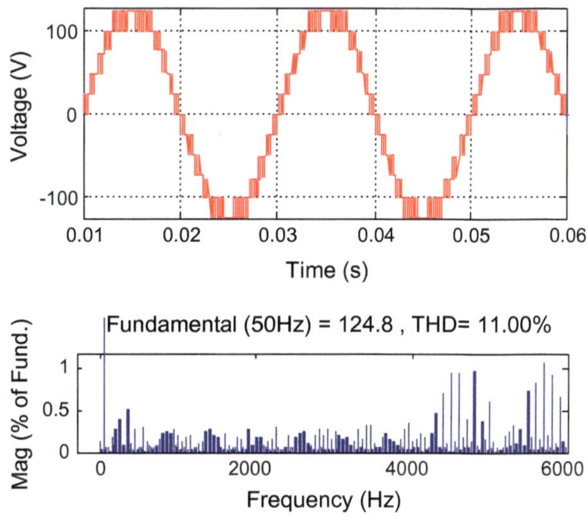

Fig. 4 Simulation result of phase voltage (V) and its corresponding THD of the proposed 11-level inverter with V_{DC} = 50 V with switching frequency F_s = 6 kHz and modulation index, M_i = 1.0

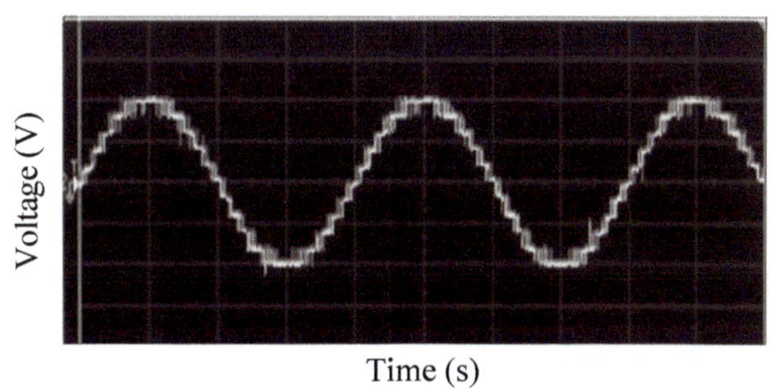

Fig. 5 Experimental results of phase voltage of the proposed 11-level MLI having V_{DC} = 50 V, (voltage scale: 75 V/Div., timescale: 10 ms/Div.)

magnitude across the voltage double circuit is taken as 25 V, respectively. Here, resistive and inductive load (RL) is considered for successful operation of the inverter having their magnitudes of $R = 25\ \Omega$, $L = 3$ mH for simulation and experimental purposes. It is observed from the above figure that the THD of the voltage waveform of the 11-level inverter is 11%, respectively; also high switching frequency of about 6 kHz is used for both simulation and hardware circuits to improve the THD and to minimize the filter components of the circuit.

A hardware setup is designed not only to develop the 11-level voltage waveform but also to confirm the working of the designed inverter under real-time condition. The three DC voltage sources are supplied from the single-phase combined transformer and rectifier connection, electrolytic capacitors. The 11-level MLI comprises 10 IGBTs and 3 DC voltage sources as shown in Fig. 5. The gate signals of the semiconductor switches are developed utilizing level-shifted pulse-width modulation technique.

6 Conclusion

The present work demonstrated a novel single-phase 11-level MLI configuration using minimum component counts and the three DC voltage sources. Addition of a bidirectional switch and one DC voltage source can increase the voltage levels to twice times. The generalized structure of the proposed inverter is possible. To validate the superiority of the proposed 11-level inverter using minimum component counts, the necessary components used for the designing of the conventional MLIs like the number of DC sources, number of switches, number of capacitors are compared with the required components of the proposed inverter to draw conclusion that this 11-level inverter works better than the traditional MLIs in terms of harmonic profile, total number of components required and shrinking of filter size. It is also observed that the proposed inverter configuration has lower conduction loss due to less number of components and the DC voltage sources used compared to the traditional MLIs. The working of the proposed 11-level inverter is confirmed using MATLAB simulation software. Experimental validation is done in a laboratory setup model just to counterpart with the simulation results.

References

1. R. H. Baker, L. H. Bannister, *Electric power converter*. U. S: Patent 3 867 643, 1975.
2. A. Nami, F. Zare, A. Ghosh et al., A hybrid cascade converter topology with series-connected symmetrical and asymmetrical diode-clamped H-bridge cells, *IEEE Trans. Power Electron.*, 26 (1) (2011) 51–65.
3. U. Choi, K. Lee, F. Blaabjerg, Diagnosis and tolerant strategy of an open-switch fault for T-type three-level inverter systems, *IEEE Trans. Ind. Appl.*, 50 (1) (2014) 495–508.
4. Y. Hinago, H. Koizumi, A single-phase multilevel inverter using switched series/parallel DC voltage sources. *IEEE Trans. Ind. Electron.*, 57 (8) (2010) 2643–2650.
5. W. K. Choi, F. Kang, H-bridge based multilevel inverter using PWM switching function, *INTELEC 2009 31st Int. Telecommunications and Energy Conf.*, (2009) 1–5.
6. S. Gui-Jia, Multilevel DC-link inverter, *Ind. Appl. IEEE Trans.*, 41 (3) (2005) 848–854.
7. E. Babaei, A cascade multilevel converter topology with reduced number of switches. Power Electron., *IEEE Trans.*, 23 (6) (2008) 2657–2664.
8. J. Ebrahimi, E. Babaei, G. B. Gharehpetian, A new multilevel converter topology with reduced number of power electronic components, *IEEE Trans. Ind. Electron.* 59 (2) (2012) 655–667.
9. B. P. McGrath, D. G. Holmes, Multicarrier PWM strategies for PWM inverters, *IEEE Trans Ind Electron*ics, 49 (4) (2002) 858–867.
10. M. Angulo, P. Lezana, S. Kouro, J. Rodriguez, Level-shifted PWM for cascaded Multilevel inverters with even power distribution, *IEEE Power Electronics Specialist Conference,* 2 (23) (2007) 2373–2378.
11. B. P. McGrath, D. G. Holmes, Multicarrier PWM strategies for multilevel inverters, *IEEE Trans. Ind. Electron.*, 49 (4) (2002) 858–867.
12. K. C. Jana, S. K. Chowdhury, S. K. Biswas, Performance evaluation of a simple and general space vector pulse-width modulation-based M-level inverter including over-modulation operation, *IET Power Electron.*, 6 (4) (2013) 809–817.
13. A. K. Gupta, A. M. Khambadkone, A general space vector PWM algorithm for multilevel inverters, including operation in over modulation range, *IEEE Trans. Power Electron.*, 22 (2) (2007) 517–526.

14. H. Lou, C. Mao, D. Wang et al., Fundamental modulation strategy with selective harmonic elimination for multilevel inverters, *IET Power Electron.*, 7 (8) (2014) 2173–2181.
15. P. Meshram, V. Borghate, A simplified nearest level control (NLC) voltage balancing method for modular multilevel converter (MMC), *IEEE Trans. Power Electron.*, 30 (1) (2014) 450–462.
16. A. Edpuganti, A. Rathore, Optimal low switching frequency pulsewidth modulation of nine-level cascade inverter, *IEEE Trans. Power Electron.*, 30 (1) (2015) 482–495.
17. B. Mahato, R. Raushan, K. C. Jana, Modulation and control of multilevel inverter for an open-end winding induction motor with constant voltage levels and harmonics, *IET Power Electron.*, 10 (1) (2017) 71–79.

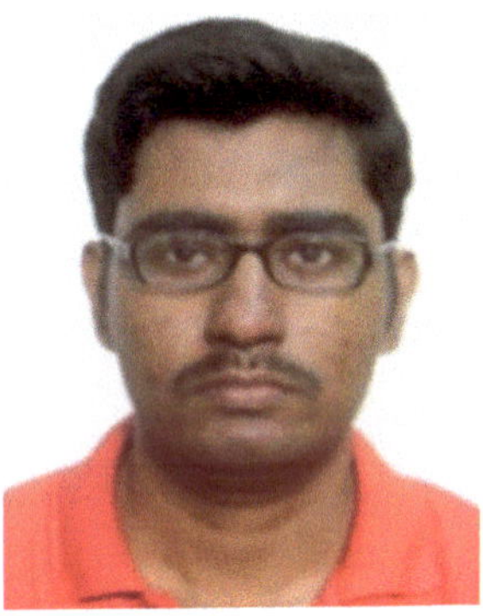

S. Majumdar received his M.Tech and B.Tech degree from **IIT (ISM)**, Dhanbad, and **AOT**, Adisaptagram, in 2016 and 2013, respectively. Since 2016, he is working towards the **Ph.D.** degree from **IIT (ISM)**, Dhanbad, in the department of electrical engineering. His main research interest includes modelling and design of multilevel inverters, renewable energy sources.

B. Mahato received his M.E. and B.Tech degree from **BIT**, Mesra, Ranchi, and **GNIT**, Kolkata, in 2014 and 2011, respectively. Since 2014, he is working towards the **Ph.D.** degree from **IIT (ISM)**, Dhanbad, in the department of electrical engineering. His main research interest includes power factor correction, phase-controlled rectifiers, multilevel inverters, converter, AC drives, and hybrid electric vehicles.

K. C. Jana received his M.Tech and Ph.D. degree from NIT Durgapur and Jadavpur University in 2003 and 2013, respectively. He is currently serving as an assistant professor in **IIT (ISM)**, Dhanbad, from June 2012. He has been working in **BIT**, Mesra, Ranchi, from July 2003 till May 2012. His research is in the area of modelling and design of multilevel inverter, real-time control of power electronics devices, design and implementation of efficient power converters.

Effect of Reinforced Soil–Structure Interaction on Foundation Settlement Characteristics of a Three-Dimensional Structure

Nayana N. Patil, H. M. Rajashekar Swamy and R. Shivashankar

1 Introduction

A limited number of studies have been conducted on soil–structure interaction effect considering three-dimensional space frames. SSI studies that take into account the yielding of structures and soil nonlinearity are scarce, especially investigating the effects of nonlinearity of SSI system on overall behaviour in terms of displacements and stresses.

King and Yao [1] and Roy and Dutta [2] were the few researchers who made use of the finite element method to consider superstructure—raft/combined footing–soil as a single compatible unit. The SSI studies conducted by Noorzaei et al. [3] and Viladkar et al. [4] clearly indicated that a two-dimensional plane frame SSI analysis might substantially overestimate or underestimate the actual interaction effect in a space frame. The interactive behaviour of the 3D frame-isolated footing–soil system was studied by Rajashekhar Swamy et al. [5]. Rajashekhar Swamy et al. [6] conducted linear and nonlinear SSI analyses of structure resting on raft foundation to find the maximum settlement as well as differential settlements in soil increase in nonlinear analysis when compared to linear analysis. They observed that maximum vertical stresses decrease in nonlinear analysis when compared to linear analysis. However, the stress resultants, in members of the frame, were found to vary (either decrease or increase) depending on location in nonlinear analysis when compared to linear analysis. SSI studies have been carried out for structures supported on unreinforced soil. But reinforced soil–structure interaction (RSSI) dealing with

N. N. Patil · H. M. Rajashekar Swamy (✉)
Department of Civil Engineering, Faculty of Engineering and Technology,
M. S. Ramaiah University of Applied Sciences, Bengaluru 560058, India
e-mail: swamyraja2005@gmail.com

R. Shivashankar
Department of Civil Engineering, National Institute of Technology,
Surathkal, Karnataka, India

J. Chattopadhyay et al. (eds.), *Innovations in Soft Computing and Information Technology*, https://doi.org/10.1007/978-981-13-3185-5_13

structures supported on reinforced soil is yet to be explored. The analysis that treats structure–foundation–reinforced soil as a single system is coined as reinforced soil–structure interaction (RSSI) analysis in this work.

2 Motivation and Objective

Numerous studies have been carried out for structures supported on unreinforced soil. But reinforced soil–structure interaction (RSSI) dealing with structures supported on reinforced soil is yet to be explored.

3 Methodology

Finite element software has been developed to carry out the SSI and RSSI analyses. The structure under consideration has been chosen from Rajashekhar Swamy et al. [6]. SSI analysis is performed on the structure with isolated footings (2 m × 2 m) supported on the soil mass of size 153 m × 95 m × 20 m and beams carrying uniformly distributed load (UDL) of 31 kN/m. In RSSI analysis, the isolated footings of structure are underlain by geogrid and the reinforced soil–structure is analysed for the same loads.

4 Linear SSI Analysis of Space Frame-Footing–Soil System

The proposed physical model has been used for the interactive analysis of a four-storey, five-bay-by-three-bay, space frame-isolated footing–soil system. Figure 1 shows the isometric view of the space frame-isolated foundation–soil system. The layout details of the frame are shown in Fig. 2. The geometrical and material properties of the frames, its components and the isolated foundation are presented in Table 1.

Finite element formulation in the SSI analysis of the frame-isolated footing–soil system is as shown in Fig. 3a–c. The soil is modelled with 43 × 10 × 27 layers in the longitudinal, vertical and transverse directions, respectively, resulting in 11,610 brick elements. Each footing of size 2 × 2 m is modelled by four plate elements of size 1 m × 1 m. The number of plate elements used is 96. The number of beam elements in the longitudinal direction (*X*-direction) is 80, 72 in transverse (*Z*-direction) and 96 in the vertical (*Y*-direction). The graphs are plotted in terms of dimensionless parameters *X*/*L* and *Z*/*B* where *L* and *B* are dimensions of the frame in *X* and *Z* directions as shown in Fig. 3a.

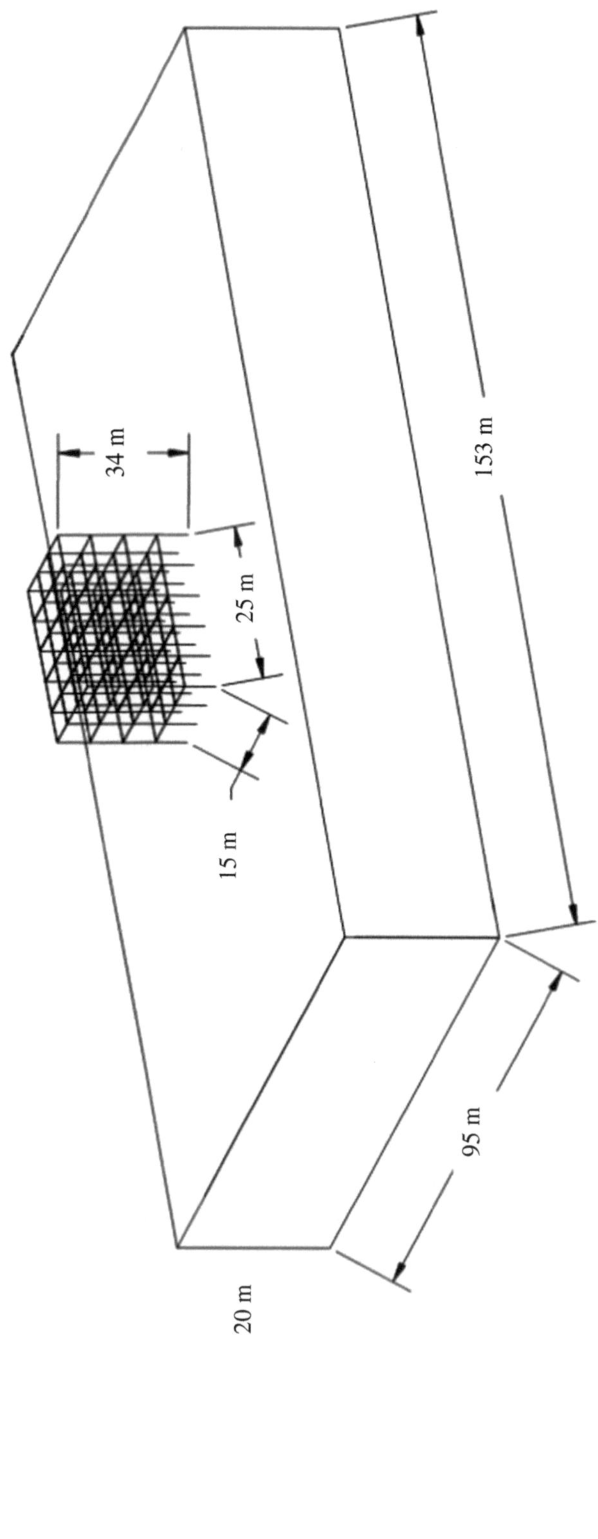

Fig. 1 Structure–footing–soil system [6]

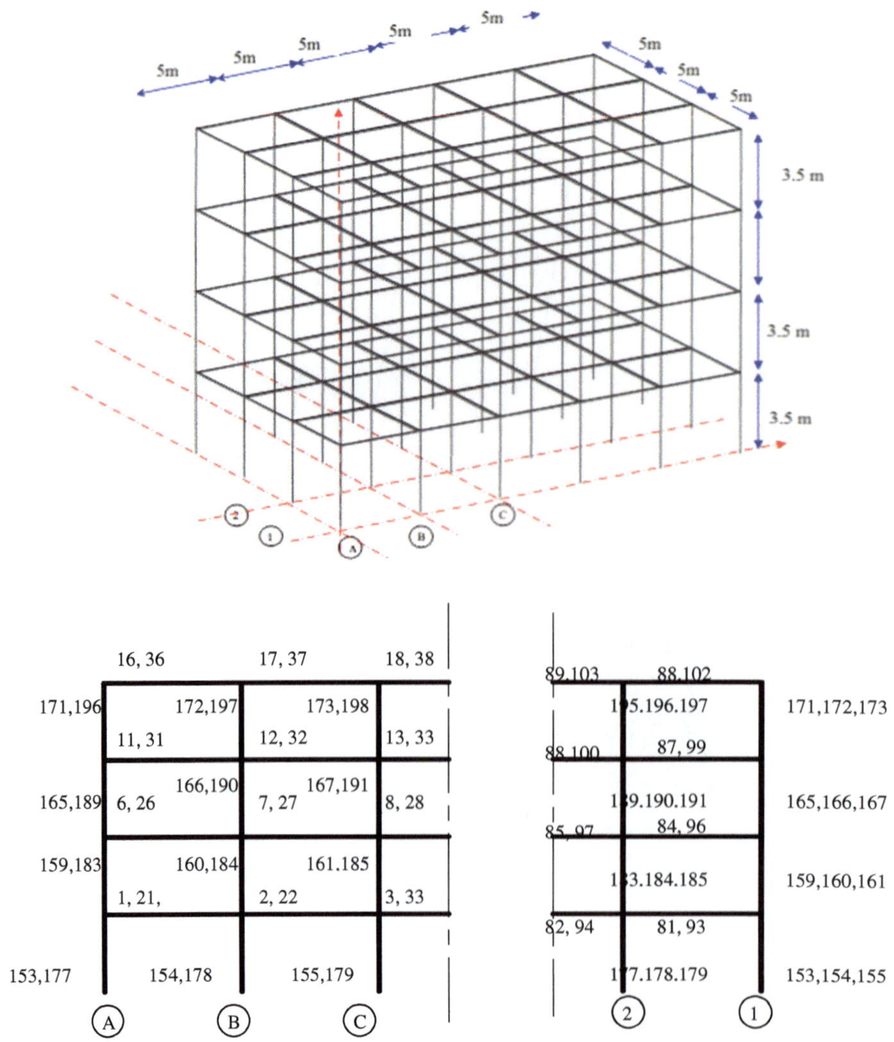

Fig. 2 Details of quarter frame [6]

The various components of the system with respective degrees of freedom are shown in Fig. 4 and are modelled as follows:

1. Columns and beams are modelled as one-dimensional beam elements with six degrees of freedom per node (three translational and three rotational degrees of freedom) as shown in Fig. 4a.
2. Soil mass is modelled as eight-node brick element with three translational degrees of freedom per node as shown in Fig. 4b.

Table 1 Details of the validation of SSI problem [6]

Sl. No.	Structure	Component	Details
1	Frame	No. of storeys	5
		No. of bays	5 × 3
		Storey height	3.5 m
		Bay width	5 m
		Beam size	0.3 m × 0.6 m
		Column size	0.4 m × 0.4 m
		Footing size	3.0 m × 3.0 m × 0.4 m
2		Soil mass	153.0 m × 95.0 m × 0.0 m
3	Elastic modulus of soil		1.33×10^7 N/m^2
4	Poisson's ratio of soil		0.45
5	Bulk modulus of concrete		6.1×10^6 N/m^2
6	Elastic modulus of concrete		1.4×10^{10} N/m^2

3. Individual footing is modelled using plate elements with five degrees of freedom per node, i.e. three translational degrees of freedom and two rotational degrees of freedom as shown in Fig. 4c.

5 Validity of the Proposed Physical Model

Figure 5 shows the settlements of the isolated footings obtained from the proposed analysis and their comparison with Rajashekhar Swamy et al. [6]. Glance at the figure suggests that there is a very good agreement between the values of settlement. This justifies the finite element mesh extent considered.

6 Linear RSSI Analysis of Space Frame-Footing–Soil System

To conduct linear RSSI analysis, the frame-footing–reinforced soil model used is as shown in Figs. 6 and 7. Under each column footing, four layers of geogrid are laid at *D*/*B* ratios of 0.25, 0.5, 0.75 and 1 as shown in Fig. 6. The size of isolated footing is 2 m × 2 m, and the sizes of geogrid are 4 m × 4 m. The geometric details of geogrid are shown in Fig. 8. Properties of geogrid are given in Table 2.

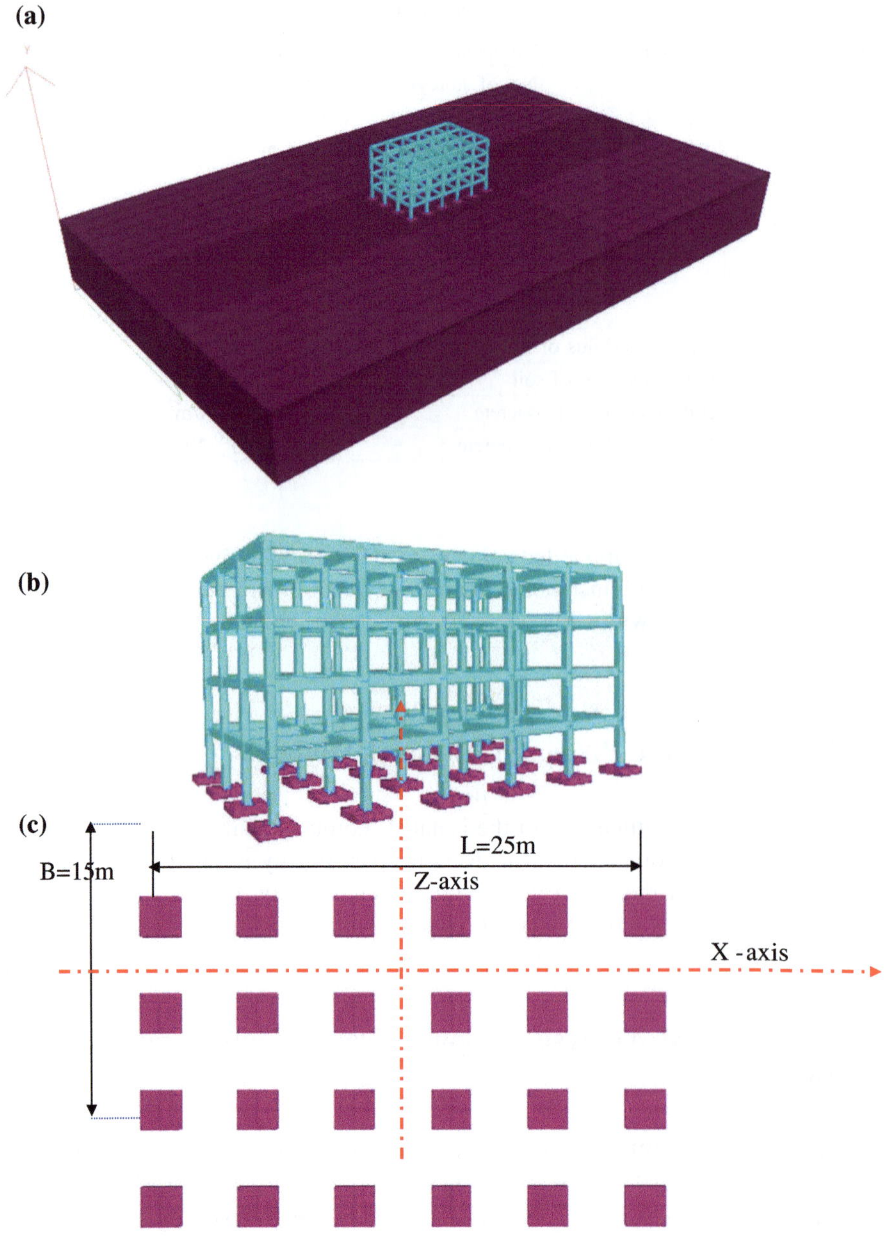

Fig. 3 **a** Frame-isolated footing–soil system, **b** structure foundation system, **c** reference axis and arrangement of isolated footings [8]

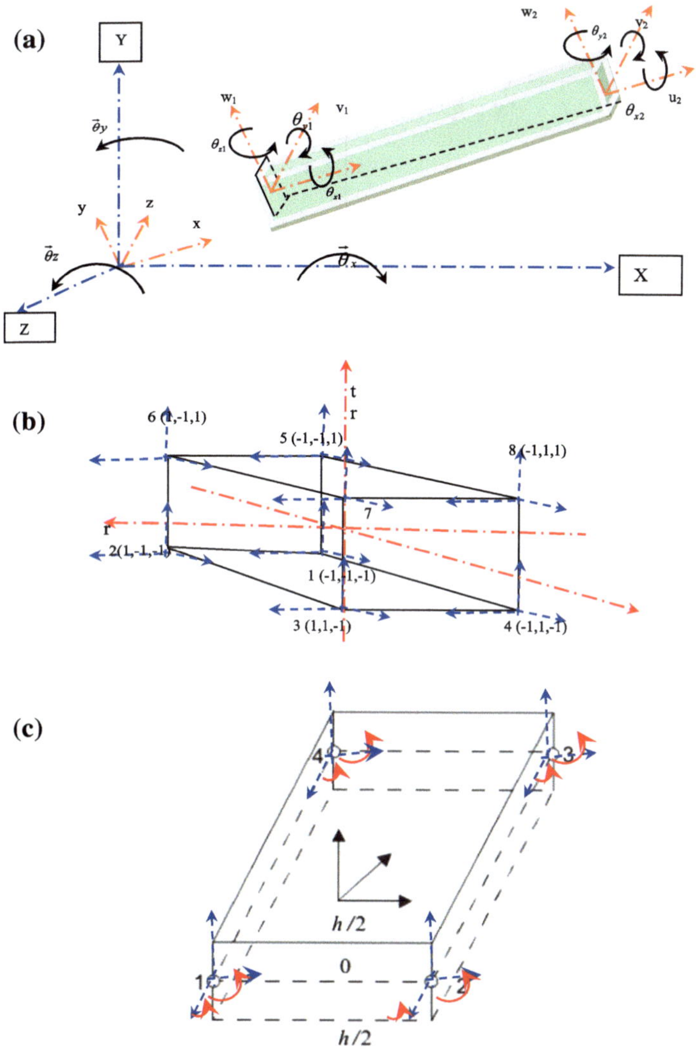

Fig. 4 Details of element types **a** Euler–Bernoulli beam element used for beams and columns, **b** brick element for soil, **c** plate element used for footing [8]

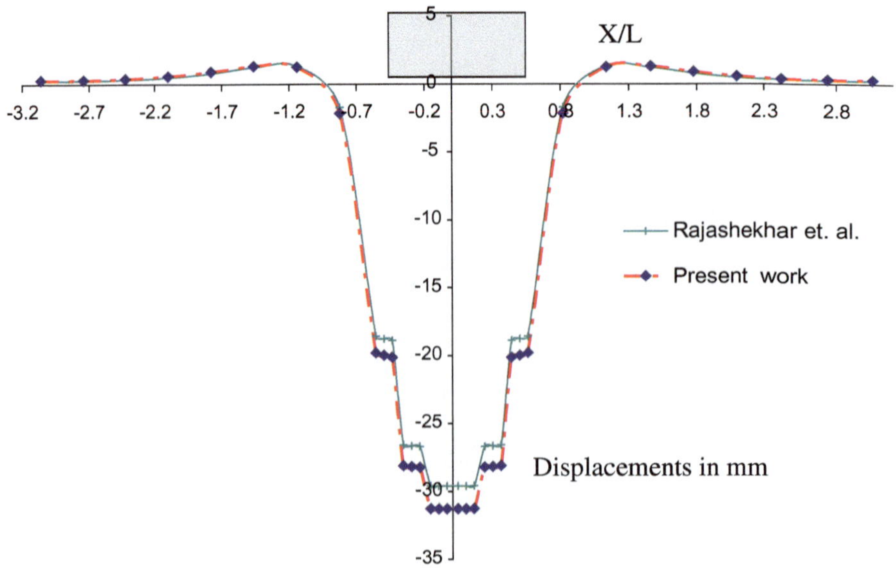

Fig. 5 Comparative settlements in mm at the centre in the present work and the referred work [6]

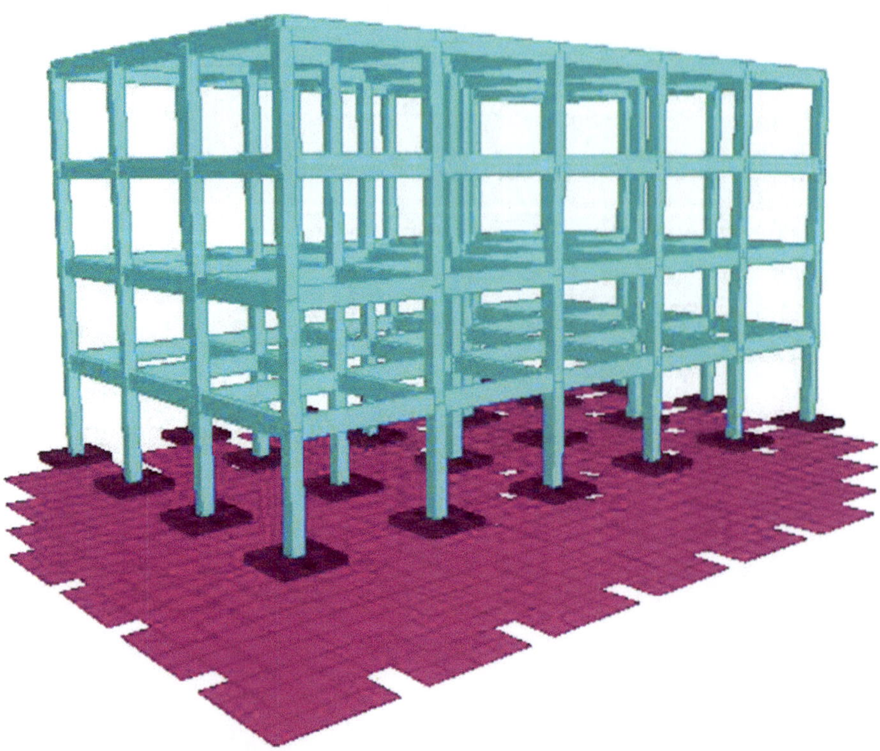

Fig. 6 Frame-footing–reinforcement module [8]

Table 2 Properties of geogrids used for the study [8]

Properties	Values
Rib thickness (mm)	0.75
Aperture size (MD/XD) (mm)	25/33
Junction thickness (mm)	2.8
Tensile strength at 5% strain (kN/m)	8.46 (MD), 13.42 (XD)
Aperture shape	Rectangular
Colour	Black
Type of polymer used	Polyethylene

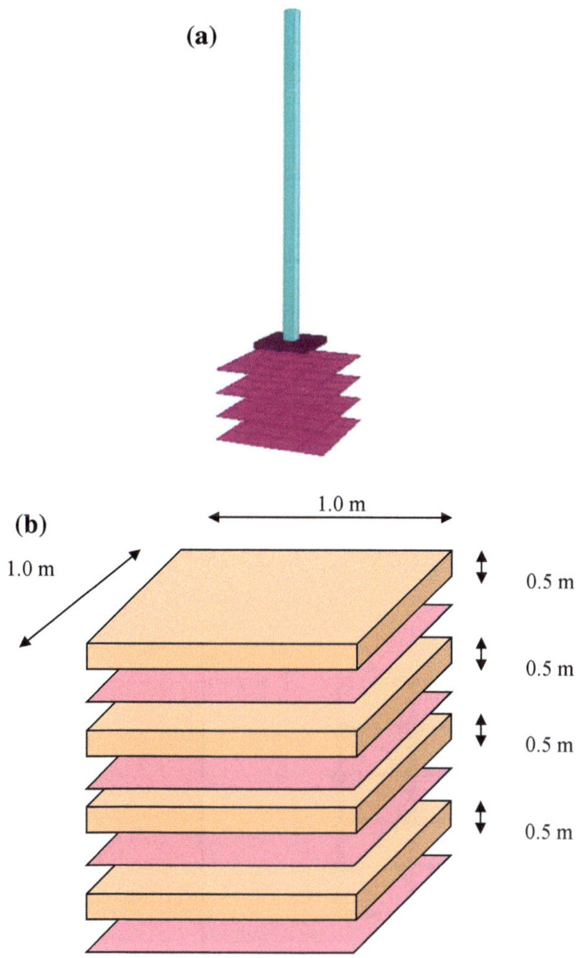

Fig. 7 Arrangement of geogrid: **a** modelling of column–foundation–geogrid and **b** soil–geogrid arrangement represented as macroelement in RSSI analysis [8]

Geogrid consists of apertures of size 25 × 34 mm as shown in Fig. 8b. For a 1 m × 1 m-size geogrid, the number of apertures are 30 × 40 in mutually perpendicular directions as shown in Fig. 8a. Since modelling the geogrid with apertures shown in Fig. 8a proves to be difficult, due to software limitation and enormous execution time, macroelement approach is adopted. This method overcomes the tedious process of modelling the geogrid with small aperture sizes and is validated elsewhere by the authors.

The geogrid of 1 m × 1 m with aperture size of 33 mm × 25 mm (shown in Fig. 8a) is modelled using two-dimensional rectangular element having four nodes with two degrees of freedom per node as shown in Fig. 8c.

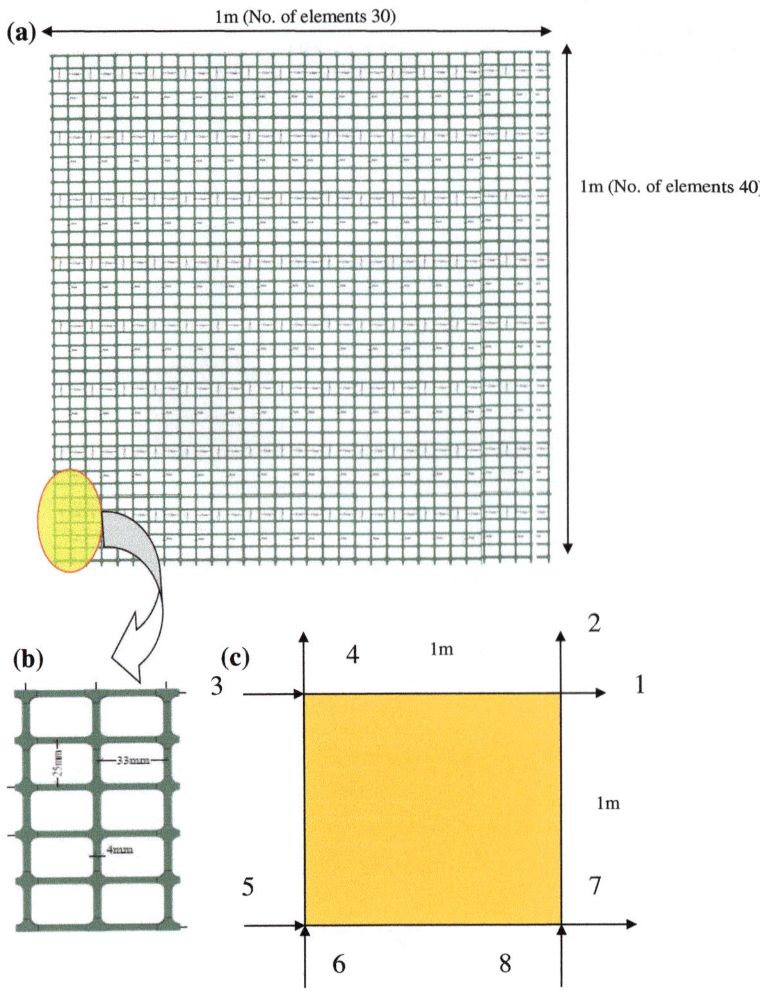

Fig. 8 Details of geogrid and macroelement: **a** geogrid of size 1 m × 1 m with apertures, **b** geometrical details of geogrid, **c** geogrid represented as macroelement of size 1 m × 1 m [8]

Fig. 9 Footing and geogrid arrangements [8]

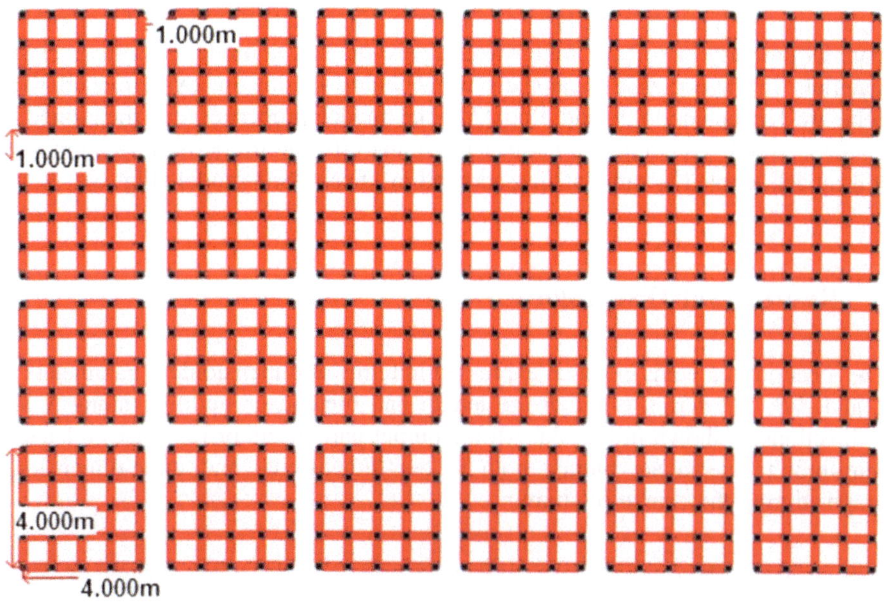

Fig. 10 FEM modelling of geogrid [8]

$$[K] = \begin{pmatrix} 83.39 & 25.017 & -60.753 & -2.082 & -17.423 & -25.018 & -5.214 & -2.082 \\ 25.017 & 88.66 & -2.083 & -5.411 & -25.017 & -17.826 & 2.082 & -65.423 \\ -60.753 & -2.083 & 83.39 & -25.017 & -5.214 & 2.082 & -17.423 & 25.018 \\ 2.082 & -5.411 & -25.017 & 88.66 & -2.082 & -65.423 & 25.017 & -17.826 \\ -17.423 & -25.018 & -5.214 & -2.052 & 83.39 & 25.017 & -60.753 & 2.082 \\ -25.017 & -17.826 & 2.082 & -65.423 & 25.017 & 88.66 & -2.083 & -5.411 \\ -5.214 & 2.082 & -17.423 & 25.019 & -60.753 & -2.083 & 83.39 & -25.017 \\ -2.082 & -65.423 & 25.017 & -17.826 & 2.082 & -5.411 & -25.017 & 88.66 \end{pmatrix} \quad (1)$$

Figure 9 shows the arrangement of footings and geogrid below foundations. Figure 10 shows the arrangement of macroelements representing geogrid in plan.

7 Nonlinear SSI and RSSI Analyses of Space Frame-Footing–Soil System

The finite element model for nonlinear SSI nonlinear analysis is similar to the model used in linear SSI analysis as shown in Figs. 1 and 2, except for nonlinear material property of soil. In nonlinear analysis, the soil is modelled as hypoelastic material. The hypoelastic parameters were obtained from the experimental work done by Krishnamoorthy and Rao [7]. The model properties are mentioned in Tables 3 and 4. To conduct nonlinear RSSI analysis, the frame-footing–reinforced soil model adopted is same as that adopted in linear RSSI model (Fig. 6).

8 Results and Discussions

The location of a point in soil is expressed in terms of dimensionless parameters *X*/*L* and *Z*/*B*. The terms *L* and *B* are length and breadth of building along *X* and *Y* directions as shown in Fig. 3a. All the output parameters or responses related to semi-infinite media are plotted both in two-dimensional and three-dimensional graphs during the course of study.

8.1 Displacements in Linear SSI Analysis

Figure 11a shows the vertical deformation at foundation level in the longitudinal direction of the soil mass. Maximum vertical displacement and horizontal displacements are presented in Table 5. Figure 11b, c shows the displacement

Table 3 Properties of soil used in nonlinear SSI and RSSI analyses (after Krishnamoorthy and Rao [7])

Properties	Values
Liquid limit	54
Plastic limit	40
Plasticity index	14
Shrinkage limit	20
Water content	28
Specific gravity	2.65
Wet density (kN/m^3)	18.18

Table 4 Hypoelastic model parameters used in nonlinear SSI and RSSI analyses (after Krishnamoorthy and Rao [7])

Model parameters		Values
K modulus	$\lambda\backslash V_i$	0.02
	$\kappa\backslash V_i$	0.003
	P'cons	21,000 Pa
J modulus	*A*	100
	N	100
G modulus	*E*	0.001
	F	0.56

contours at foundation level and along the longitudinal section. Figure 12a, b shows the longitudinal displacements, at foundation levels, and Fig. 12c, d shows the transverse displacements at foundation levels.

8.2 *Displacements in Linear RSSI Analysis*

Figure 13a shows the vertical deformation at foundation level along the longitudinal direction of the soil mass. Maximum vertical displacement and horizontal displacements are presented in Table 5. Figure 13b, c shows the displacement contours at foundation level and along longitudinal section taken at centre. Figure 14a, b shows the longitudinal displacements at foundation levels, and Fig. 14c, d shows the transverse displacements at foundation levels.

Figure 15a, b shows lateral displacements at foundation level along *X* and *Z* directions found to reduce by 42 and 45.8%, respectively, in linear RSSI analysis with respect to linear SSI analysis. Figure 16 shows the vertical displacements in linear SSI and RSSI analyses at foundation level along longitudinal line at $Z/B = 0.0$. It is observed that the vertical displacements are reduced merely by 3.72% in RSSI analysis when compared to linear SSI.

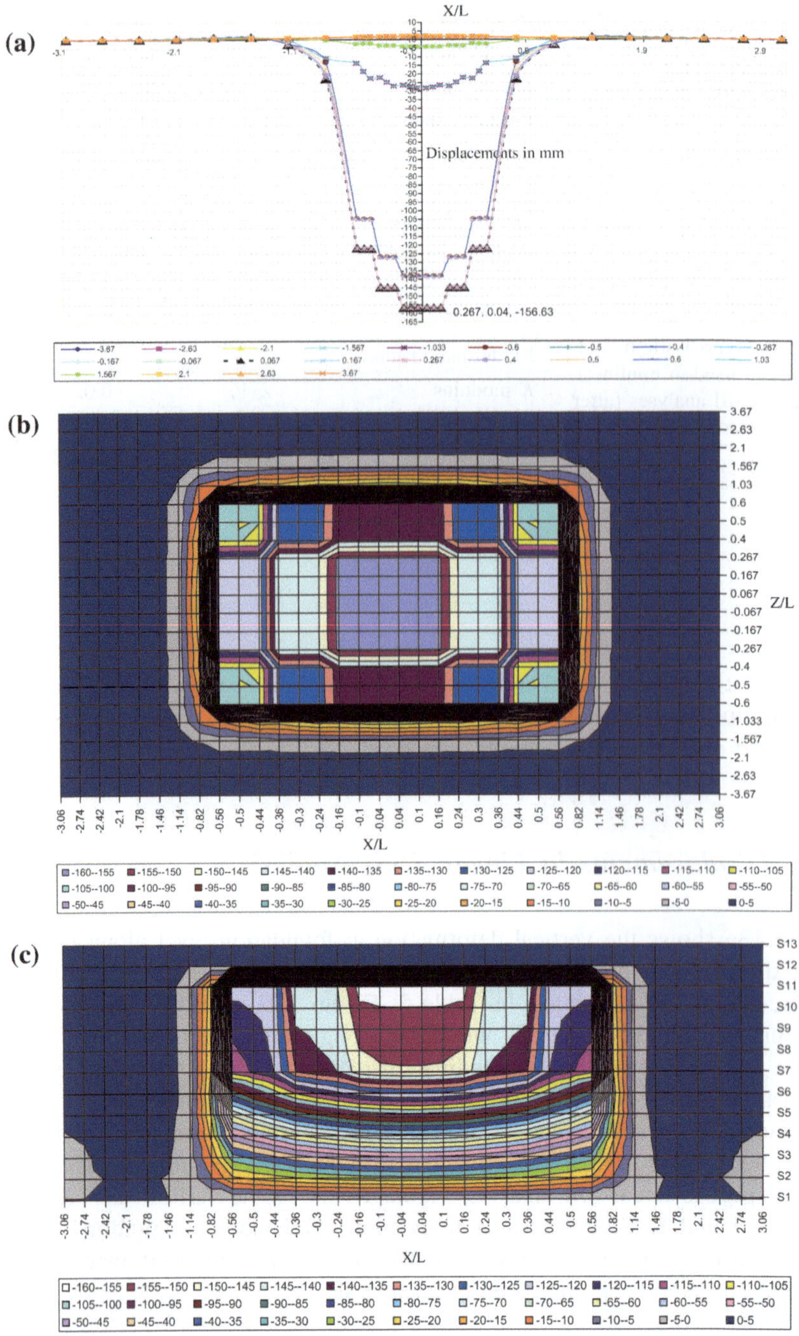

Fig. 11 Vertical displacements in mm in linear SSI analysis: **a** vertical displacements at foundation level, **b** contours of vertical displacements at footing level, **c** vertical displacements along longitudinal section [8]

Table 5 Vertical and horizontal displacements in reinforced soil for different analyses [8]

Type of analysis	Axis	*X*/*L*	*Z*/*B*	Maximum displacement
Linear SSI	*X*-axis	0.267	0.04	−156.63 mm vertical
		0.82	0.167	9.69 mm horizontal
	Z-axis	1.03	0.04	12.98 mm horizontal
Linear RSSI	*X*-axis	0.267	0.16	−150.8 mm vertical
		0.82	0.167	5.62 mm horizontal
	Z-axis	1.03	0.04	7.03 mm horizontal
Nonlinear SSI	*X*-axis	0.16	0.267	−185.3 mm vertical
		0.82	0.167	11.32 mm horizontal
	Z-axis	0.04	1.03	12.925 mm horizontal
Nonlinear RSSI	*X*-axis	0.16	0.267	−173.8 mm vertical
		0.04	1.03	9.72 mm horizontal
	Z-axis	0.04	1.03	10.98 mm horizontal

8.3 Deformation and Settlements in Nonlinear SSI Analysis

Figure 17a shows the vertical deformation at foundation level in the longitudinal direction of the soil mass. Figure 17b shows displacement contours at foundation level. Figure 17c shows displacements along longitudinal section. Figure 18 shows horizontal displacements along longitudinal and transverse directions. In Fig. 18a, b, longitudinal displacements at foundation levels are shown. Similarly, in Fig. 18c, d, transverse displacements at foundation levels are shown.

8.4 Deformation and Settlements in Nonlinear RSSI Analysis

Figure 19a shows the vertical deformation at foundation level in the longitudinal direction of the soil mass. Figure 19b shows displacement contour at foundation level. Figure 19c shows displacements along longitudinal section. Maximum lateral displacements along *X* and *Z* directions are reduced by 14 and 15% in nonlinear

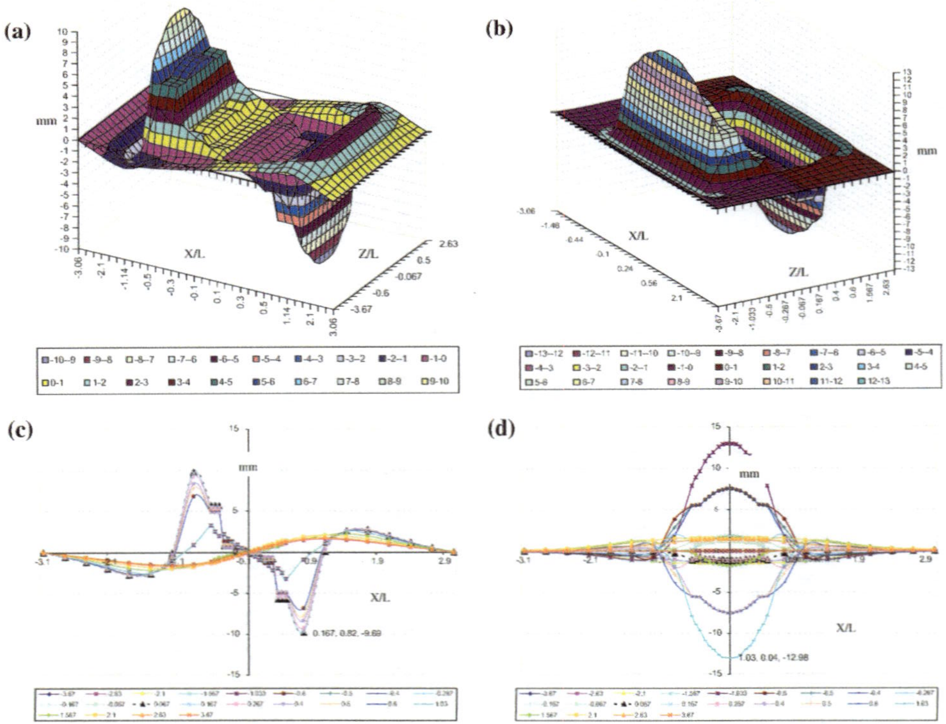

Fig. 12 Horizontal displacements in mm in linear SSI analysis: **a** and **b** longitudinal displacements at foundation level; **c** and **d** transverse displacements at foundation level [8]

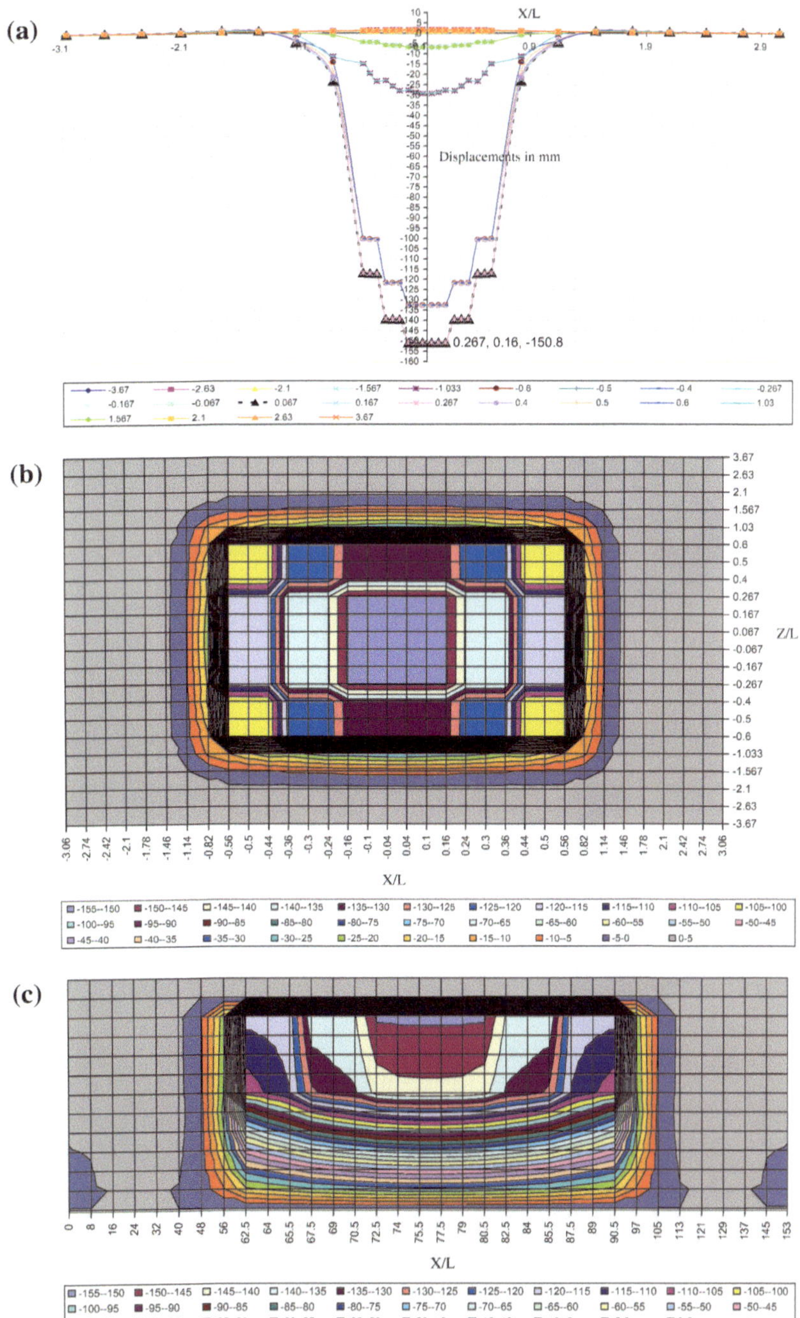

Fig. 13 Vertical displacements in mm in linear RSSI analysis: **a** vertical displacements at foundation level, **b** contours of vertical displacements at footing level, **c** vertical displacements along longitudinal section at centre [8]

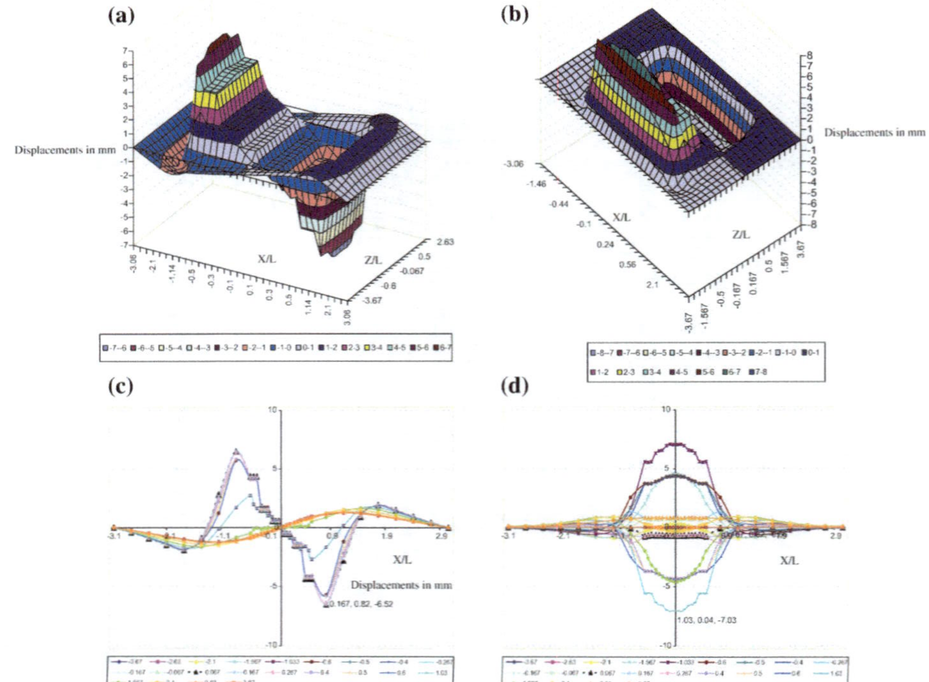

Fig. 14 Horizontal displacements in mm in linear RSSI analysis: **a** and **b** longitudinal displacements at foundation level; **c** and **d** transverse displacements at foundation level [8]

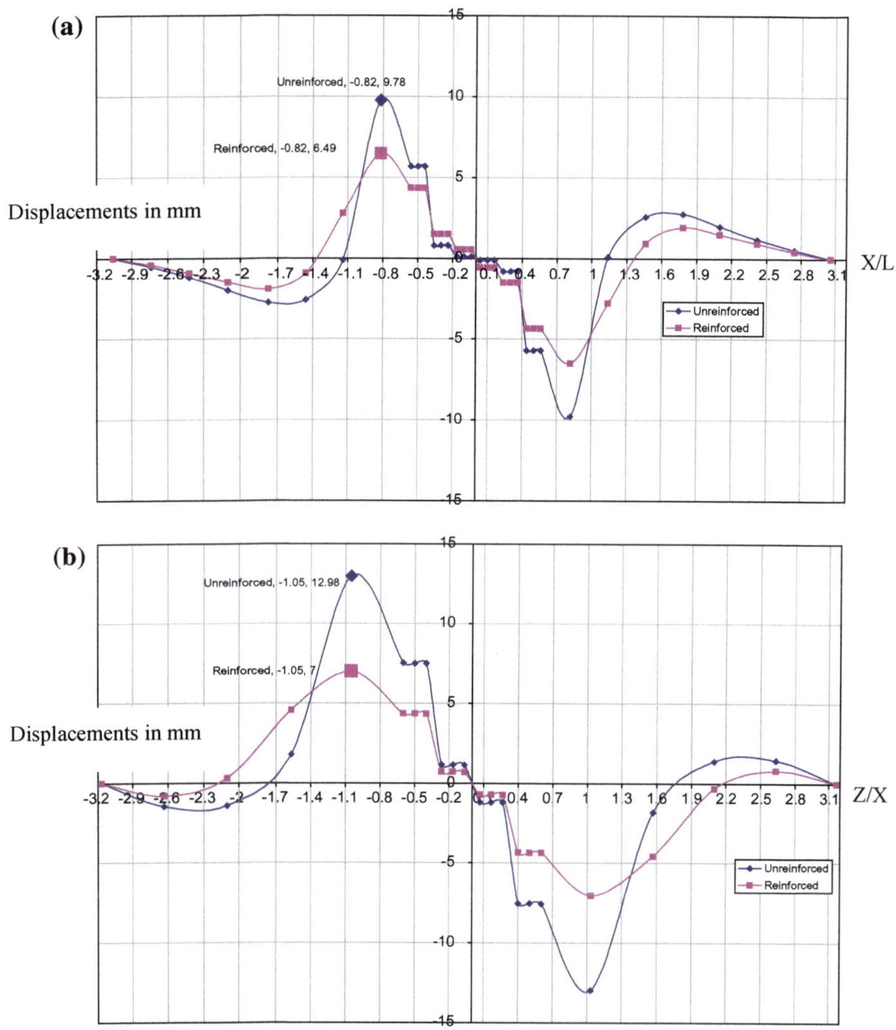

Fig. 15 Horizontal displacements in mm in linear RSSI analysis: **a** longitudinal displacements at foundation level and **b** transverse displacements at foundation level [8]

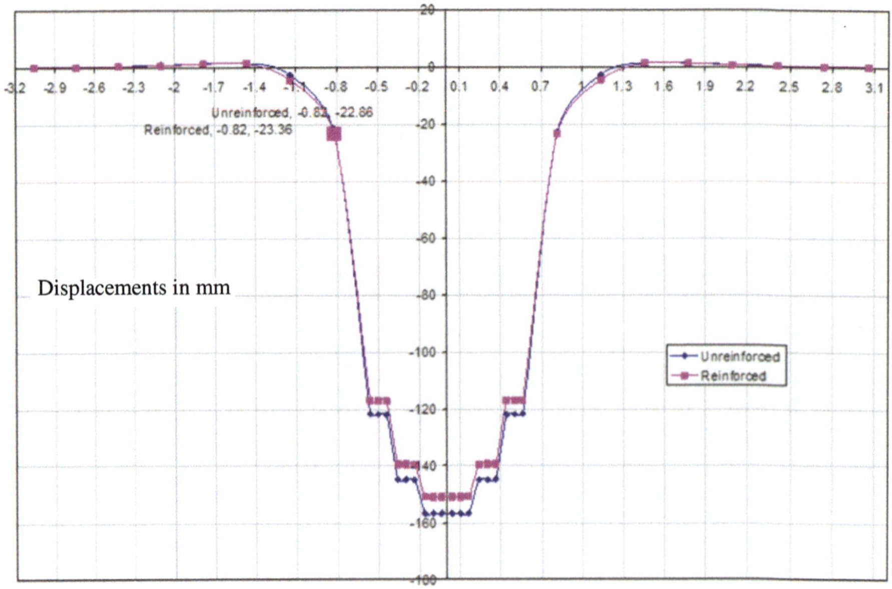

Fig. 16 Vertical displacements in mm in linear SSI and RSSI analyses at foundation level [8]

RSSI analysis when compared to nonlinear SSI analysis. But the vertical displacements are reduced merely by 6.2%.

Figure 20 shows horizontal displacements along longitudinal and transverse directions. In Fig. 20a, b, longitudinal displacements at foundation levels are shown. Similarly, in Fig. 20c, d, transverse displacements at foundation levels are shown.

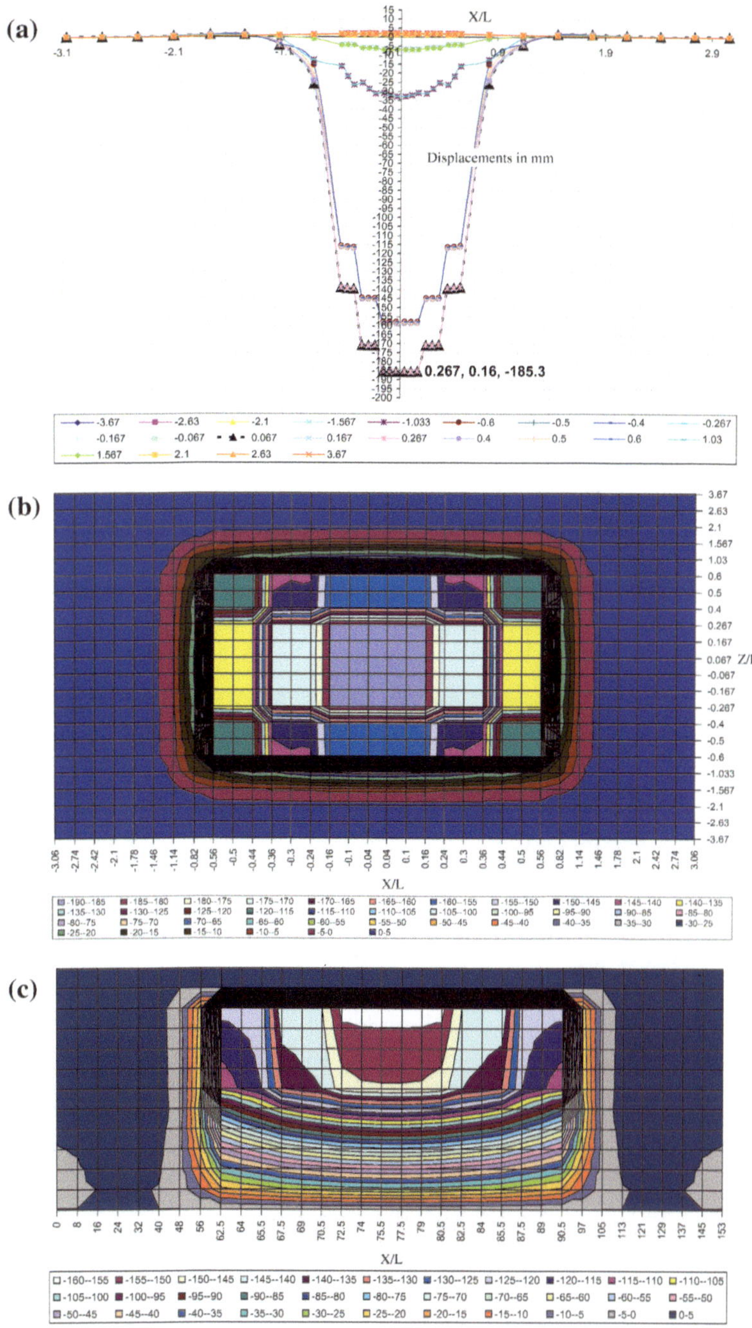

Fig. 17 Vertical displacements in mm in nonlinear SSI analysis: **a** vertical displacements at foundation level, **b** contours of vertical displacements at footing level, **c** vertical displacements along longitudinal section at centre [8]

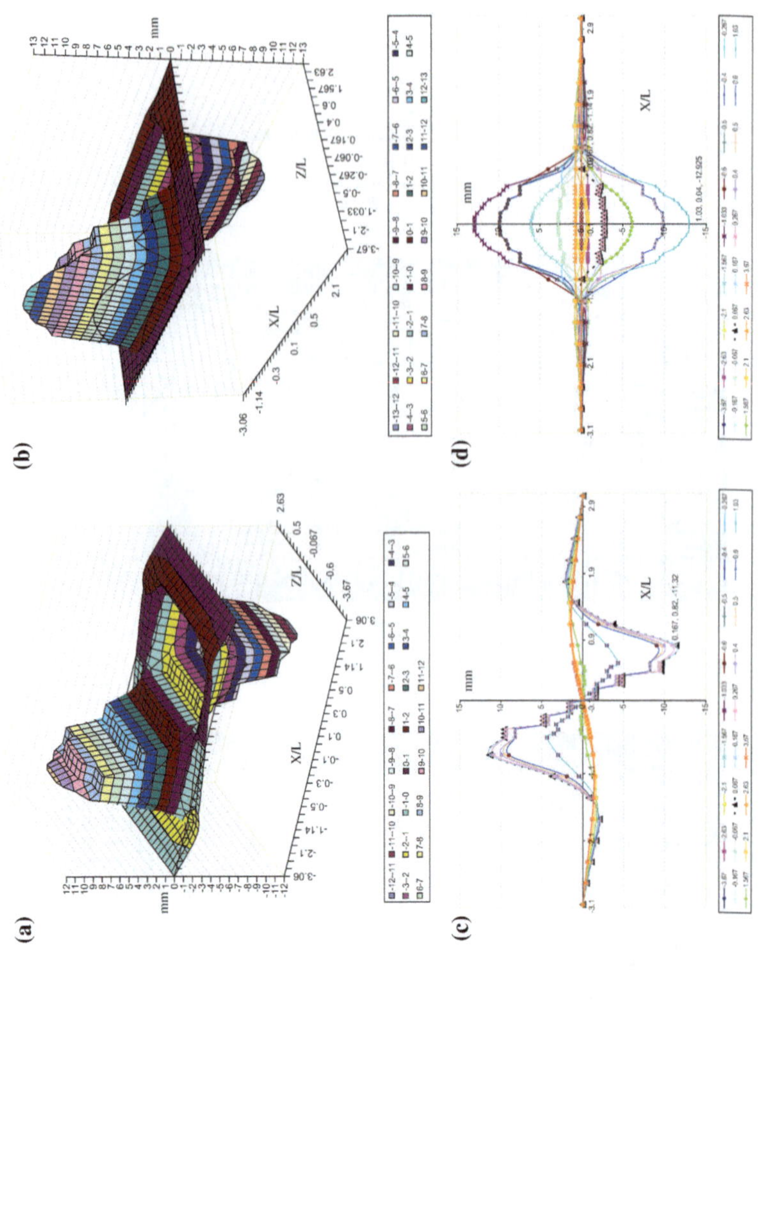

Fig. 18 Horizontal displacements in mm in nonlinear SSI analysis: **a** and **b** longitudinal displacements at foundation level; **c** and **d** transverse displacements at foundation level [8]

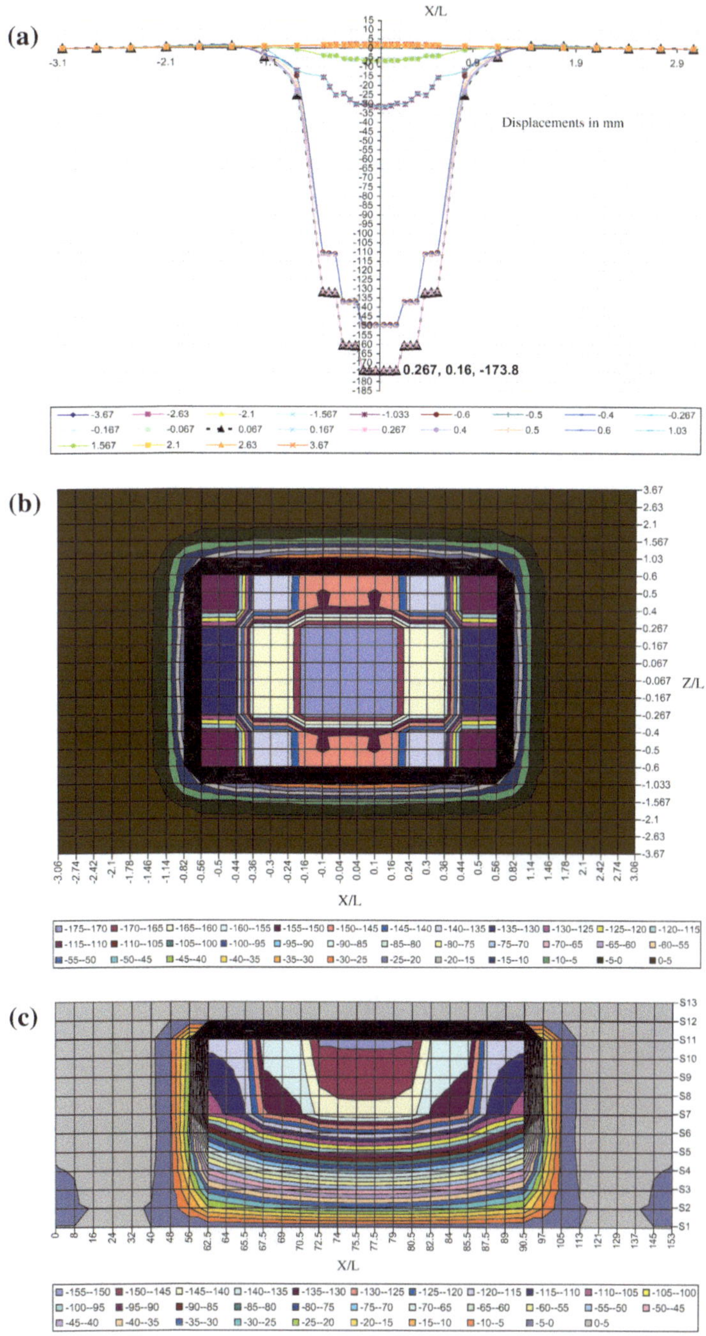

Fig. 19 Vertical displacements in mm in nonlinear RSSI analysis: **a** vertical displacements at foundation level along longitudinal directions, **b** contours of vertical displacements at footing level, **c** vertical displacements along longitudinal section at centre [8]

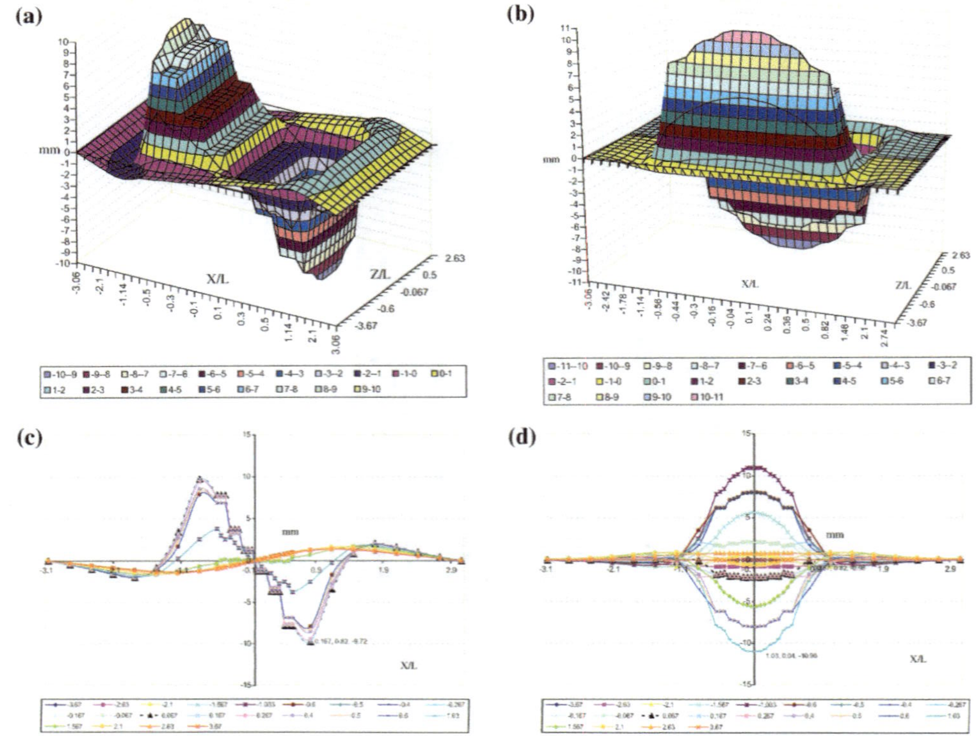

Fig. 20 Horizontal displacements in mm in nonlinear RSSI analysis: **a** and **b** longitudinal displacements at foundation level; **c** and **d** transverse displacements at foundation level [8]

9 Conclusions

1. Maximum lateral displacements along *X* and *Z* directions are reduced by 42 and 45.8% in linear RSSI analysis when compared to linear SSI analysis.
2. The vertical displacements are reduced merely by 3.72%. Maximum lateral displacements along *X* and *Z* directions are reduced by 14 and 15% in nonlinear RSSI analysis when compared to nonlinear SSI analysis.
3. The vertical displacements are reduced merely by 6.2%. As a result, the axial forces in beams have been increased in ground floor and have been reduced in higher floors.
4. Any change in the differential settlement, contact pressure and foundation stiffness results in significant changes in the moments and forces in the superstructure. In RSSI analysis, horizontal displacements and horizontal stresses have been reduced compared to SSI analysis as a consequence of shear forces which have been reduced in columns. Under RSSI analysis, the performance of the structure results in reduced displacements and member end actions resulting in economic structures.

References

1. King GJW and Yao ZE. Simplified Interactive Analysis of Long Framed Building on raft Foundation. Jnl. Structural Engineer. 1983; Vol.61, B (3) pp. 62–67.
2. Roy R and Dutta SC. Differential settlement among isolated footings of building frames: the problem its estimation and possible measures. Int J Appl Mech Eng. 2001; 6(1) pp 165– 86.
3. Noorzaei J, Viladkar MN and Godbole PN. Nonlinear soil-structure interaction in plane frames. Eng. Comput. 1994; 11, 303–316.
4. Viladkar MN, Godbole PN and Noorzaei J. Soil–structure interaction in plane frames using coupled finite infinite elements. *Computers and Structures* 1991; 39(5): pp 535–46.
5. Rajashekhar Swamy HM. Non-linear Dynamic analysis of Soil Structure Interaction of Three Dimensional Structure for Varied Soil conditions. 2012; PhD Thesis. Manipal University.
6. Rajashekhar Swamy HM, Krishnamoorthy A, Prabakhara DL and Bhavikatti SS. Evaluation of the Influence of Interface Elements for Structure—Isolated Footing—Soil Interaction Analysis. Interaction and Multiscale Mechanics. 2011; Vol. 4 No. 1 (2011) 65–83.
7. Krishnamoorthy and Rao NBS. Prediction of stress strain behaviour of soil using hypoelasticity constitutive model. Indian Geotechnical Journal. 2001;31(3).
8. Nayana NP. NonLinear Analysis and Behaviour of Reinforced Soil Structres. 2019; PhD Thesis, NIT-K, Surathkal.

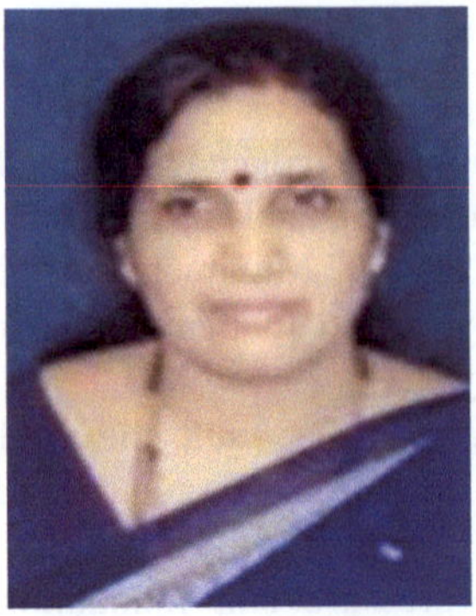

Dr. Nayana N. Patil obtained her doctoral degree from the National Institute of Technology (NIT-K) at Surathkal, India, in the year 2018 in the area of Geotechnical Engineering. She obtained her M.Tech degree in Civil Engineering (Geotechnical Engineering) from Basaveshwar Engineering College, Bagalkot in the year 2006. She obtained her BE degree in Civil Engineering, with a first class distinction, from Karnataka University Dharwad, in the year 1988. She obtained IIIrd rank to the University. She has around 20 years of teaching experience and has served as a Consultant too. Dr. Nayana N.Patil is currently an Asisstant Professor of Civil (Geotechnical) Engineering at the MS Ramaiah University of Applied Sciences (MSRUAS) since 2014. Prior to joining MSRUAS, Dr. Nayana N.Patil served in various academic and research positions at Bellary Engineering College, Bellary and Sahyadri College of Engineering and Management, Mangalore. Her areas of research interest include soil behaviour, soil reinforcement techniques, fibre reinforced concrete, green construction, soil-structure interaction etc. She has guided over 25 M.Tech and 40 B.Tech students in various areas of Civil Engineering. She has published over 25 technical papers in national and international journals and conferences.

Dr. H. M. Rajashekar Swamy obtained his Doctoral degree from Manipal Institute of Technology, Manipal, India in the year 2012 in structural dynamics. He obtained his M.Tech degree in General Programming (Structures and Geotechnical Engineering) from Indian Institute of Technology, Powai, Bombay. He obtained his B.Tech in Civil Engineering from NIT-K, Surathkal, India. He has more than 32 years of teaching experience and has worked in various capacities of administration, consultancy and construction. Currently he is doing research in the fields of structural dynamics, nonlinear Soil-Structure interaction under transient and earthquake loads. His fields of interest include static and dynamic analysis of SSI problems, frame and wall interaction under static and dynamic loads, structural health monitoring of structures, static and dynamic FEM applications, fluid-structure interaction etc. He is presently serving as Professor and the Head of the Department of Civil Engineering at MS Ramaiah University of Applied Sciences, Bangalore. He has guided over 50 M.Tech and 100 B.Tech students in various areas of Civil Engineering. He is guiding 4 PhD scholars and has published over 35 technical papers in national and international journals and conferences.

Prof. R. Shivashankar obtained his doctoral degree from the Asian Institute of Technology (AIT) at Bangkok, Thailand, in the year 1991 in the area of Geotechnical Engineering. He obtained his M.Tech degree in Civil Engineering (Soil Mechanics and Foundation Engineering) from IIT Madras in the year 1980. He obtained his BE degree in Civil Engineering, with a first class distinction, from University Visvesvaraya College of Engineering (UVCE), Bangalore University, in the year 1978. Prof. R. Shivashankar is currently a Professor of Civil (Geotechnical) Engineering at the National Institute of Technology Karnataka (NITK) at Surathkal, Mangalore, since 2002. Prior to joining NITK, Prof. R. Shivashankar served in various academic and research positions at Bangalore University,

Asian Institute of Technology (AIT) at Bangkok, Thailand; Saga University, Japan; KREC/NITK. His areas of research interest include soil behaviour, soil reinforcement techniques, ground improvement, pile foundations, soil-structure interaction etc. He has been a principal coordinator/co-coordinator for five major sponsored research projects on soil reinforcement techniques. He has guided 12 PhD students so far and over 75 M.Tech students in various areas of geotechnical engineering. He has published over 200 technical papers in national and international journals and conferences, given keynote lectures and chaired sessions. He has presented papers in several conferences in India and abroad. He is a member of several professional organizations. He is currently a member of the Technical committee TC107 on 'Laterites and Lateritic soils' of the International Society of Soil Mechanics and Geotechnical Engineering (ISSMGE).

A Robust Framework for Socially Responsive Services: A Constraint-Based Social Network Representation

Kamakhya Narain Singh, Chinmaya Misra, Soumita Seth, Sumanta Kumar Mandal and Biresh Kumar

1 Introduction

NGOs can be defined as non-profit organizations and a group of members who work in some ways for serving the individuals or a community [1]. A well-established literature shows that the service provider and the service receiver influence the behavior of each other and which is highly influenced by social network structure. Therefore, NGOs can be defined as private organizations who work in local level or national level or international levels whose objective is to help or share the resources among needy parts of the society. It is a non-profit community-based organization.

The main goal of NGOs is to share resources [2]. Here resource signifies any types of help that can be provided by the individuals or community. It may be idea, some solution, physical resource, or some kinds of interdependency, and it will satisfy the demands of the user. It should be feasible in true sense; that is, it should

K. N. Singh (✉) · C. Misra
Kalinga Institute of Industrial Technology, A Deemed to be University, Bhubaneswar, India
e-mail: kamakhya.vphcu@gmail.com

C. Misra
e-mail: cmisra@yahoo.com

S. Seth
Computer Science and Engineering Department, Aliah University, Kolkata, India
e-mail: soumita.seth@gmail.com

S. K. Mandal
Department of Computer Science and Engineering, CEB, Bhubaneswar, India
e-mail: sumantab4u2@gmail.com

B. Kumar
Department of Computer Science and Engineering, Amity University, Ranchi, India
e-mail: bkumar@rnc.amity.edu

J. Chattopadhyay et al. (eds.), *Innovations in Soft Computing and Information Technology*, https://doi.org/10.1007/978-981-13-3185-5_14

be technologically, economically, and culturally feasible [3]. Here our main focus is that the distribution should be fair and unbiased. The needy people should get the help. The help may be educational like a fellowship for a poor student, medical help for a patient, or some help for a disabled [4]. Otherwise, we may face the following major problems:

- Overall crisis.
- Requirement of service of needy people.
- Development of poorer section of society.

Galvin et al. [5] propose the concept of resource management and the usage of resources. Resources will be supplied by the assigned agency. Better collaboration and interaction are possible with the advancement of Internet with time. The social network paved its way into popularity by superseding the Internet. Social network analysis is a vast area of research which deals with the understanding and analysis of the behavior of individuals or community. It predicts the prospective existence of the relationships between community members.

Socially Responsive Resource Usage Protocol (SRRUP) defined to standardized social ethics and utilization of resources [6, 7]. This protocol helps to emerge in an improvised environment by providing the way of accessing the resources to needy people. To allocate the resources, more socially distant individual is given more priority than less socially distant individual. Social distance of the user represents the defined logical distance between user and resource [8–10]. A resource has three parameters: place, cost, and constraints, causing prevailing consequences in its usages. Place can be defined where the resource can be availed [11–13]. A person who needs resources pay the price to avail the resources is known as cost. A set of affirmation and considerations measure to a resource usage are the constraints.

In our proposed framework, we have made the following contributions:

- The distribution of resources will be impartial and those resources can be avail for those people who genuinely have the requirements.
- The resource will be available to the persons based on their financial, social, disadvantageous positions, and their locations (remote or rural).

In operating system, we termed user as a process. Process enters in waiting state if it does not get requested resources. Some time waiting, process does not come out from waiting state because requested resources are held by some other waiting process and so on then it becomes deadlock. Such situation occurs for inquiry table distribution of resources. For this work, we have given attention to fair resource allocation. To describe it precisely, best option is to use a directed graph; this is called resource distribution graph. The graph has nodes V and edges E. V is partitioned into $\{P1, P2, \ldots, Pn\}$, the set of active processes and $R = \{R1, R2, \ldots, Rm\}$, the set resource types in the system. Figure 1 depicts the graph where a directed age from process Pi to resource type Rj is referred by $Pi \rightarrow Rj$; it shows that process Pi has requested an instance resource type Rj and is currently waiting for that resource. A directed edge from resource type Rj to process Pi is referred by $Rj \rightarrow Pi$; it means that an instance of resource type Rj has been allocated to process

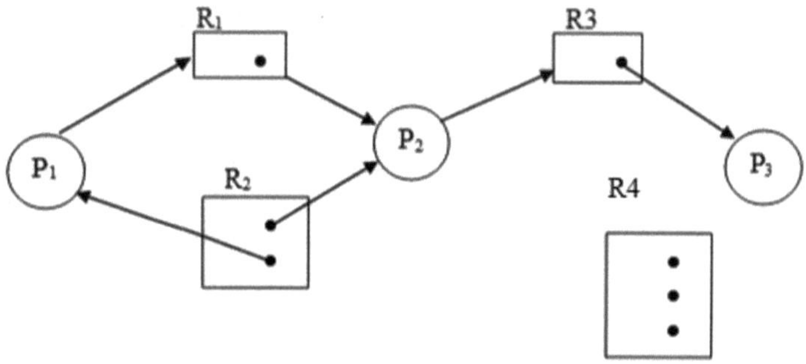

Fig. 1 Resource allocation graph

Pi. A directed edge $Pi \rightarrow Rj$ is named a request edge; a directed edge $Rj \rightarrow Pi$ is named an assignment edge.

Though different resource allocation graph is used in many applications like job processing, hospital management, and real-time scheduling, this is mainly used for avoidance of deadlock where same resources are demanded by multiple processes and others are not releasing the resources [14]. In our proposed research work, multiple instances of resource are possible. Hence, this situation will not raise.

Section 2 presents the working mechanism of the proposed model. The experimental result and comparison studies are presented in Sect. 3. In the last section, future work and conclusions has been provided.

2 Proposed Model

2.1 Concept of the Proposed Model

In our earlier work [1], if a volunteer is not able to provide a support due to lack of resources then he will wait for required resources. To improve the performance by reducing the waiting time of our earlier work, we have proposed a new scheme.

If two volunteers have the less resources than receiver's requirement like volunteers $v_1 < r$ and $v_2 < r$. If $v_1 + v_2 \geq r$ then both volunteers jointly provide resources to that receiver for completion of his job. This way it is reducing the waiting time and utilizing the resources effectively.

In this model, we have designed a network comprised of two layers: one layer represents the service provided by the service provider and other layer represents the members who will receive the services. The services will be received by the weaker sections of the society such as orphans, disabled, poor or needy, and old peoples. The service will be given voluntarily by individual or by a community. Here the volunteers mainly provide two types of services. One is to identify the deserving persons who are willing to take services. Second responsibility is to accumulate the resources to provide services without interruption to the needy, old, disabled, orphan, or persons staying in remote areas. The services required vary from user to user. Some users may require financial help for continuing their study. Some may require physical assistance or medical supervision for their agony and pain. Volunteer also has different resources with different capacities. At a time, more than one service can be provided by volunteer. If a user requires services a volunteer alone is not able to help, then services can be provided by more than one volunteer to the same user.

So the communication is one-to-one (volunteer to receiver), one-to-many (receiver to volunteers), and many-to-one (receivers to volunteer). The volunteers may approach external sources to serve the users because requirements of the user vary. For approaching the external sources, volunteers have to follow system rules: Volunteer cannot take more than user's requirement. Based on resource availability the needed service will be given. Volunteers of a specific group of users communicate with other volunteer if required. User quits the organization or concern volunteer may stop his support to user after fulfilling the requirement.

2.2 *Mathematical Representation of the Proposed Model*

We have introduced user by u and volunteers by v with appropriate suffixes. Users seek their requirements through $u(x_1, x_2, \ldots x_k)$ and volunteer represents services through $v(y_1, y_2, \ldots, y_p)$ where x_i and y_i represent direct amount that should be non-negative real numbers. It is used to find the strength of the organization on the basis of user need and volunteer service.

Using community detection in social network analysis [5, 6], we organize and find various categories of person with different need and support to run the organization smoothly.

We prefer two types of sorting:

Case I. User-based category.
Case II. Volunteer-based category.

This can be classified to get a real scenario.

Case I Suppose that more than one service can be provided by each volunteer but he has only one service to provide. Then we seize an estimable amount of users to him.

Let $v_1(y_1)$ represent a volunteer with a type of support s_1 (like disables, orphans).

Let a category of users $\{u_{11}, u_{12}, \ldots, u_{1k}\}$ are associated with v_1 with demand of $y_{11}, y_{12}, \ldots, y_{1k}$ where $u_{1i}(y_{1i})$ represents a typical member of this category.

Now there are two possibilities:

Lemma 1 $s_1 \geq \sum_{j=1}^{k} y_{1j}$

(Sum of required is less than service quantum)

Here, either the volunteer or organization v_1 themselves asks for more users to extend this task.

Now we can extend $u_{1k+1}(y_{1k+1}), u_{1k+2}(y_{1k+2}), \ldots, u_{1p}(y_{1p})$ where $p > k$ and $s_1 \geq \sum_{j=1}^{p} y_{1j}$.

This process goes on till $\sum_{j=1}^{p} y_{1j} \leq s_1 < \sum_{j=1}^{p+1} y_{1j}$.

This improve the performance of v_1. Improvement of v_1 is evaluated on the basis of more users (u_{1j}'s).

Lemma 2 $s_1 \geq \sum_{j=1}^{k} y_{1j}$

In this case, the support of v_1 is not sufficient to satisfy the needs of u_{1k}'s. So either the volunteer himself (v_1) or NGO contacts some other volunteer v_2 of similar task as demanded by u_{1k}'s. Now here options are:

(a) if $\sum_{j=1}^{l} y_{1j} \leq s1 < \sum_{j=1}^{k} y_{1j}$ where $l < k$, we can detach those $\{u_{1l+1}, u_{1l+2}, \ldots, u_{1k}\}$ from v_1 and associate them to v_2 directly. Here v_2 is an existing volunteer of NGO.

or (b) v_2 cannot approach directly to u_1's. NGO will link up v_2 to v_1 so now: $s_1 + s_2 \geq \sum_{j=1}^{k} y_{1j}$

where s_1 and s_2 are the support of v_1 and v_2, respectively. In this case, for purely temporary basis either NGO or volunteer gets a support from outside.

Case II If each u_i needs many services and no one volunteer alone is sufficient to fulfill their needs, then each u will be assigned to a group of v's. Here user represents his needs through $u_1(y_{11}, y_{12}, \ldots, y_{1k})$ and group of volunteer represent their strength through $\{v_{11}(s_{11}), v_{12}(s_{12}), \ldots, v_{1k}(s_{1k})\}$. We expect $y_{1j} \leq s_{1j}$, where $y = 1, 2, \ldots, k$ else NGO has to include more v's to fulfill the needs of u_1.

Lemma 1 Suppose $y_{1j} = s_{1j} \; \forall \; j = 1, 2, \ldots k$
Then all the needs of u_1 are fulfilled and u_1 may leave.

Lemma 2 Suppose $y_{1j} < s_{1j}$ for some $y = p \leq k$.
Then the $v_{1p}(s_{1p})$ can support some other users 'u' also. NGO will utilize $v_{1p}(s_{1p})$ remaining services to fulfill the needs of other user u_2 like $u_2(y_{2p})$ for which $y_{1p} + y_{2p} \leq s_{1p}$.

This way such v_{1p} are associated with more users. This process repeats until $\sum_{j=1}^{k} y_{1j} \leq s_{1j}$.

Here, the public presentation of the organization possibly is assessed by number of users associated with a volunteer and the number of users leaving the system with satisfaction.

2.3 *Algorithms of the Proposed Model*

Here, we declare all attributes which exist in this proposed algorithm.

$u \leftarrow$ User.
$y \leftarrow$ Requirement.
$u_{1j} \leftarrow y_{1j}$ Users $(u_{11}, u_{12}, \ldots u_{1j})$ with requirements $(y_{11}, y_{12}, \ldots y_{1j})$.
$v \leftarrow$ Volunteer.
s $\leftarrow$ Service.
$v \leftarrow s$ Volunteer (v) provides service (s).
$v_1 \leftarrow s_1$, v_1 provides s_1-type service, $v_2 \leftarrow s_2$, v_2 provides s_2-type service etc.

In this algorithm, our aim is to provide services (s) to a user (u), who are registered ($u \leftarrow y$) with requirements (y). Volunteer (v) provides service to user (u) to fulfill their needs. In the first case, the organization is robust, while in the second case the organization asks for improvement.

Algorithm 1: Algorithm for Case I

Begin
$u_{1j} \leftarrow r_{1j}, \forall j = 1, 2, \ldots k$;
$v_1 \leftarrow s_1$, $v_2 \leftarrow s_2$
if $s_1 > r_{11}$,
  then v_1 supports r_{11}
end
for l=2 to k do
  if $s1 > \sum_{j=1}^{l} r_{1j}$ then
    v_1 supports $\{x_{11}, x_{12}, \ldots\ldots, x_{1l}\}$
  end
  if $\sum_{j=1}^{l} r_{1j} \leq s1 \leq \sum_{j=1}^{l+1} r_{1j}$ then
    v_1 supports $\{x_{11}, x_{12}, \ldots\ldots, x_{1l}\}$ but not x_{1l+1}
  end
end
select $v_2(s_2)$ such that
$s_1+s_2 \geq \sum_{j=1}^{k} r_{1j}$ or $s_2 \geq \sum_{j=1}^{k} r_{1j} - s_1$ then
$\{v_1, v_2\}$ supports $\{x_{11}, x_{12}, \ldots\ldots, x_{1k}\}$
End

Algorithm 2: Algorithm for Case II

Begin
$u_1 \leftarrow \{r_{11}, r_{12}, \ldots, r_{1k}\}$;
for l= 1 to k do
  $v_l \leftarrow s_{1l}$
end
for l= 1 to k do
  if $r_{1l} \leq s_{1l}$ then
    u_1 leaves the system as satisfied customer
    $\{v_{11}, v_{12}, \ldots v_{1k}\}$ supports u_1.
  end
end
Now $u_2 \leftarrow \{r_{21}, r_{22}, \ldots r_{2k}\}$
for l = 1 to k do
  if $r_{2l} \leq s_{1l} - r_{1l}$
    then u_2 leaves the system as satisfied customer
      $\{v_{11}, v_{12}, \ldots, v_{1k}\}$ supports $\{u_1, u_2\}$
  end
end
for p = 1 to k do
  if $r_{2p} \leq s_{1p} - r_{1p}$
    then some v'_{1p} enter the system with $v'_{1p} \leftarrow s'_{1p}$ and
    $r_{2p} \leq s'_{1p} + s_{1p} - r_{1p}$
    u_2 leaves the system as satisfied manner.
      $\{v_{11}, v_{12}, \ldots v_{1k}, v'_{1p}\}$ supports $\{u_1, u_2\}$
  end
end
This process continues until all u_i's and v_k's are exhausted.
End

3 Experimental Result Discussion

Here, we elaborated our program output. We have taken all possible kinds of inputs and elaborated desired outputs of concerned inputs.

3.1 Experimental Environments

We have implemented these algorithms and tested using all possible desired inputs using a Java program for the preliminary study and executed it on a Lenovo Miix 510-121SK laptop with 2.3 GHz Intel Core i5-6200U 6th Gen processor, 8 GB DDR3 RAM and 256 GB hard disk, Windows 10 with 64-bit operating system and Java1.8. In this program, our objective is to make a communication among various methods. Program works on the basis of client and server program. Client represents the user and server represents the providers. There are three types of user-defined communication methods:

- Volunteer_to_receiver—This method represents the task like one volunteer is alone quite sufficient to fulfill the needs of a user of all type of services.
- Receiver_to_volunteers—This method provides the dependency of volunteers to fulfill the requirements of receiver.
- Receivers_to_volunteers—This method works as services of more than one user can be fulfilled by more than one volunteer. Above communications can be initialized in various ways like users represented by '*u*' and volunteers by '*v*'. The needs of user are represented by y_i, $i = 1, 2, \ldots, n$ and services of volunteers are represented by s_j, $j = 1, 2, \ldots, n$. Table 1 represents the input provided to the system and Table 2 reflects the desired output.

Table 1 User needs table

User/need	r_1	r_2	r_3	…	…	r_n
$u^1{}_{1000}$	1	0	0	0	0	0
$u^1{}_{0100}$	0	1	0	0	0	0
…	…	…	…	…	…	…
$u^2{}_{1100}$	1	1	0	0	0	0
$u^2{}_{1010}$	1	0	1	0	0	0
…	…	…	…	…	…	…
$u^3{}_{1110}$	1	1	1	0	0	0
…	…	…	…	…	…	…
Sum	=4	=3	=	=	=	=
Total =						

Table 2 Volunteer support table

Provider/support	s_1	s_2	s_3	…	…	s_n
${v_1}^1$	k_1	0	0	0	0	0
${v_2}^1$	0	k_2	0	0	0	0
…	…	…	…	…	…	…
${v_1}^2{}_2{}^2$	p_1	p_2	0	0	0	0
${v_1}^3{}_3{}^2$	p_1	0	p_3	0	0	0
…	…	…	…	…	…	…
${v_1}^5{}_2{}^4{}_3{}^3$	q_1	q_2	q_3	0	0	0
…	…	…	…	…	…	…
Sum	=	=	=	=	=	=

Notation: u^i is a user (i = 1 to n)

$u^1{}_{1000}$ denotes user wants single resource of y_1.
$u^2{}_{0100}$ denotes user wants single resource of y_2.
$u^1{}_{1100} = u^2{}_{1100}$ denotes user wants two resources y_1 and y_2, respectively.
$u^1{}_{1110} = u^2{}_{1110} = u^3{}_{1110} = \ldots$ represents a user needs three resources y_1, y_2, and y_3.
$u^1{}_{1111} = u^2{}_{1111} = u^3{}_{1111} = u^4{}_{1111} \ldots$ represents a user with four resources y_1, y_2, y_3, and y_4.
${v_1}^1$ has a strength of single resource r_1 for type s_1.
${v_2}^1$ has a strength of single resource r_2 for type s_2.
${v_1}^5{}_2{}^6$ has a strength of five resources r_1 for type s_1 and six resources r_3 for type s_3.
…… ……
${v_i}^j$ has a strength of j for type s_i where i represents types of service and j represents for some resources.

Simple Case: Assume each user 'u' needs one resource of size '1' only. That is, $y_1 = y_2 = \ldots = y_n = 1$, users with variable requirements, e.g., $u^1{}_{1000}(y_1)\ y_1 \geq 1$.

3.2 *Result of Various Communication Method*

- In this function, we have supplied different kinds of inputs for all the above-mentioned methods. For volunteer_to_receiver communication method, we represent through Table 3. Here $u^1{}_{3000}, u^2{}_{0500}$, and $u^3{}_{0020}$ denotes corresponding user and ${v_1}^5, {v_1}^5{}_2{}^6$, and ${v_3}^8$ the volunteer. Here u^1 wants three resources of type s_1, u^2 wants five resources of type s_2, and u^3 requires two resources of type s_3, while (Table 4) v_1, v_2, and v_3 contain 5, (5 of s_1, 6 of s_2),

Table 3 User needs table of receiver

User/need	r_1	r_2	r_3	r_4
$u^1{}_{3000}$	3	0	0	0
$u^2{}_{0500}$	0	5	0	0
$u^3{}_{0020}$	0	0	2	0

Table 4 Provider support table of volunteer

Provider/support	s_1	s_2	s_3
$v_1{}^5$	5	0	0
$v_1{}^5{}_2{}^6$	5	6	0
$v_3{}^8$	0	0	8

and 8 resources, respectively. So, v_1 supports three resources to u^1, v_2 supports five resources of s_2 to u^2, and v_3 gives two resources to u^3. Now, v_1 has two, v_2 has (5 of s_1, 1 of s_2), and v_3 has six resources.

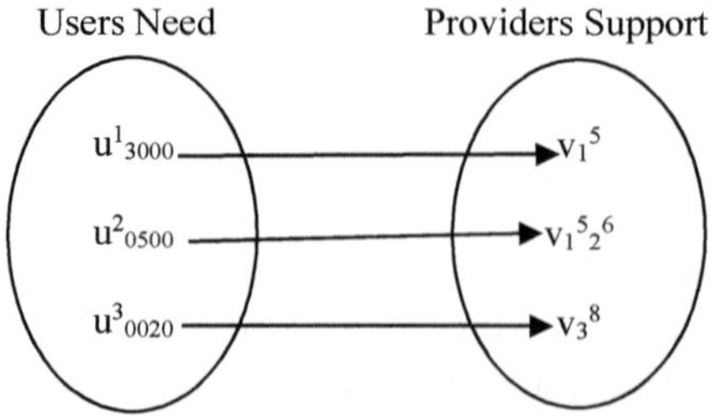

Volunteer_to_receiver Communication Method

- For receiver_to_volunteers, we represent through Table 5. Here $u^4{}_{4020}$ is denoted as a user which needs four resources of service s_1 and two resources of s_3. Since (Table 6) v_1 alone is not able to fulfill user need, because it has only two resources of s_1, so v_1 needs help from another volunteer to satisfy user. Here v_1, and v_2 support two and two resources of type s_1, respectively, and v_3 supports two resources of s_3. Now, v_1 has exhausted, v_2 has (3 of s_1, 1 of s_2), and v_3 has 4 of s_3-type resources.

Table 5 User needs table of receiver_to_volunteers

User/requirement	r_1	r_2	r_3	r_4
$u^4{}_{4020}$	4	0	2	0

Table 6 Provider support table of receiver_to_volunteers

Provider/strength	s_1	s_2	s_3
$v_1{}^2$	2	0	0
$v_1{}^5{}_2{}^1$	5	1	0
$v_3{}^6$	0	0	6

Table 7 User needs table of receiver_to_volunteers

User/requirement	r_1	r_2	r_3	r_4
$u^5{}_{4080}$	4	0	8	0
$u^6{}_{0605}$	0	6	0	5
$u^7{}_{5002}$	5	0	0	2

Table 8 Provider supports table of receivers_to_ volunteers

Provider/strength	s_1	s_2	s_3	s_4
v_1	Exhausted	0	0	0
$v_1{}^3{}_2{}^1$	3	1	0	
$v_3{}^4$	0	0	4	0
$v_1{}^5{}_2{}^4{}_3{}^5{}_4{}^3$	5	4	5	3
$v_1{}^5{}_2{}^5{}_3{}^5{}_4{}^5$	5	5	5	5

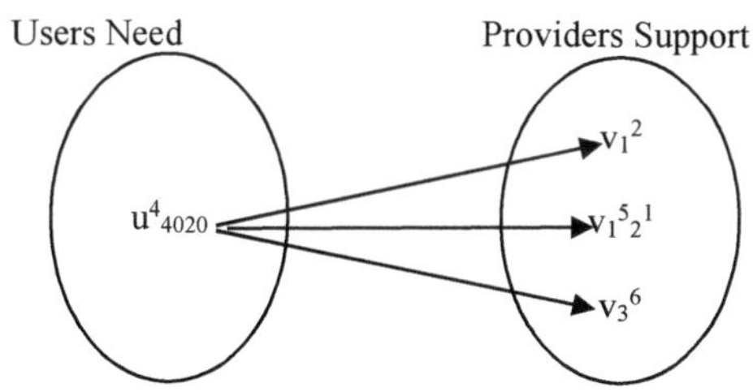

Receiver_to_volunteers Communication method

- For receivers_to_volunteers, we represent through Table 7. Here, $u^5{}_{4080}, u^6{}_{0605}$, and $u^7{}_{5002}$ represent respective users. In Table 8, two more donors ($v_1{}^5{}_2{}^4{}_3{}^5{}_4{}^3$ and $v_1{}^5{}_2{}^5{}_3{}^5{}_4{}^5$) have been added from external sources. Here $u^5{}_{4080}$ wants four resources of s_1 and eight resources of s_3. From Table 8, $v_1{}^3{}_2{}^1$ supports three resources of s_1, $v_1{}^5{}_2{}^4{}_3{}^5{}_4{}^3$ supports one resource of s_1, and $v_3{}^4$ supports four resources of s_3 and $v_1{}^5{}_2{}^4{}_3{}^5{}_4{}^3$ supports four resources of s_3. Now, $v_1{}^3{}_2{}^1$ has $v_2{}^1$ only, $v_3{}^6$ has exhausted, and $v_1{}^5{}_2{}^4{}_3{}^5{}_4{}^3$ has $v_1{}^5{}_2{}^4{}_3{}^5{}_4{}^3$. User $u^6{}_{0605}$ wants six resources of s_2 and five resources of s_4, but $v_2{}^1$ does not have sufficient resources to support. So it requires support from $v_1{}^4{}_2{}^4{}_3{}^3{}_4{}^3$ and $v_1{}^5{}_2{}^5{}_3{}^5{}_4{}^5$. After giving the resources to $u^6{}_{0605}, v_2{}^1$ exhausted, $v_1{}^4{}_2{}^4{}_3{}^3{}_4{}^3$ has $v_1{}^4{}_3{}^1$ and $v_1{}^5{}_2{}^5{}_3{}^5{}_4{}^5$ has $v_1{}^5{}_2{}^4{}_3{}^5{}_4{}^3$. User $u^7{}_{5002}$ needs five resources of s_1 and two resources of s_4. $v_1{}^4{}_3{}^1$ supports four resources of s_1 and $v_1{}^5{}_2{}^4{}_3{}^5{}_4{}^3$ supports one resource of s_1 and two resources of s_4. After giving the resources to $u^7{}_{5002}, v_1{}^4{}_3{}^1$ has $v_3{}^1$ and $v_1{}^5{}_2{}^4{}_3{}^5{}_4{}^3$ has $v_1{}^4{}_2{}^4{}_3{}^5{}_4{}^1$.

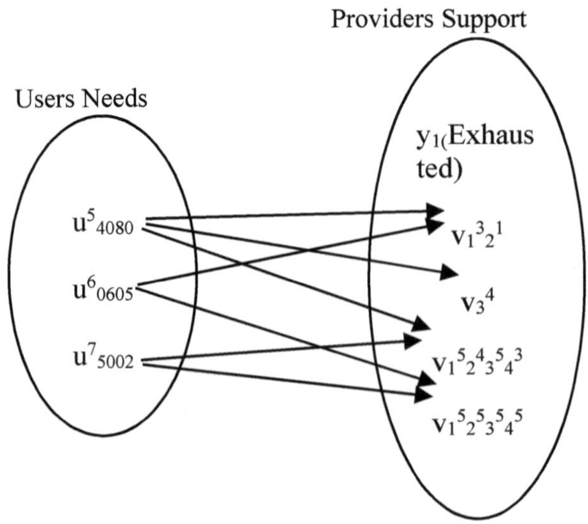

Receivers_to_volunteers Communication Method

The number of satisfied user (u_i's) and number of registering new users (u_i's) measure the strength of the system.

4 Conclusions and Future Work

This framework suggests to provide services to various sections of the society and helps to improve the efficiency, effectiveness, and reliability of the system. It suggests couple of systematic methodologies like one volunteer to one receiver, one receiver to many volunteers, and many receivers to many volunteers to allocate the resources. In this chapter, we have formalized the holistic approach for fair distribution of service from one network to another network. Here one network represents the list of newly registered users and other represents the volunteers with types of services. Services have a broad range of latitude which leads the challenges for the system approaches to a service provider. There are two constraint matters to support a kind of services: identify the needy and supply the services to fulfill their requirements. For identification, either volunteer identifies or user has to approach to NGO. System provides the kind of services with the help of three communication methods. Once the user's needs are fulfilled, then either user might leave or volunteer stops his service. In both the cases, user has to give their feedback to the system. If the user leaves the system with satisfaction, then system is robust else system requires support. In implementation, we tried to assign providers to users on the basis of user's choice (preferential allocation in a prescribed order, say u_1 to v_1 like that).

In future work, user's feedback can be considered for better improvement of the system. We have to make sure best utilization of resources by keeping providers non-idle and consistent use of resources. We need to check the nearness between the resource providers. We should make the user as a temporary provider like increasing the demand of the user like $u^1{}_{6000}$ which represents a user with need of $y_1 = 6$. Here u_1 should be a supplier on some conditions but there is no provision for y_1 in the existing system.

References

1. Singh K.N, Misra C, Seth, S., Mantri J.K, and Dash S. R., "A Framework for Social Service Volunteers: A Social Network Representation". International Journal of Pure and Applied Mathematics Volume 114 No. 10, 11–23, 2017.
2. Eric D. Werker and Faisal Z. Ahmed.: What do Non-Governmental Organization do?. Journal of Economic Perspectives, 2007.
3. Rout, A.,: A UML Framework for Socially Responsive Resource Usage Protocol, IJCCT, Vol-3, Issue-2. 0975–7449, 2012.
4. Mohanty, H., "Socially responsive resource usage: a protocol". In International Conference on Distributed Computing and Internet Technology, pp. 243–254. Springer Berlin Heidelberg, 2011.
5. Silberschatz, P. B. Galvin and G. Gagne, Operating System Priciples, WSE, Wiley India Pvt. Ltd, 2006.
6. Bhattacharyya, D., Seth, S., and Tai-hoon K., "Social network analysis to detect inherent communities based on constraints". Appl. Math 8, no. 1L, 385–396. 2014.
7. Seth, S., Bhattacharyya, D., and Tai-hoon K., "CBACCN: Constraint Based Community discovery in Complex Networks". International Journal of Applied Engineering Research 9, no. 23 18115–18127, 2014.
8. Yamakami, T., "Servicenics approach: A social service engineering framework." In Digital Information Management (ICDIM), 2013 Eighth International Conference on, pp. 358–362. IEEE, 2013.
9. Nepusz, T., A. Petróczi, and F. Bazsó. "Multigraph Approach to Social Network Analysis." 2012.
10. Yamakami, T., "A two-layer view model of service engineering: Implications based on service engineering in mobile social games in Japan." In Information Science and Digital Content Technology (ICIDT), 2012 8th International Conference on, vol. 3, pp. 562–566. IEEE, 2012.
11. Fogg, B. J., Cathy S., David R. D., Leslie M., Julianne S., and Ellen R. T., "How do users evaluate the credibility of Web sites?: a study with over 2,500 participants." In Proceedings of the 2003 conference on Designing for user experiences, pp. 1–15. ACM, 2003.
12. Barik, R. K., Dubey, H., Samaddar, A. B., Gupta, R. D., & Ray, P. K. FogGIS: Fog Computing for Geospatial Big Data Analytics. arXiv preprint arXiv:1701.02601, 2016.
13. Goswami, V., and Misra C., "Discrete-time modelling for performance analysis and optimisation of uplink traffic in IEEE 802.16 networks." International Journal of Communication Networks and Distributed Systems 10, no. 3: 243–257, 2013.
14. Das, Satya Ranjan, and Sanjay Mohapatra. "Social and public impact of ICT enabled education." Information Technology, 2008. ICIT'08. International Conference on. IEEE, 2008.

Kamakhya Narain Singh He is a life member of Indian Society for Technical Education. He is Assistant Professor in School of Computer Applications, KIIT Deemed to be University, Bhubaneswar. He was born in [Jehanabad] in 1980. He did MCA from University of Hyderabad in [2007], M.Tech from JNTU Hyderabad in [2014] and pursuing PhD from North Odisha University, Baripada. He has 2 years of industry and 10 years of Teaching experience. His research area is [IoT, Sensor Network, Social Network and Web Technology]. In 2007, He started his professional career as a [Software Engineer] in Sofmen Technology Pvt Ltd, Indore. He Joined as a lecturer in College of Engineering, Bhubaneswar In [2009]. He has also worked as a guest faculty in Sambalpur University Institute of Information Technology in 2009-2010. In [2013], joined as Assistant Professor in School of Computer Application, KIIT Deemed to be University, Bhubaneswar. He has published many papers in reputed journals and international conferences.

Dr. Chinmaya Misra He is a Life Member of the Indian science Congress. He is the Associate Professor in the School Of Computer Applications, KIIT Deem to be University, Bhubaneswar, India. He is working in Energy Saving Mechanism in WiMaX and Image Processing and Queuing System, Cloud Computing and Social Networking. He has completed his MCA from North Bengal University in 1999 and received his MS degree from School of Information Technology, IIT Kharagpur in [2008]. He completed his PhD from KIIT Deemed to be University from School of Computer Science and Enginnering in [2014]. In [1999] he started his professional career as [Programming Assistant] in IIT Kharagpur and joined as lecturer in KIIT Deemed to be university as [Lecturer] in 2008. He has published many research works in reputed journals and conferences.

Soumita Seth She is Visiting Lecturer in the Department of Computer Science and Engineering, Aliah University, Kolkata, West Bengal (India) as well as the Department of Computer Science and Engineering, Guru Nanak Institute of Technology, Kolkata, West Bengal (India). More specifically, her work examines in the field of Machine learning, Bioinformatic. Soumita Seth was born in [Kolkata, India], in 1987 and, received her Bachelor's degree B.Tech in [IT] from [Department of Information Technology, WBUT(Now MAKAUT)], in [2009]. She has completed M.Tech. in [CSE] from [Department of CSE, WBUT(Now MAKAUT)], in [2011]. In [2011], She started her professional career as [Assistant Professor] in [Computer Application Department, The Heritage Academy, Kolkata]. She also worked as [Academic Associate] in [VGSOM, IIT Kharagpur]. She also worked as [Research Fellow] in [CSCR, ISI Kolkata]. Presently, She excelled in teaching since last 3 years. Her current research interests include, [Machine Learning, Datamining, Bioinformatic].

Sumanta Kumar Mandal He is Software Engineer in Accenture Services Pvt. Ltd, Bangalore (India). More specifically, his work examines in the field of Software developer using Java, Cloud Computing (Azure DevOps). Sumanta Kumar Mandal was born in [Baripada, India], in 1982 and, received his Master's degree in [MCA] from [Department of CSE, Biju Patnaik University, Rourkela], in [2007]. He has completed M.Tech. in [CSE] from [Department of CSE, Biju Patnaik University, Rourkela], in [2010], He is Pursuing Ph.D Programme in the Department of CSE, KIIT University, Bhubaneswar with major in Machine Learning. In [2011], he started his professional career as [Software Engineer] in [IBM, Pune]. He excelled in corporate since last 7 years and worked 2 years as a Junier Lecturer in CEB, Bhubaneswar. His current research interests include, [Machine Learning Sensor Networks].

Biresh Kumar He is a Life Member of the Computer Society of India. He is Assistant Professor in the Department of Computer Science and Engineering, Amity university, Jharkhand, Ranchi (India). More specifically, his work examines in the field of Machine learning, Software Engineering, IoT. Biresh Kumar was born in [Ranchi, India], in 1978 and, received his Master's degree in [MCA] from [Department of CSE, Burdwan University], in [2004]. He has completed M.Tech. in [CSE] from [Department of CSE, BIT Mesra], in [2009], He is Pursuing Ph.D Programme in the Department of CSE Usha Martin University, Ranchi with major in Machine Learning. In [2004], he started his professional career as [lecturer] in [CSE Department, CIT, Ranchi]. He excelled in teaching since last 14 years. His current research interests include, [Machine Learning, IoT, Swarm Intelligence].

Security Enhancement Approach in WSN to Recovering from Various Wormhole-Based DDoS Attacks

Sandip Mandal and Rama Sushil

1 Introduction

WSN have lots of some limitation in compare to different cable or without-cable network, even communication system is also different in some cases. Lots of constraints are available in WSN which prevents to establish some securities methods directly [1]. Quite important and tough matter is to secure the sensor networks because of common features like not secured packet communication, data constraints, local coordination, and hidden challenges before placing the nodes, manual level of attacks, and hostile environmental factors. DDoS may be transferred maximum repeating access to a main station or server thus the main station collapsed due to heavy pressure. The target entities might fetch or sometimes focusing to remove an admin's data or client's information and comprised in a way so that the handler is pretending. Such way it's slow down the sensor network or harms the actual services of sensors. In security terms, wormhole attack is a quite high level of issues where two affected sensors create dummy path in the wireless sensor network, which is mainly for the packet transfer between both edges [1, 2]. The attackers have the capacity to create patterns of attacks, like blocked the data messages, that truly affects various network-based ideas like power balanced routing scheme, local data gathering etc. The main approach to the wormhole attack is that hackers can quietly start a virtual concept of wormhole by not fetching the methods moreover any security methods focused on the sensor network. Several of approaches of wormhole-based security issues are already identified in WSN. Next, we are going to show these prototypes of wormhole types of issues.

Like data spoofing types of methods, hacker comprises the original face of any other in present WSN, after that maximum transmitted packets fetched by hackers.

S. Mandal (✉) · R. Sushil
Department of Information Technology, DIT University Dehradun, Dehradun 248001, Uttarakhand, India
e-mail: sandy06.gcect@gmail.com

J. Chattopadhyay et al. (eds.), *Innovations in Soft Computing and Information Technology*, https://doi.org/10.1007/978-981-13-3185-5_15

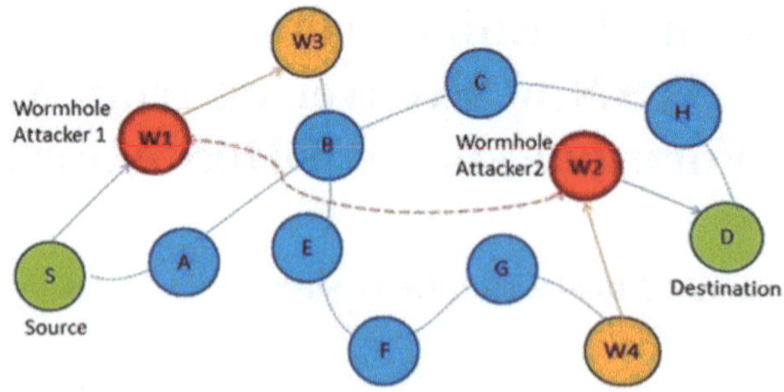

Fig. 1 A scenario of wormhole-based issues in WSN

There are many techniques about data spoofing targets, like Sybil, tracking identity, and hidden sensor attacks. Spoofing is quite same as background wormhole attacks in a system and might be tracked by these logics like all parameter validation approach or packet-based fingerprinting.

Very popular and multiple patterns of Wormhole injection, [3] key as sinkhole, gray-hole and black-hole issues. Such a variant of security issues targets as to store the sink data from fetching data by misleading packets transmission or to change wrong information to sink. Like a black-hole security issue, an invalid sensor considers owning a member of the route and next release [4] all data messages which would be transmitted. Dropdown the probability of tracking, the affected node can create a good system architecture, named as gray-hole; it releases the data packets in a selective way. Such types of attacks effects on various topologies of network protocol like a relay, deviation, data transmission, and delay and network lifetimes. Actual features of wormhole attack is that it's not tracked by conventional security approaches even easy to start without any broad changes in the sensor network. Wormhole attacks can be applied through various strategies like information encapsulation, maximum transmission voltage, broad-quality communication paths, message relaying, and routing algorithm distortion approach [5] (Fig. 1).

1.1 Wormhole Attacks Based on Packet Encapsulating

Such types of approaches are very broad in WSN because wormhole not required any prior security knowledge, or focused features, like a good quality transmission wire route or a gain initiator. Most frequent method to discard some security-based issues in which wireless sensor points select the speedy path swing rather than another one asked to perform smaller unit sharing.

1.2 Wormhole Issues in High-Density Route

The limited route wormhole is defined on relaying packets that are actually for the nearest region in another area. Due to manual factors on propagation by the channel path, the actual time for the data message to propagate to sensor's next adjacent will be lower to the phase acquired by packets to communicate to an attacker, restructure the security path, and next communicate to member sensors on immediate next to comprised route.

1.3 Wormhole Issues by Message Relay

Data message relay-based issues may be occurred by single or major affected nodes. Such types of attack, when a single affected member considers as the neighbors for carrying the relay data. Such variation of comprising are named as "Message-Relay" attack in the published work.

1.4 Wormhole Based on Transmission Protocol Variation

If member nodes want to transfer the message to the distant area in the covered zone, it may impact with positive energy. The main issue is when a hacker going to receive the data from the malicious comprised node. Sometimes, this launches Siebel attack or linear cryptanalytic attacks.

In local networks, a basically null effect is there or any algorithm which depends via lots of environmental-based issues that are not effective. Transmit the data from one node and expand to full domain, the gain of communication is higher in network and power radio will lapse for each node because energy dissipation has pointed to point linking role with transmitter between two members.

If energy exhausted, transmission effected and it's problematic and it occurred as DDoS-Distributed Denial of Service types attack.

2 Related Work

Sometimes wormhole-based issues are quite flexible to occur but hard to gain, and DDoS security attack [4] identifying and tracking has been an excellent research idea. Maximum established methods for finding the wormhole uses locating hardware, time-based devices, toolkit, or IDS devices Next part, some wormhole-based identification and possible creativeness process are defined.

2.1 Distance Related Concept Depend on Consistency Vector

The distance-vector process might be accrued on data propagation time-based knowledge, locating devices, or positioning devices. These processes actually require advance devices and for such things, it's not related to such work.

A wormhole identification process is that where its dependent on adjacent members and CVT approach. The top approach depends on issues that by adopting route through communication are, the attacker joined the new neighbors of member sensors on the boundary of area coverage. In that case, a message sent timing between two affected nodes are quite large than actual data delivery time of regular and adjacent nodes. Such types of system approach do not require any specialized hardware setup.

The second approach is a delivery-to-time protocol (DTT) [4] to check security breaches at the starting time of initiator set up the mechanism by tracking communication phase between twice member sensors with the selected route. The main motto is that such types of nodes are comparatively higher than not-affected and nearby nodes, through this approaches any person may be checked the possible wormhole attack in WSN. In that case, both nodes are within mention transmission coverage. Like in [6], no extensive hardware required for DTT mechanism.

2.2 A Special Hardware-Based Approach

Moving devices are fitted for fetch tracking [3] and nearby member discovery (MD) [7] in sensor areas. In [7], the adjacent member function is tracked as areas thus every area is termed through high-frequency antenna devices. Each cluster or area is named according to numeric no. like 1 to *M*. At a first when a signal fetched via any new sensors, it can identify the direction of signals and redirect to all neighbors. Next step sensor nodes help their surrounding nodes to check the validity, as an approach, by searching that such away untracked nodes information is appeared to all path or not.

2.3 Synchronized Clock-Based Solution

Synchronized clock-based results measure that network points in the area are closely flipped and single messages even the recorded timing as its triggered. The logic for such results through message is stored, by destination nodes checks the transmitting phase duration with the actual duration when the message initiated. The ending member has the information of phase path value and total duration, quite possible via tracking if data traversed through long paths. If the transmission path is so long that the actually allowed route distance, we can say the wireless network may be affected by security issues. To discard these issues by extensive types of devices by duration catching, a CVT method was established by Haeselmann [8]. The CVT mean actual duration which retrieved from the path-based packet request PBRQ tracking duration by sending a message to A and path-based packet reply (PBRP) fetching data through a member sensor B compare to member sensor A. As sensor node B fetches a PBRQ, the calculation for the CVT done. Anyway, if CVT crossed a breakpoint, the PBRQ might be dropped down. Such away this understood that network algorithm topology messages should not change and every sensor nodes automatically updated based on flipped, and two common security challenges present on any sensor combinations. Sensor node A may fetch the CVT through all surroundings nodes and itself. The main reason is CVT of such nodes is comparatively higher than not-affected and adjacent nodes; node A may identify the actual and dummy nodes. In such an approach, the single node calculates the CVT between in it and other adjacent nodes. No special hardware setup is needed for such implementation and also it's quite feasible to launch; thought it not tracked the exposed types of attacks, the main reason is that nodes created through exposed types of attack.

2.4 N-*axis Mapping Projection-Based Solution*

N-Axis Mapping Projection of the wormhole (NAMP-VOW) is presented in [1] to identify the DDoS wormhole attacks. This method is based on an important stand that network with affected nodes must have some deviation to the original network. We structured the model with all nodes by NAMP-VOW. Continuously it's restructured and designed further as per requirement. Previously wormhole can be identified through differentiation in WSN. In our method, the single node assumes the distance to its nearby nodes using the fetched data signal rate. At first, all nodes send their location information to sink or Base Station only, which analyzed the basic network structure depend on distance vector calculation method. Otherwise, the original path fetched through the base station for all possible pair of sensor nodes (with more energy and capacity level).

If any wormhole present, the recreated image of the WSN will show different anomalies and identify security issues through projection the loopholes which occurred through targeting. If no sign of wormholes, path architecture is touching the baseline (Fig. 2b), any issues may consider as "tape" crossing both corners of the area in the whole sole. A surface smoothing approach may be adopted over a distance measurement technique. NAMP-VOW [9] then identify the loopholes in wormhole attacks.

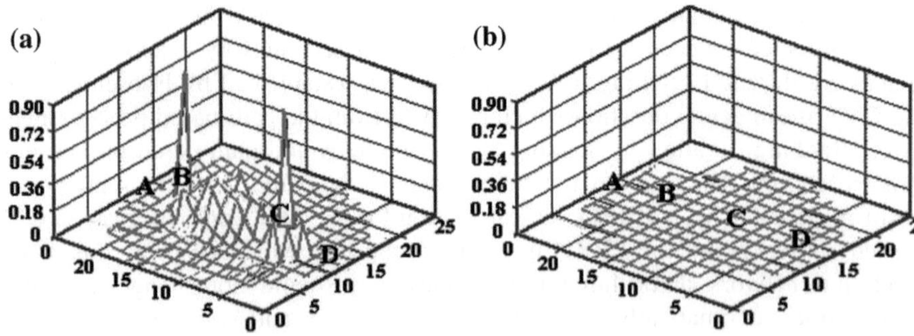

Fig. 2 Graphical representation of wormhole in WSN. **a** Security breaches via sensor B–C. **b** WSN without any wormhole

2.5 Radio Fingerprinting Approach

In [3] Researchers elaborate that hardware reconfiguring can be easily happened on wireless points devices, like Chipcon 1000, 433 MHz gain values through all identification values are covered. By this researchers just fetched this wave signals and its applicability in the sensor network. Lots of issues are always open to world forum, to design some good fingerprints to the importance of hazard and dimension by fingerprinting methods in details.

3 Proposed System Model

This chapter focuses on the Out-Band as well as In-Band modes of a wormhole security attack. Some detailed discussion of these modes of attack is carried out. Finally, a planning and implementation of modeling also preventing approach for In-Band wormhole is presented. The designed model has a direct impact on In-Band wormholes, and also has logical confidence by prevention, through network relay, flow, and delay methods. In this proposed methodologies three main fields were identified, namely flow gathering by sensors, time delay recording scheme by inaccurate issues, and proper prevention approaches established in the sensor communication.

3.1 Out-Band Approach to Wormhole Designing

Out-Band wormhole designing, [10] any attacker creates a less-density path (wormhole path) via two endpoints node on the coverage area. Sometimes it's created through wire connection which isn't accessible to sensor nodes, or by any extensive potency free air devices. When the attacker creates gain over the sensor networks for maximum packets transferring through the wormhole-based path, an attacker can distract and delay the process by wormhole nodes. For outbound wormhole creation, there is no need of any compromise node or hacking attack by an attacker.

3.2 *In-Band Approach to Wormhole Designing*

Like the In-Band wormhole design process, [10] an attacker targets end members in edge points on the area and wrongly creates a channel via latency, this path may be created by communication or routing paths. Likely in out of the bound process, this

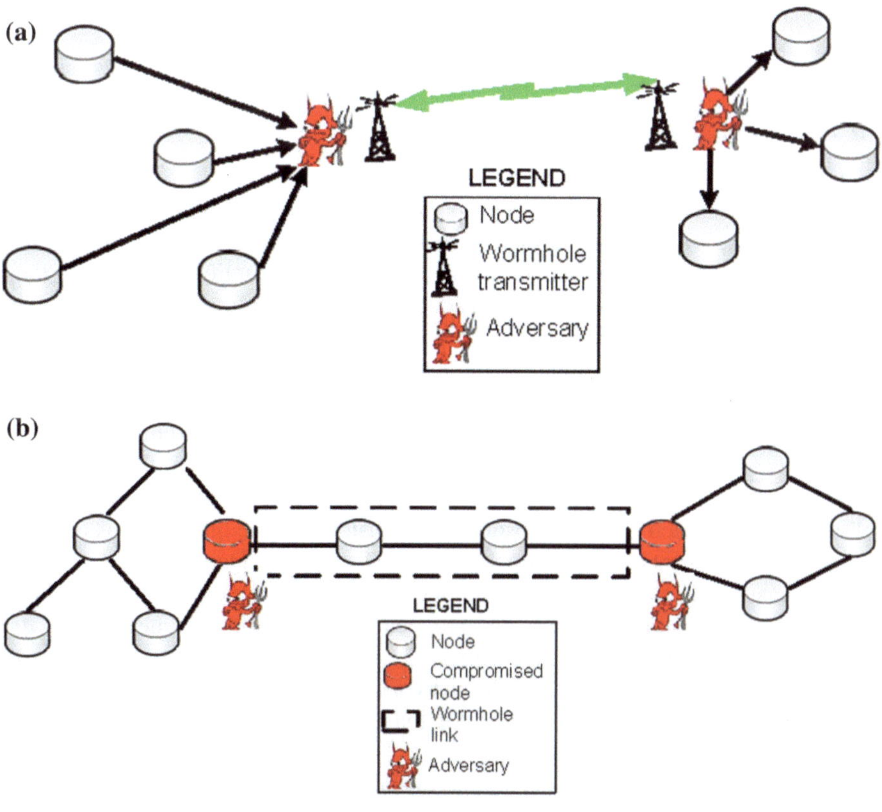

Fig. 3 Briefing of various tunneling. **a** An Out-Band wormhole network, the attacker constructs a low magnitude link between to nodes with high flexibility, such as any directional devices or another medium. **b** Case of in-band wormhole attack, the hacker hacked the sensor network members in other areas and presents a wrong single-flow path by two targeted member nodes. The path truly manages a link through untracked member points

types of the situation flow high amount of data packets between nodes. Next attacker selects a path, containing valid and affected nodes, between the end nodes which create wormhole loops. In-band attack, the attacker must require two affected sensor nodes but not required any requirements of advanced devices (Fig. 3).

3.3 In-Band Wormhole Mitigation Methods

Inbound wormhole attacks adding with affected nodes and security techniques, protection against the outbound wormhole is not feasible in case of In-Band security attacks. In case of In-band challenges, creates long delays in comprise to Out-Band wormhole path. This thing depends on the multiple flow communication protocols to send the data. Through the mathematical process, sensor nodes sometimes find one channel paths through unexpected losses in data rate and/or packet-loss relays, which are identified as wormhole paths and discarded.

3.4 Detection Technique and Effects

This approach mainly depends on two popular parameters such as the rate of data and delay time, when a data packet passes through a link. The main motto is to launch a recovery system to launch from such types of situations. The main target of the attacker is to focus on wormhole attack and try to fetch these in on track. The data packets are experiencing the network delay when passed through the affected wormhole, which is proportional to the actual nos. of nodes in the affected zone. The tracking of affected wormhole is quite depended on the probability that if a transmission channels unnecessary gather delay in the data rate and maximum flows maybe it is identified as a wormhole, because the packet delay is quite more than the actuals. Even some path have high network relay rate compare to others, date delivery between such links are absorbed as considered its edge nodes as comprise points, i.e., the points are not tracked. The Distributed Denial of Service [11] pushed because of lots of wormhole are identified in the region is shown in Fig. 4 shows how communication results stopped when there is unexpected data increasing in the communication channel.

Initially, the magnitude [12] of the network transmission is quite higher at the early phase of wormhole, but in later for maximum energy catching due to relay thoroughly effect in magnitude level.

As we applying the proposed mitigation and detection approach it is clearly understood from Fig. 5 that a stable network response recorded, the networks total outcome (from Fig. 4) are passed through a managed network system that protects the WSN from several attacks and outside activities.

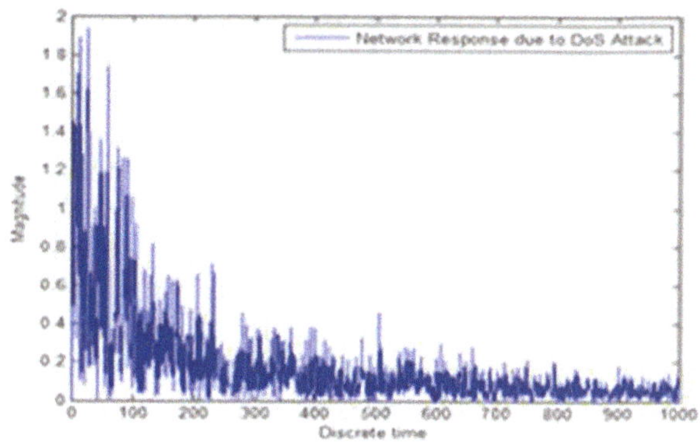

Fig. 4 Presenting an interruption in data rate due to wormhole

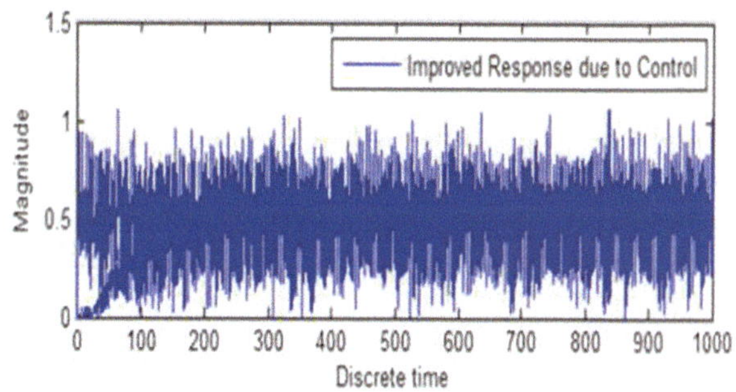

Fig. 5 Presenting the rectified data rate level after applying the recovering method

4 Simulation Results

As we checked in this proposed method, simulation works have been performed in MATLAB tools. Accrued values by Figs. 4 and 5 show the whole impact of wormhole issues [13] on area performance whatever mitigation concepts applied or not applied, both.

Easily it can be checked from Fig. 3 that the network shows fluctuating behavior when passed through an infected wormhole. The overall WSN data transmission takes place with broad magnitudes standards in the initial stages and then sensor network evenly starts to disjunction due to unavailability of resources or energies. Figure 5 indicates how an actual transmission pattern is gathered after adding the elaborated preventing approach on the wormhole attacked WSN.

5 Conclusion and Future Work

So, this briefed and presented topic, the researcher focused the actual implication of security attack in the wireless-based sensor network, in proposed method an attacker creates a path among two different regions, maybe using a side channel, or by focusing the sensor points, such as In-Band wormhole cases. Through depend on the wormhole channel, the attacker may be avoiding the time issues in compare to real-time data system values, by drop down some unassessed message in a comprised path. Through a briefer note, a genuine and user run framework for modeling and detect all attacks in a sensor network. Under the mentioned approach, transmission flow in a data network, and some relay gathered in focus, and prevention algorithm is striated, interconnected passive dynamical systems. In case of In-Band wormhole, some logical functions used to measure actual relay by the wormhole path or tunnel considered a mathematical logic for comprised sensor nodes. In the broad view, we verified that the In-Band wormhole causes large disturbances in the physical system by replaying packets, unnecessarily consuming the energy of nodes. Our simulation achieved as the network defense consider that unique flow of flow holding among nodes, no importance of wormhole.

Future work is planned on detecting other types of wormhole related attacks and presenting preventive and defensive methods to prevent such attacks and also try to design an easy detention scheme also. To prevent different DDoS attacks or issues via practical happening within the sensor communication network. In this way, it represents a broad and multiple-layer-based problems of current loopholes and monitors its applying and proper protocol choosing moreover the present scope with more challenging issues in the digital world.

References

1. H. Yang, H. Luo, F. Ye, S. Lu, and L. Zhang. "Security in mobile ad-hoc networks: Challenges and solution," IEEE Wireless Communications, vol. 11, no. 1, pp. 38–47, Feb. 2004.
2. M. Meghdadi, S. Özdemir, and Ý. Güler. "Security wireless sensor networks: Problems and solutions," (manuscript in Turkish), Journal of Informatica Technologies, Gazi University, Ankara, Turkey, vol. 1, no. 1, pp. 35–42, Jan. 2008.
3. F. Hu, D. Dong, and Y. Xiao. "Attacks and countermeasures in multi-hop cognitive radio networks," International Journal of Security and Networks, vol. 4, no. 4, pp. 263–71, 2009.
4. Najma Farooq, Irwa Zahoor, Sandip Mandal," Recovering from In-Band Wormhole Based Denial of Service in Wireless Sensor Networks" Vol.4, No.3 (June 2014) International Journal of Current Engineering and Technology E-ISSN 2277 – 4106, P-ISSN 2347-5161.
5. V. Karyotis, S. Papavassiliou, M. Grammatikou, and V. Maglaris. "A novel framework for mobile attack strategy modeling and vulnerability analysis in wireless ad-hoc networks," International Journal of Security and Networks, vol. 1, no. 3/4, pp. 255–65, 2006.
6. R. Poovendran and L. Lazos. "A graph-theoretic framework for preventing the wormhole attack in wireless ad-hoc networks," Wireless Networks, vol. 13, no. 1, pp. 27–59, Jan. 2007.
7. G.C. Dalton II, K. S. Edge, R. F. Mills, and R.A. Raines. "Analysing security risks in computer and radio frequency identification (RFID) networks using attack and protection trees," International Journal of Security and Networks, vol. 5, no. 2/3, pp. 87–95, 2010.
8. T. Haenselmann. Sensor networks. Published under the Free Documentation License (GND FDL). Available from: http://www.pi4.informatik.uni-mannheim.de/~haensel/sn_book/ [last cited in 2006].

9. K. Xing, S. Srinivasan, M. Rivera, J. Li, and X. Cheng. "Attacks and countermeasures in sensor networks: A survey," The George Washington University Technical Report GWU-CS-TR-010-05, 2005.
10. Y.C. Hu, A. Perrig, and D.B. Johnson. "Packet leashes: A defense against wormhole attacks in wireless ad-hoc networks," Proceedings of 22nd IEEE INFOCOM, pp. 1976–86, Apr. 2003.
11. R. Ramanathan. "On the performance of ad hoc networks with beamforming antennas," MobiHoc'01: Proceedings of the 2nd ACM international symposium on Mobile ad-hoc networking and computing: ACM Press, pp. 95–105, 2001.
12. A.A. Pirzada and C.S. McDonald. "Circumventing sinkholes and wormholes in ad-hoc wireless networks," Proceedings of International Workshop on Wireless Ad-hoc Networks, London, England, Kings College, London, 2005.
13. J. Zhen and S. Srinivas. "Preventing replay attacks for secure routing in ad-hoc networks," Proc. of 2nd Ad Hoc Networks and Wireless, pp. 140–50, 2003.
14. A. Akyildiz, I.F. Su, W. Sankarasubramaniam, and E. Cayirci. "A survey on sensor networks," IEEE Communications Magazine, vol. 40, no. 8, pp. 102–114, Aug. 2002.
15. S. Hadim and S.N. Mohamed. "Middleware challenges and approaches for wireless sensor networks," IEEE Distributed Systems, vol. 7, no. 3, pp. 1–23, Mar. 2006.
16. H. Mohammadi, E.N. Oskoee, M. Afsharchi, N. Yazdani, and M. Sahimi. "A percolation model of mobile ad-hoc networks," International Journal of Modern Physics C (IJMPC), vol. 20, no. 12, pp. 1871–902, 2009.
17. Y. Xiao, V.K. Rayi, B. Sun, X. Du, F. Hu, and M. Galloway. "A Survey of key management schemes in wireless sensor networks," Computer Communications Journal, Special issue on security on wireless ad-hoc and sensor networks, vol. 30, no. 11–12, pp. 2314–41, Sep. 2007.
18. M. Hutter, T. Plos, and M. Feldhofer. "On the security of RFID devices against implementation attacks," International Journal of Security and Networks, vol. 5, no. 2/3, pp. 106–18, 2010.
19. Y.C. Hu, A. Perrig, and D.B. Johnson, "Wormhole detection in wireless ad-hoc networks," Department of Computer Science, Rice University, Technical Report TR01-384, Jun. 2002.
20. L. Lazos, R. Poovendran, C. Meadows, P. Syverson, and L.W. Chang. "Preventing wormhole attacks on wireless ad-hoc networks: A graph-theoretic approach," in Preceding of Wireless Communications and Networking Conference, vol. 2, pp. 1193–9, Mar. 2005.
21. Najma Farooq, Irwa Zahoor, Sandip Mandal, "Recovering from In-Band Wormhole Based Denial of Service in Wireless Sensor Networks" in International Journal of Current Engineering and Technology, Vol 4, PP. 1604–1607, 2014.
22. R.R. Choudhury, X. Yang, N.H. Vaidya, and R. Ramanathan. "Using directional antennas for medium access control in ad-hoc networks," MobiCom'02: Proceedings of the 8th annual international conference on Mobile computing and networking, pp. 59–70, 2002.
23. S. Yi, Y. Pei, and S. Kalyanaraman. "On the capacity improvement of ad-hoc wireless networks using directional antennas," MobiHoc 2003, Proceedings of the 4th ACM international symposium on Mobile ad-hoc networking and computing. New York, NY, USA: ACM Press; pp. 108–16, 2003.
24. M. Takai, J. Martin, R. Bagrodia, and A. Ren. "Directional virtual carrier sensing for directional antennas in mobile ad-hoc networks," MobiHoc'02, Proceedings of the 3rd ACM international symposium on Mobile ad- hoc networking and computing, pp. 183–93, 2002.
25. S. Brands and D. Chaum. "Distance-bounding protocols," In Theory and Application of Cryptographic Techniques, pp. 344–59, 1993.
26. W. Wang and B. Bhargava. "Visualization of wormholes in sensor networks," WiSe'04, Proceedings of the 2004 ACM workshop on Wireless security. ACM Press, pp. 51–60, 2004.
27. S. Özdemir, M. Meghdadi, and Ý. Güler. "A time and trust based wormhole detection algorithm for wireless sensor networks," (manuscript in Turkish), in 3rd Information Security and Cryptology Conference (ISC'08), pp. 139–4, 2008.
28. H. Chen, W. Lou, X. Sun, and Z. Wang. "A secure localization approach against wormhole attacks using distance consistency," EURASIP Journal on Wireless Communication and

Networking- Special Issue on Wireless Network Algorithms, Systems, and Applications, pp. 22–32, 2010.
29. H. Chen, W. Lou, and Z. Wang. "Conflicting-set-based wormhole attack resistant localization in wireless sensor networks," Book Chapter Lecture Notes in Computer Science – Ubiquitous Intelligence and Computing, vol. 5585/2009, pp. 296—309, 2009.
30. L. Lazos and R. Poovendran, "Serloc: Secure range-independent localization for wireless sensor networks," Proceedings of the ACM Workshop on Wireless Security, pp. 21–30, Oct. 2004.
31. R. Shokri, M. Poturalski, G. Ravot, P. Papadimitratos, and J.P. Hubaux. "A practical secure neighbor verification protocol for wireless sensor networks," ACM WiSec, 2009.
32. I. Khalil, S. Bagchi, and N.B. Shroff. "MOBIWORP: Mitigation of the wormhole attack in mobile multi-hop wireless networks," Elsevier Ad Hoc Networks, vol. 6, no. 3, pp. 344–62, 2008.
33. P. Papadimitratos, M. Poturalski, P. Schaller, P. Lafourcade, D. Basin, S. Capkun, and J. P. Hubaux. "Secure neighborhood discovery: A fundamental element for mobile ad-hoc networking," IEEE Communications Magazine, Feb. 2008.
34. M. Poturalski, P. Papadimitratos, and J.P. Hubaux. "Secure neighbor discovery in wireless networks: Formal investigation of possibility," ACM ASIACCS2008, pp. 189–200, 2008.
35. Y.C. Hu, A. Perrig, and D.B. Johnson. "Wormhole attacks in wireless networks," IEEE Journal on Selected Areas of Communications, vol. 24, no. 2, pp. 370–80, 2006.

Sandip Mandal obtained his Master of Engineering degree in Computer Science and Engineering from IIEST Shibpur. Before that he completed his Bachelor in Technology from Govt. Engineering College Kolkata. He has more than 7 year experience in teaching and research. He presented no. of quality research papers in conferences as well as published some papers in indexed bjournals. His research area is Wireless Sensor Network, Telecommunications, Networking and Security Analysis.

Prof. Rama Sushil obtained her PhD degree from IIT Roorkee. She has 19+ years of teaching and research experience, Published a book on "Artificial Intelligence", good number of research papers published in peer reviewed int. Journals of repute. She has presented and published her research work in International & National conferences like in ieee xplore & springer. Two best paper awards are in her credit. She has presented her research work in Japan, Italy, USA & Taiwan. Published two book chapters titled "Application of Cloud Computing in Library Information Service and "M-Commerce" by renowned publishers. She is Editor in chief & reviewer of two international journals of repute. Currently the PhD research scholars under her guidance are working in the field of Cloud, Big Data & cryptography.

Brain Mapping of Topological Images Using reBUT

Priyanka Srivastava and K. S. Patnaik

1 Introduction

Earlier research on the basic science of sensation examines different types of information the brain receives from the external world. To elucidate its classical view, various theories and concepts exist. They work on the concept that the retinal receptors are sensitive to simple signals which are related to the external world. The messages from those signals are then transferred to the primary visual cortex of the brain, where specific traits of vision such as form, motion, and color are isolated in multiple parallel paths and their associative areas are characterized as "unimodal," because they are influenced by a sole sensory modality. The messages from unimodal are then sent to another area known as "multisensory neurons or heteromodal," which contains more than one sensory modality (auditory, somatosensory, visual, and many more). This classical approach implies hierarchical processing of information where low-level inputs are transformed into various representations and multisensory integration occurs at multiple processing cortical stages [1].

Latest developments in human neurosciences relay the conceptual states of the brain with its physical states. Multisensory neurons accept responses from numerous senses modalities (independent sources of information called "cues") and combine that information to process complex brain activities, such as object categorization. Neurons involved in multisensory integration determine itself to treat stimuli as either integrated or independent events. Multisensory integration increases by initial sensory capability, broad involvement with heteromodal cues which requires cortical influences. Neurons of the brain apply the specifics of

P. Srivastava (✉) · K. S. Patnaik
Department of Computer Science and Engineering,
Birla Institute of Technology, Mesra, Ranchi, India
e-mail: priyanka12_1987@rediffmail.com

K. S. Patnaik
e-mail: kspatnaik@bitmesra.ac.in

J. Chattopadhyay et al. (eds.), *Innovations in Soft Computing and Information Technology*, https://doi.org/10.1007/978-981-13-3185-5_16

experience to novel incoming stimuli [2] based on heteromodal multisensory principle [2]. Without the help of cortical activity, neurons become receptive to several sensory modalities, but are unable to integrate different sensory inputs.

In such a multilayered framework, the Borsuk–Ulam theorem (BUT) from topology comes into action. This theorem states that the two opposite points on a sphere, when projected on a one-dimension lower circumference, provide a sole point showing a matching description, i.e., two antipodal points that produce similar signs.

Borsuk–Ulam Theorem, developed by K. Borsuk, is used for continuous mappings from an n-dimensional hypersphere S^n to a feature space that is an n-dimensional Euclidean space R^n. According to BUT, every continuous map $f: S^n \rightarrow R^n$ must classify a pair of antipodal points. Borsuk was the first person to associate the geometric concept of antipodal shapes and mappings, and termed it as homotopies. BUT is useful for the assessment of brain activity and quantum entanglement.

2 Examples

Figure 1a shows a simple sketch of BUT in which two antipodal points are denoted by black circles. Figure 1b displays an example which reflects that a viewer is noticing a guitar player from front. The ecological and exterior inputs from different sensory modalities (as sound and sight) meet on a single set of multisensory neurons in the cortical layers of observer's brain.

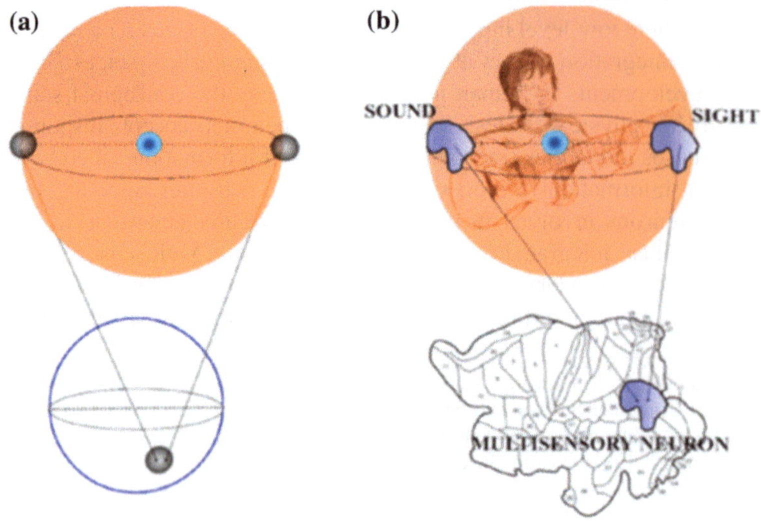

Fig. 1 Overview of **a** Borsuk–Ulam Theorem and **b** overview of multisensory modalities

The two antipodal points of Fig. 1a on an n-sphere can be redirected as R^n using BUT. Besides the antipodal points, a broad-spectrum occurrence of either two antipodal shapes, or functions, or signals also exists [3]. To gauge the probable use of BUT in brain signal analysis, we consider the brain surface as an n-sphere and the feature spaces for brain signals as finite Euclidean topological spaces. Brain activity is characterized by different feature vectors, where a particular feature vector $x \in R^n$ models the explanation of a brain signal. The BUT states that, for description of $f(x)$ of a brain signal x, we anticipate to find an antipodal feature vector $f(-x)$ which defines the brain signal from two opposite sides of the brain. Thus, a pair of antipodal or opposite brain regions having matching descriptions forms a particular brain signal.

In Fig. 2, the sets of vertices (P_1, P_1'), (P_2, P_2'), (P_3, P_3'), (P_4, P_4'), and (P_5, P_5') are antipodal points. Antipodal points are the endpoints of line segments which pass through the center of a circle or a sphere and are opposite to each other [4]. The antipode (P') is the echo of the point P as it is drawn by joining the point P and center of the circle.

Let us consider S^1, a 1-sphere real line segment of a circle with radius r, in which each point (x) on the segment corresponds to a real number, and it can be defined as

$$S^1 = \{x \in R: x^2 = r^2, \quad r \in R\} \tag{1}$$

Similarly, a 2-sphere S^2 can be defined as

$$S^2 = \{(x, y) \in R \times R: x^2 + y^2 = r^2, \quad r \in R\} \tag{2}$$

An n-sphere S^n with radius r (hypersphere) is denoted as

$$S^n = \{(x_1, \ldots, x_n) \in R^n: x_1^2 + \cdots + x_n^2 = r^2\} \tag{3}$$

Commonly BUT is a continuous map denoted as $f\colon S^n \to R^n$. There exists a point $x \in S^n \subseteq R^{n+1}$ such that $f(x) = f(-x)$. The point $-x$ is the antipode of x.

Let us assume that X, Y are two topological spaces. Then the function or map f: $X \to Y$ on a set X to a set Y is defined as a subset $X \times Y$, for each $x \in X$ there is a unique $y \in Y$ such that $(x, y) \in f(y = f(x))$. The mapping f is distinct by a rule.

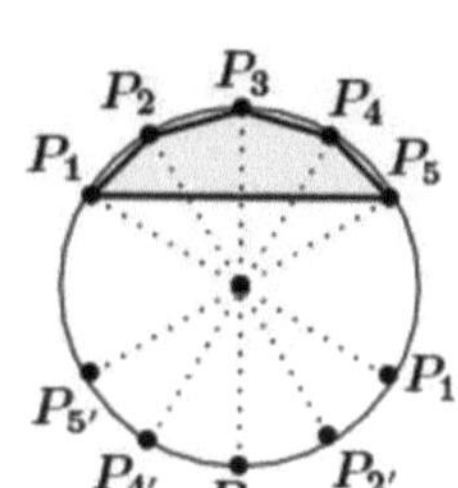

Fig. 2 Sample view of antipodal points

Each subimage of a digital image X is a sample of a region. Image regions A, $B \subset X$ have features such as connectedness, area, and diameter. Each region of the image is having particular feature vector. For example, Phi(A) = (connected$_A$, area$_A$, diameter$_A$) is a feature vector that describes A and Phi(B) = (connected$_B$, area$_B$, diameter$_B$), as feature vector of B [5].

Pixels of a digital image are termed as points, and each point has a specified location with varied features. The pixel features comprise cornerness, gradient orientation, gradient magnitude in the horizontal and vertical directions, and edge strength from vertical. All the pixel features are the base for feature vectors' descriptions.

Figure 3 denotes two-point-based variations of the Borsuk–Ulam Theorem correlated with topological space equipped with a descriptive proximity.

reBUT is a region-based extension form of BUT. Such regions can be spatially far or descriptively near [4]. Assume that Re1 and Re2 as two regions of the brain with matching activity such as word recognition. It has been observed that the same word motivates the regions of the left (Re1) and right (Re2) hemispheres of the brain. This implies an instantiation of the reBUT model of cortical activity.

A concept of the inferior prefrontal cortex (IPFC) areas of the left and right hemispheres of the brain is presented in Fig. 4a, b, respectively. Region 10 of left hemisphere and region 9 of right hemisphere are there for visual, physical, and numeric conceptions for cortical action of the brain.

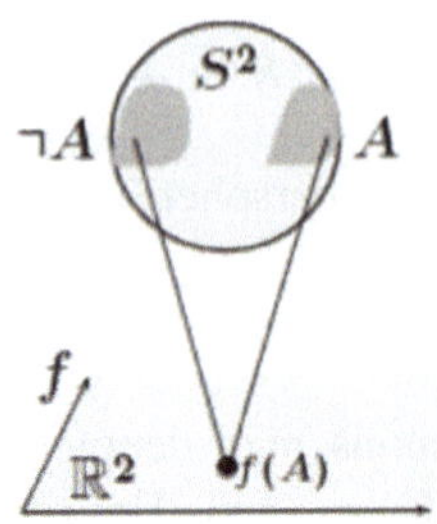

Fig. 3 $2S^2 \rightarrow R^2$

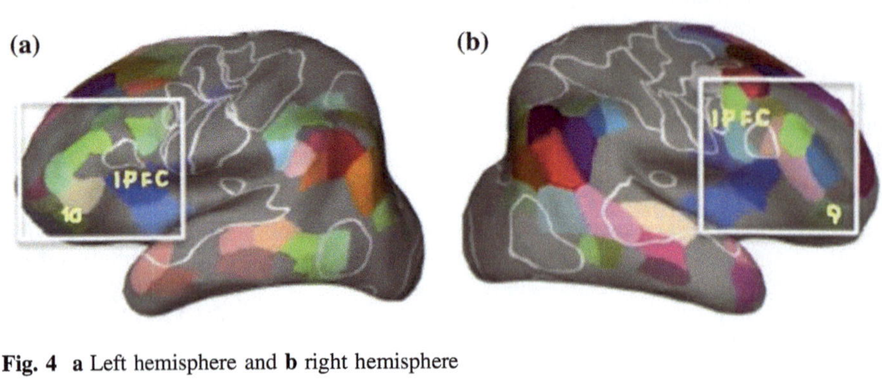

Fig. 4 **a** Left hemisphere and **b** right hemisphere

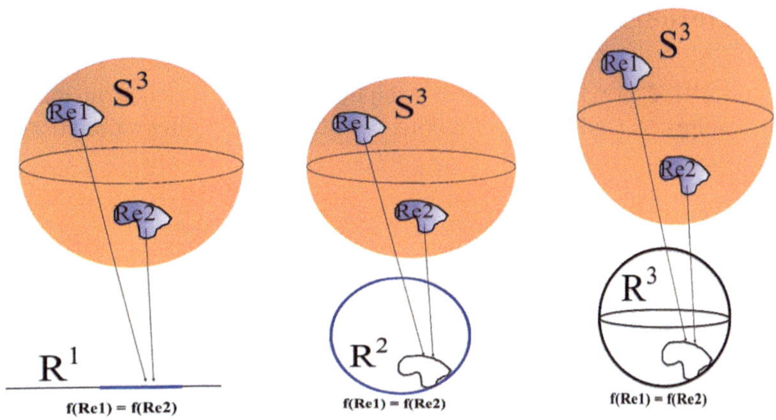

Fig. 5 Antipodal cortical regions with matching descriptions

The hemispheres of the brain are represented by a 3-sphere S^3 and let f: $2S^3 \rightarrow 2R^k$, for $k = 1, 2, 3$ and the cortical regions Re1, Re2 $\in 2S^3$, have two features (tactile and numeric selectivity) with equivalent brain signal values that are real numbers.

Figure 5 shows the following three cases.

Case 1: f: $2^{S3} \rightarrow 2^{R1}$ defined by f(Re1) = f(Re2) $\in 2^{R1}$, defined by
$f(\text{Re1}) = \{\emptyset(\text{Re1}) \in R^1\text{: Re1} \in 2^{S3}\}$, such that
$\emptyset(R1) = \emptyset(R2)$ = cortical signal for visual or tactile or numeric selectivity.

Case 2: f: $2^{S3} \rightarrow 2^{R2}$ defined by f(Re1) = f(Re2) $\in 2^{R2}$, defined by
$f(\text{Re1}) = \{\emptyset(\text{Re1}) \in R^2\text{:Re1} \in 2^{S3}\}$, such that
$\emptyset(R1) = \emptyset(R2)$ = cortical signal for visual/tactile or tactile or numeric/tactile selectivity.

Case 3: f: $2^{S3} \rightarrow 2^{R3}$ defined by f(Re1) = f(Re2) $\in 2^{R3}$, defined by
$f(\text{Re1}) = \{\emptyset(\text{Re1}) \in R^3\text{: Re1} \in 2^{S3}\}$, such that
$\emptyset(R1) = \emptyset(R2)$ = cortical signal for visual, tactile and numeric selectivity.

For every cases, let f(Re1) be a set of signals of the brain allied with the selection of a visual or tangible or numeric concept. In Fig. 4a, b, the region 10 in lPFC of the left hemisphere and region 9 (indicated by Re2) of the right hemisphere would be mapped to the same region of the feature space in $R1$, i.e., f(Re1) = f(Re2). Figure 5 denotes the mapping on 2^{S3} into 2^{R1}. Likewise, the brain mapping is done as 2^{S3} into 2^{R2} and the mapping on 2^{S3} into 2^{R3} respectively in three different parts.

3 Conclusions

BUT and reBUT provide a uniform computational experiments on multisensory neuron features to study the projections of multisensory integration from real to abstract phase.

References

1. Tozzi A, Peters JF. A symmetric approach elucidates multisensory information integration. Information 2016; 8(1):2–14.
2. Tozzi A, Peters JF. A topological approach unveils system invariances and broken symmetries in the brain. Journal of Neuroscience Research. 2016; 94(5):351–365.
3. Matousek J. Using the Borsuk–Ulam Theorem. Lectures on Topological Methods in Combinatorics and Geometry. Springer Science & Business Media. 2008.
4. Peters JF, Tozzi A. Region-based Borsuk-Ulam theorem. arXiv 2016;02987(1):1–15.
5. Borsuk K. Fundamental retracts and extensions of fundamental sequences. Fund. Math. 1969 64(1):55–85.

Priyanka Srivastava is a research scholar in department of Computer Science Engineering, BIT Mesra, Ranchi. She is also serving as an Assistant Professor, Dept. of CS & IT, Jharkhand Rai University, Ranchi with a teaching experience of 9 years. After graduating from Ranchi Women's College, she did her M.Sc.(IT) from BIT, Mesra. Priyanka conducts research mainly in the areas of Digital Image Processing, Design of Algorithms, Programming paradigms.

Dr. K. S. Patnaik is an Associate Professor, Dept. of Computer Science & Engineering, BIT Mesra, India with a research experience of 14 years. He has received Bachelor's degree in Electrical Engineering from NIT, Silchar, Master of Engineering in Software Engineering from BIT Mesra, Ranchi and PhD from BIT Mesra, Ranchi. Dr Patnaik conducts research mainly in the areas of Digital Image Processing, Software Engineering, Soft Computing, Internet-of-Things.

Iterative Decoding of LDPC-RS-Coded Multiple Description Image

Saikat Majumder, Shrish Verma and Bibhudendra Acharya

1 Introduction

A new paradigm of terrestrial broadcasting has been envisioned in [1]. The objective of this research is efficient use of radio frequency spectrum in order to eliminate TV white space and avoid interference from unlicensed devices [2–4]. Such a system is envisioned to be resistant to co-channel interference and robust to multipath distortion. In [1], the authors proposed a 1/4 rate LDPC code for DVB system to obtain negative SNR threshold. Product codes consist of two or more component codes interconnected with random interleavers and offer a powerful error correction method substituting schemes where a large block length is required for correcting mixed types of errors. A hybrid iterative decoding scheme, consisting of both soft and hard decision decoding, was proposed for low-density parity check-Reed–Solomon (LDPC-RS) product code in [5] for cloud transmission system. In [2], authors proposed another two-dimensional (2-D) code based on LDPC and RS code for cloud transmission. Next, we highlight other significant works in the area of image transmission using product codes. We are specifically interested in FEC-based multiple description coding (MDC) which utilize such 2-D product code structure.

To improve performance against deep fades, FEC-based MDC has been considered as a smart scheme for multimedia communication over wireless channel [6–12]. MDC enables generation of descriptions from a single source such that each

S. Majumder · S. Verma · B. Acharya (✉)
Department of Electronics and Telecommunication, National Institute of Technology, Raipur 492010, Chhattisgarh, India
e-mail: bacharya.etc@nitrr.ac.in

S. Majumder
e-mail: smajumder.etc@nitrr.ac.in

S. Verma
e-mail: shrishverma@nitrr.ac.in

J. Chattopadhyay et al. (eds.), *Innovations in Soft Computing and Information Technology*, https://doi.org/10.1007/978-981-13-3185-5_17

description can be decoded to reconstruct the source, but with reconstruction fidelity less than if all the descriptions were available. When certain minimum numbers of descriptions are available at the receiver, the descriptions can be combined to enhance the quality of reconstructed source. Because of independent decoding capability of descriptions in MDC, the loss of some of the descriptions does not hamper the decoding of other correctly received descriptions.

Orthogonal frequency division multiplexing (OFDM) has been applied to many applications in the field of telecommunication systems for its ability to combat inter-symbol interference over frequency selective fading channel, compared to single carrier communication. OFDM-based transmission of progressive image using space–time block code was proposed in [13]. A 2-D channel coding scheme for image transmission using OFDM which employ time and frequency diversities simultaneously was proposed in [7, 14]. A hard-decision iterative decoding algorithm for FEC-based MDC (based on RS and convolutional code) transmitted over OFDM channel was proposed in [8]. In this scheme, correctly decoded symbols obtained from RS codewords are utilized to improve the decoding performance of the Viterbi algorithm at the next iteration of decoding. The performance of the algorithm was evaluated over OFDM system demonstrating significantly improved performance over other schemes employing product code structures. A different approach to iterative decoding of FEC-based multiple description codes was proposed in [10], where component decoders exchange soft extrinsic information between RS decoder and convolutional decoder. This is different from the existing schemes where hard decoding is performed on the output of the inner channel code and applied to the outer RS decoder.

In this paper, we propose a scheme for iterative decoding of LDPC-RS-based MDC for transmission of progressively coded image. For this purpose, LDPC-RS product code proposed in cloud transmission literature [1–5] is applied. The proposed algorithm takes advantage of correctly decoded codeword output of RS decoder to augment the error correction ability of the LDPC decoder at the next iteration of sum–product algorithm (SPA). The algorithm works by selectively modifying the log-likelihood ratios of the identified reliable bits. We show that the proposed iterative LDPC-RS product code-based progressive image transmission scheme performs significantly better than baseline LDPC-RS coding scheme without iterative decoding. We also compare the proposed system to a similar iterative design for transmission of progressive image using RSCC-RS product code in [8] and demonstrate improvement in performance.

Rest of the paper is organized as follows. In Sect. 2, we describe the preliminaries of FEC-based multiple description coding and explain the encoder structure in the proposed system which uses LDPC-RS product code. In Sect. 3, we present the system model and proposed decoding scheme. The simulation results are presented in Sect. 4. Finally, conclusion is presented in Sect. 5.

2 Encoding in LDPC-RS-Based MDC

We briefly describe the proposed encoding scheme for FEC-based MDC where RS code act as FEC code and LDPC code as channel code. An (*n*, *k*) RS code is a maximum distance separable (MDS) code, if $d_{\min} = n - k + 1$, where $d_{\min}$ is referred as minimum distance of the code. An (*n*, *k*) MDS code is a construction where *k* information bits or symbols are encoded into *n* channel bits or symbols and reception of any $n - d_{\min} + 1$ of the *n* symbols allows *k* information symbols to be recovered. RS codes are typically decoded by Berlekamp–Massey (BM) algorithm [15]. In RS code, each symbol is of *m* bits and source symbols for RS code are formed by grouping *m* bits into a symbol. Let $\mathbf{S} = [S_1, S_2, \ldots]$, where $S_i \in \mathrm{GF}(2^m)$, be such a stream of symbols from a progressively coded source. It is possible to reconstruct the source from the prefixes of **S** as symbol packets are received progressively. Presence of even a single error renders the subsequent symbols useless and source reconstruction stops at that resolution. For example, if there is an error in symbol *i*, source can be decoded using the preceding symbols $[S_1, S_2, \ldots, S_{i-1}]$ and symbols subsequent to S_i cannot be utilized in the process of decoding.

A FEC-based multiple description code is obtained by applying FECs of different code rates to different parts of symbol stream with the number of parity bits or code strength depending on the relative importance of the information symbols and producing *n* equally important descriptions. The realization of *n*-channel FEC-based multiple description code using RS code is shown in Fig. 1. There are *p* RS codewords (columns), each of length *n* but containing different number of information symbols. Let the number of information symbols in codeword *l* be denoted as k_l, where $l = 1, \ldots, p$. The information symbols are provided with unequal error protection using (n, k_l) systematic RS codes, where $1 \le k_l < n$. Thus, each RS code can correct up to $n - k_l$ erasures. Since in a progressively coded bitstream, information symbols at the beginning of the stream **S** are more important than the symbols appearing after them, error protection decreases for RS codewords as $k_1 \le k_2 \le \ldots \le k_p$. As shown in Fig. 1, first RS codeword $l = 1$ consists of first information symbol 1 and rest of the symbols being RS parity. For RS codeword at $l = 2$, information symbols are 2, 3 and the rest being parity symbols. Similarly, the information symbols 4, 5 and parity symbols present in RS codeword are distributed among *n* different descriptions. Hence, conversion of progressive symbol stream into multiple descriptions (rows) spreads the contiguous information symbols across packets and any RS codeword *l* can be recovered completely if number of lost descriptions (lost rows) is not more than $n - k_l$. LDPC code is applied on each description for transmission over wireless channel. Finally, MDC descriptions are mapped to *n* subcarriers in the OFDM frame. At the receiver, after LDPC decoding, descriptions are either decoded error-free or there is a decoding failure. In case of failure, descriptions are declared as being 'erased'. Thus, depending on the quality of channel, decoding of some of the descriptions may fail. This will result in decrease in quality of image or video and will depend on the probability of receiving a description correctly.

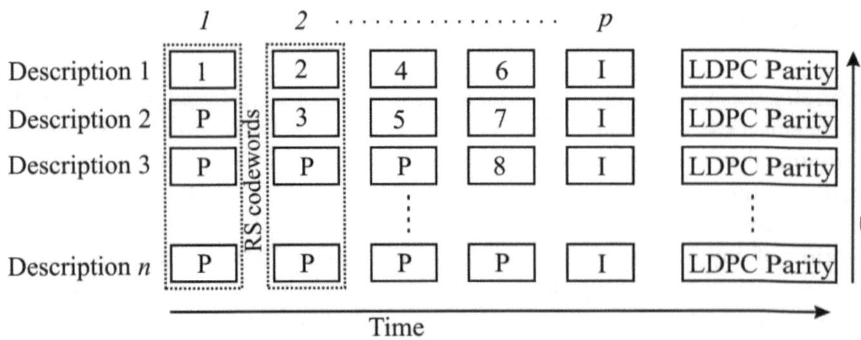

Fig. 1 Construction of FEC-based MDC from a progressively coded source stream

3 Proposed Methodology

The model of proposed system is shown in Fig. 2. In FEC-based MDC literature, encoding process consists of RS codes along the column and convolutional code as rows [11, 12]. Proposed application of LDPC code, instead of systematic convolutional code, makes this coding scheme relevant to newer research in the field of cloud transmission and terrestrial video broadcasting [2, 5]. We next describe the proposed scheme in detail for the system model considered in this paper.

The encoder, as depicted in Fig. 2, consists of a progressive source encoder (e.g. SPIHT) followed by RS encoder. The RS code-based multiple description encoder, described in earlier section, generates n equally important descriptions $\mathbf{c}_1, \ldots, \mathbf{c}_n$ or taken together as $\mathbf{C} = [\mathbf{c}_1, \ldots, \mathbf{c}_n]^{\mathrm{T}}$. Outer interleaver Π_1 is applied on each description $\mathbf{c}_i$ so as to distribute the bits in a symbol throughout a packet or description. Π_1 is a bit-level interleaver in contrast to the symbol-level interleavers

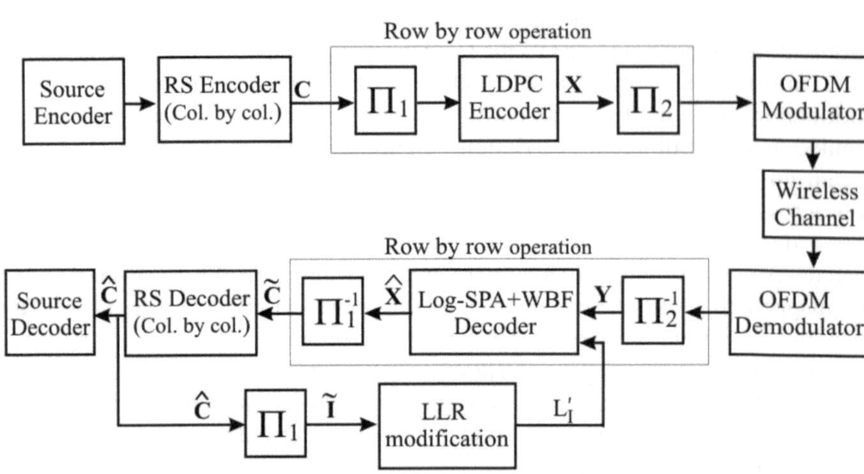

Fig. 2 Encoding and iterative decoding of LDPC-RS-coded FEC-based MDC

used in FEC-based multiple description coding literature. Each interleaved description ($\Pi_1(\mathbf{c}_i)$) is then individually encoded with systematic LDPC code to get $\mathbf{x}_{i,}$ and the consolidated output is given as $\mathbf{X} = [\mathbf{x}_1, \ldots, \mathbf{x}_n]^{\mathrm{T}}$. Another bit-level random inner interleaver Π_2 is applied on the LDPC codewords, row by row. Π_2 is a bit-level random channel interleaver whose function is to distribute the effect of fading throughout the packet. The encoded descriptions constitute a 2-D arrangement of symbols as shown in Fig. 1, and at the last stage, symbols are OFDM modulated and transmitted.

We next describe the proposed iterative decoding scheme depicted in the lower part of Fig. 2. OFDM demodulator provides the bit reliabilities, which after deinterleaving with Π_2^{-1} is applied to LDPC decoder. This input signal is modelled as $\mathbf{y}_i = \mathbf{h}_i\mathbf{x}_i + \mathbf{w}_i$, where $\mathbf{h}_i$ is the fading coefficient for combined channel consisting of OFDM channel and interleaver, and $\mathbf{w}_i$ is additive white Gaussian noise. Input to LDPC decoder is LLR matrix $L^{\mathrm{ch}}(\mathbf{X})$, where $\mathbf{X} = [\mathbf{x}_1, \ldots, \mathbf{x}_n]^{\mathrm{T}}$. LLR corresponding to tth bit in row i of $\mathbf{X}$ is

$$L^{\mathrm{ch}}(x_{i,t}) = \log\frac{P(x_{i,t} = 0|\mathbf{y}_i)}{P(x_{i,t} = 1|\mathbf{y}_i)} \tag{1}$$

for $t = 1, \ldots, mp/R$ and R is the LDPC code rate. Standard log-SPA is used in this paper for decoding of LDPC code. Given high computational complexity of SPA decoder, mixed set of decoding technique shall be applied for LDPC code [5]. After each step of SPA, if the results do not satisfy the parity constraints, parallel weighted bit-flipping decoding (PWBF) [16] is applied on the a posteriori LLR obtained from SPA. One step of SPA includes several iterations of SPA decoder. PWBF algorithm, on the other hand, offers a transaction between decoding performance and computational complexity. PWBF allocates each bit in the hard-decision output vector with a number indicating its reliability. Unreliable bits are then identified and flipped. If the flipped hard-decision vector satisfies the parity check matrix, the algorithm stops; otherwise, bit-flipping function is recalculated for each bit and bit-flipping process is continued. Though SPA is the best performing decoder for LDPC code over continuous output channels, PWBF has the advantage of faster convergence. It has been demonstrated in [16] that PWBF algorithm converges in about 5–10 iterations with bit error rate performance close to that of standard sum–product algorithm. We next explain the calculations associated with a SPA decoder.

Let $L = L^{\mathrm{ch}}(\mathbf{x}_i)$ be the initial LLR required for decoding of ith LDPC codeword, and H is the LDPC parity check matrix. Then, for each $H(s,t) = 1$, set $L(t \to s) = L(t)$, where $L(t \to s)$ is the message sent from variable node t to check node s. Outgoing message from check node s transmitted towards variable nodes is computed as

$$L(s \to t) = 2\tanh^{-1}\left(\prod_{t'=N(s)-\{t\}} \tanh\left\{\frac{1}{2}L(t' \to s)\right\}\right) \tag{2}$$

where $N(s)$ are neighbouring variable nodes of s. Each variable node computes the outgoing message $L(t \rightarrow s)$ as

$$L(t \rightarrow s) = L(t) + \sum_{s' \in N(t) - \{s\}} L(s' \rightarrow t) \tag{3}$$

where $N(t)$ are neighbouring check nodes of t. For each variable nodes $t \in [1, n]$, a posteriori LLR $L^{\mathrm{p}}(t)$ is calculated as

$$L^{\mathrm{p}}(t) = L(t) + \sum_{s \in N(t)} L(s \rightarrow t) \tag{4}$$

When hard decision is performed on a posteriori LLRs of the bits in ith description, the estimate of bits of LDPC codeword in ith description is

$$\hat{x}_{i,t} = \begin{cases} 1, & L^{\mathrm{p}}(t) < 0 \\ 0, & \text{otherwise} \end{cases} \tag{5}$$

The decoded LDPC codeword $\widehat{\mathbf{x}}_{\mathrm{i}} = \left[\hat{x}_{i,1}, \hat{x}_{i,2}, \ldots, \hat{x}_{i,pmn^{(1)}/k^{(1)}}\right]$. If $\widehat{\mathbf{x}}_{\mathrm{i}}$ satisfies the parity check constraints, any further decoding for description i is halted; else, PWBF is applied on a posteriori LLR $L^{\mathrm{p}}(t)$ obtained from the previous SPA stage. If PWBF succeeds in decoding, the hard-decision decoded codeword is saved as $\hat{x}_i$.

Irrespective of the correctness of the decoded description, the bits within each description are deinterleaved by Π_1^{-1}. As already mentioned, Π_1 is a bit-level intradescription interleaver and has no contribution in improving the number of correctly decoded descriptions. But in an iterative system like the one shown in Fig. 2, the interleaver has special function in the subsequent iterations. Let c be the number of correctly decoded RS codewords after one iteration of decoding. Because of unequal error protection, the codewords with lower code rate (or more parity bits) are more likely to be decoded correctly. Since error protection decreases for RS codewords as $k_1 \leq k_2 \leq \cdots \leq k_p$, as shown in Fig. 1, there will be c contiguous correct symbols at the beginning of each packet or description. This makes us sure of correctness of mc bits in a LDPC codeword and can be utilized to improve the reliability of the remaining $m(p-c)$ bits in the description. In the subsequent iterations, mc reliable information bits of LDPC code are spread using Π_1 and are utilized to modify the channel LLR input to the SPA decoder. The iteration loop is run until all the codewords are successfully decoded or maximum number of iterations is reached.

4 Simulation Results

Performance of the proposed iterative decoding scheme is evaluated in this section. The images are encoded using SPIHT algorithm [17] to produce a progressively coded bitstream. The bitstream is converted into a stream of $GF(2^8)$ symbols by combining $m = 8$ bits. The multiple description encoder uses (n, k_i) RS code, where $n = 128$ and the description length is 36 GF(2^8) symbols (i.e. $p = 36$). Serial data stream is converted into 128 parallel bitstreams using the FEC-based multiple description encoder. Specifically, the algorithm for encoding was proposed in Mohr et al. [6] which provide the values of k_i for each RS codeword. Each packet is then encoded with $(576, 288)$ LDPC code from IEEE 802.16e standard. The 128 parallel streams QPSK modulated [2, 8] and are mapped to the OFDM system with 128 subcarriers. It is assumed that channel fading coefficients are known at the receiver.

To evaluate the advantage of iterative decoding, we compare the proposed decoder also with a baseline system consisting of FEC-based multiple description decoder, but with LDPC and RS code decoded in tandem (LDPC decoder does not receive any feedback from RS decoder). Performance of the decoding method proposed in this paper is also compared to a similar decoding method of FEC-based multiple description coding in [8], which consists of soft-decision Viterbi decoder as inner decoder and BM algorithm as outer decoder. In contrast to proposed scheme, which uses inherent property of parity check matrix for detection of errors, [8] uses 32-bit cyclic redundancy check (CRC) for detection of errors in descriptions. Thus in [8], number of information symbols transmitted per inner codeword (convolutional code) is 32 bytes, whereas remaining four bytes are dedicated for CRC. Because of this, maximum PSNR achievable by scheme in [8] will be less compared to the proposed scheme. In this paper, simulation results were obtained for transmission over AWGN, Rayleigh fading channels and frequency selective fading channels.

In Figs. 3, 4 and 5, we show results of the simulations carried out on SPIHT-coded Lena image. Standard 128 × 128 Lena image is encoded to rate of 2 bpp and

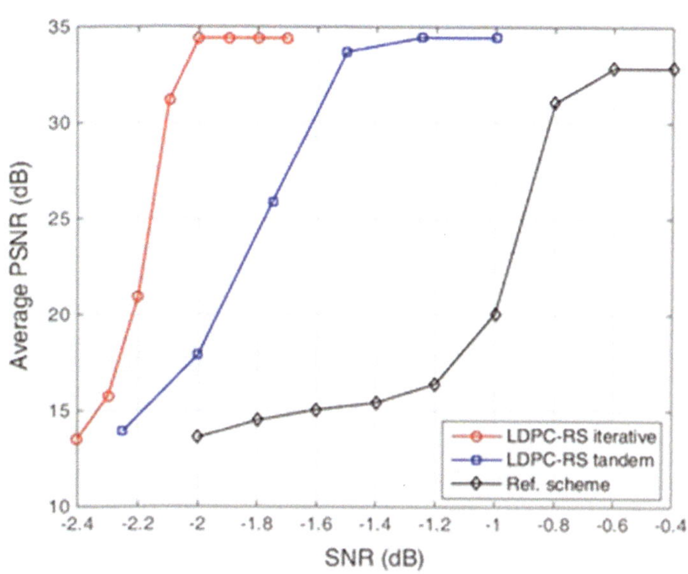

Fig. 3 Comparison of PSNR performance as function of channel SNR for 128 × 128 Lena image transmitted over AWGN channel

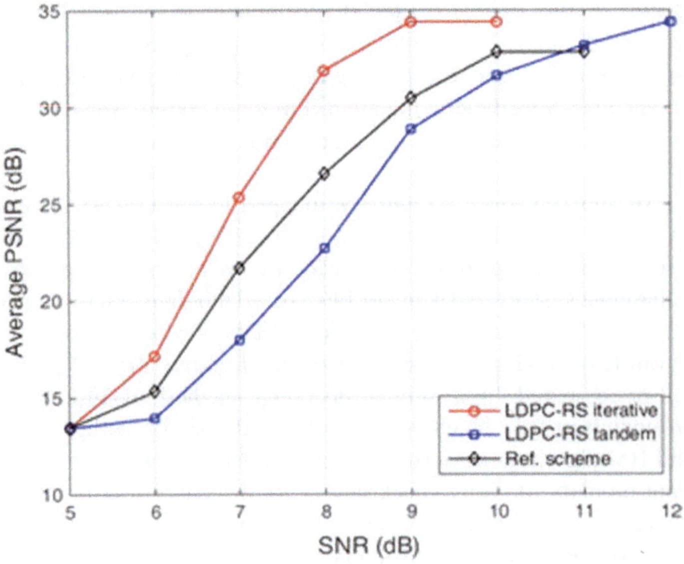

Fig. 4 Performance comparison of the proposed scheme with two reference schemes for Lena image transmitted over Rayleigh fading channel with $f_D T_s = 0.01$

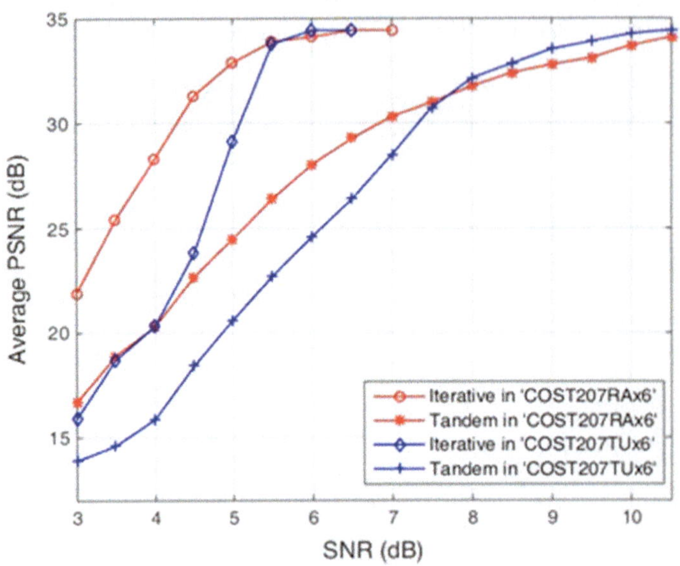

Fig. 5 Performance comparison of the proposed scheme with LDPC-RS tandem scheme for Lena image transmitted over two COST 207 frequency selective channels

transmitted. The algorithm is run for maximum of $N_{\text{iter}} = 5$ number of iterations, whereas each step of LDPC decoder consists of 25 SPA iterations. In the simulations, peak signal-to-noise ratio (PSNR) is used as a measure of reconstruction quality and is plotted against average received channel SNR. For each simulation point, 300 trials have been performed in order to average statistical variations.

Figure 3 compares PSNR performance of the proposed scheme for transmission over AWGN channel. With the proposed scheme, maximum PSNR of 34.4 dB is obtained at channel SNR of −2 dB. In contrast to this, the tandem scheme achieves this maximum at channel SNR of −1.25 dB. For AWGN channel, the reference scheme achieves maximum PSNR of 32.84 dB at channel SNR of −0.6 dB. Thus, the proposed scheme achieves maximum PSNR performance for 0.75 dB less received power compared to tandem scheme and 1.4 dB less compared to the reference scheme.

Next we present the simulations performed over Rayleigh fading channels. Figure 4 compares the PSNR performance of the proposed scheme with tandem and reference scheme with Rayleigh channel parameter $f_D T_s = 0.01$. It is evident that the proposed system outperforms both the reference schemes significantly in terms of average PSNR performance. LDPC-RS iterative scheme achieves maximum PSNR at channel SNR of 9 dB, which is 3 dB less compared to tandem scheme. In this case, reference scheme reaches maximum PSNR of 32.84 dB at channel SNR of 10 dB. The gap with respect to proposed LDPC-RS iterative scheme is now 1 dB. It can be observed that performance of reference scheme outperforms tandem scheme for transmission over fading channel. This is in contrast to AWGN channel where LDPC-RS tandem scheme performs significantly better compared to the reference scheme.

In Fig. 5, results of simulations performed over frequency selective fading channel are shown. For these simulations, we adopted 6-tap COST 207 channel models [18] available in Matlab. Performance was tested on two specific COST 207 channels, namely typical rural area (RA or COST207RAx6) and typical urban (TU or COST207TUx6). In case of RA model, the iterative scheme has advantage in channel SNR of 4.5 dB in achieving maximum PSNR value compared to tandem scheme. In case of TU also, iterative LDPC-RS scheme has advantage of 4.5 dB in achieving maximum PSNR. Though both the scheme achieve maximum PSNR values at the same channel SNR values, their performance are completely different at low SNR region. Both the schemes perform poor in TU channel. This can be attributed to the stronger multipath components in TU channel compared to RA.

5 Conclusion

In this article, a novel scheme to iteratively decode LDPC-RS-based multiple description code transmitted over fading channel has been presented. The proposed scheme employs RS codes to form multiple descriptions, and each description is channel coded with LDPC code. At the receiver, a combination of SPA and weighted bit-flipping-based LDPC decoder is applied. The proposed decoding scheme uses the successfully decoded bits in RS codeword to improve the decoding performance of inner LDPC code. Successfully decoded LDPC codewords in turn improve the decoding of RS code. Experimental results showed the superiority of the proposed scheme in comparison with similar decoding schemes in the literature.

References

1. Wu Y, Rong B, Salehian K, Gagnon G. Cloud Transmission: A New Spectrum-Reuse Friendly Digital Terrestrial Broadcasting Transmission System. IEEE T Broadcast 2012; 58:329–337.
2. Park S lk, et al. Two dimensional code based on low density parity check and Reed-Solomon codes for the terrestrial cloud transmission system. IEEE T Broadcast 2015; 61: 75–83.
3. Wu Y, et al. Comparison of terrestrial DTV transmission systems: The ATSC 8-VSB, the DVB-T COFDM, and the ISDB-T BST-OFDM. IEEE T Broadcast 2000; 46:101–113.
4. Montalban J, et al. Cloud transmission: system performance and application scenarios. IEEE T Broadcast 2014; 60:170–184.
5. Liu B, et al. LDPC-RS product codes for digital terrestrial broadcasting transmission system. IEEE T Broadcast 2014; 60:38–48.
6. Mohr A E, Riskin E A, Ladner R E. Graceful degradation over packet erasure channels through forward error correction. In: IEEE Data Compression Conference; 29–31 Mar 1999; pp. 92–101.
7. Toni L, Chan Y S, Cosman P C, Milstein L B. Channel Coding for Progressive Images in a 2-D Time-Frequency OFDM Block With Channel Estimation Errors. IEEE T Image Process 2007; 18:2476–2490.
8. Chang SH, Cosman P C, Milstein L B. Iterative Channel Decoding of FEC-Based Multiple-Description Codes. IEEE T Image Process 2012; 21:1138–1152.
9. Bais A, Dey T, Sarshar N. Unequal channel protection of multiple description codes for wireless broadcast applications. Signal Process-Image 2012; 27.
10. Majumder S, Verma S. Iterative decoding of FEC based multiple description codes for image transmission over wireless channel. In: Twenty First IEEE National Conference on Communications. pp. 1–6.
11. Sachs D G, Raghavan A, Ramchandran K. Wireless image transmission using multiple-description-based concatenated codes. In: Proceedings of SPIE 3974, Image and Video Communications and Processing 2000; Apr. 19, 2000.
12. Puri R, Ramchandran K. Multiple description source coding using forward error correction codes. In: IEEE Thirty-Third Asilomar Conference on Signals, Systems, and Computers 1999.
13. Song J, Liu K J R. Robust progressive image transmission over OFDM systems using space-time block code. IEEE T Multimedia 2002; 4:394–406.
14. Chang S -H, Cosman P C, MilsteinL B. Performance Analysis of n-Channel Symmetric FEC-Based Multiple Description Coding for OFDM Networks. IEEE T Image Process 2011; 20:1061–1076.
15. Lin S, Costello D J. Error control coding. Pearson Education India, 2004.
16. Wu X, Zhao C, You X. Parallel weighted bit-flipping decoding. IEEE Commun Lett 2007; 11: 671–673.
17. Said A, Pearlman W A. A new, fast, and efficient image codec based on set partitioning in hierarchical trees. IEEE T Circ Syst Vid 1996; 6:243–250.
18. Cho Y S, et al. MIMO-OFDM wireless communications with MATLAB. John Wiley & Sons, 2010.

Saikat Majumder received the Post Graduation degree in Electronics and Communication Engineering from the National Institute of Technology, Calicut, India and the Ph.D. degree in Electronics and Telecommunication Engineering from National Institute of Technology, Raipur, India.

He is currently an Assistant Professor in the Electronics and Telecommunication Engineering Department, National Institute of Technology, Raipur, India. His research interests include center around signal processing for wireless communication, cognitive radio and software defined radio. He has more than 20 research publications in National/International Journals and conferences.

Shrish Verma received the Post Graduation degree in Electronics and Electrical Communication Engineering from the Indian Institute of Technology, Kharagpur, India with specialization in Computer Engineering and the Ph.D. degree in Engineering from Pt. Ravi Shankar Shukla University, Raipur, India.

He is a Professor of the Department of Electronics and Telecommunication, National Institute of Technology, Raipur, India. He has published over 40 referred articles and served as a Reviewer of several journals. His current research interests include digital system design, data mining and its applications, and software fault prediction.

Bibhudendra Acharya received Post Graduation and Ph.D. degree in Electronics and Communication Engineering from National Institute of Technology, Rourkela, India.

He is currently Head and Assistant Professor in the Electronics and Telecommunication Engineering Department, National Institute of Technology, Raipur, India. His research interests include Cryptography and Network Security, Microcontroller and Embedded system, Signal Processing, Mobile Communication. He has more than 60 research publications in National/International Journals and conferences. He also served as a reviewer of several journals, including IEEE.

Dr. Acharya is a member of Computer Society of India, The Institution of Electronics and Telecommunication Engineers (India), Institute of Engineers (India), Indian Society for Technical Education (India), Cryptology Research Society of India, International Association of Engineers, Universal Association of Computer and Electronics Engineers, International Association of Computer Science and Information Technology, and The Indian Science Congress Association.

Multi-objective Scheduling Model for Reconfigurable Assembly Systems

Avinash Kumar, L. N. Pattanaik and Rajeev Agrawal

1 Introduction

Reconfigurable manufacturing systems are a current manufacturing archetype where machine apparatuses, machine compartments or material handling components can be combined, detached, altered or exchanged as needed to reply speedily to the changeable requirements [1]. Such a completely reconfigurable system does not until now exist [2], but it is the issue of dynamic investigation efforts. RAS is a part of the whole manufacturing system, and it is for the assembly processes [3]. At present, the research on the assembly line system is a significant issue at global level as automobile industry is facing lively and changeable environments. RASs are that type of assembly system which can swiftly change their capacity and functionality within a product family in order to acclimatize the fluctuating and uncertainty in the market demand [4]. RMS becomes one of the most effective answers to handle the fluctuation in the order stream since they reply to changes by offering adapted flexibility on demand by substantially reconfiguring the structure of the system [5].

Assembly comprises all essential assembly, sub-assembly processes, and apparatus mandatory to bring collectively, organize, line up, turn, and fine-tune components and constituents to produce the end product, physically append segments and parts, ingredients, and apparatuses, such as riveting, screwing, stapling, nailing, gluing, wrapping, intertwining, keying, fusing, soldering, sewing, welding.

A. Kumar (✉) · L. N. Pattanaik · R. Agrawal
Department of Production Engineering, Birla Institute of Technology,
Mesra, Ranchi 835215, Jharkhand, India
e-mail: 461.avinash@gmail.com

L. N. Pattanaik
e-mail: lnpattanaik@bitmesra.ac.in

R. Agrawal
e-mail: rajeevag1@gmail.com

J. Chattopadhyay et al. (eds.), *Innovations in Soft Computing and Information Technology*, https://doi.org/10.1007/978-981-13-3185-5_18

Assembly costs are stereotypically 26–51% of the overall cost of manufacturing. The percentage of workers tangled in assembly operations ranges from 21 to 61%. For an illustration of computer electronics manufacturing, 41–61% of total earnings are paid to assembly workforces [6]. The collected processing value of the product is consequently excessive compared to other manufacturing processes at preceding phases [7]. Assembly systems encompassing many stations for assembling a product are exploited in manufacturing virtually all types of robust goods, for example automobile and office equipment and furniture. The part is secured by immobilizers and transported on the fixed object through the assembly system.

RASs are those that can swiftly transform their capacity (numbers assembled) and functionality (product type, within a product family) to familiarize to marketplace mandate. RAS should hold the features of customization, convertibility, and scalability. Customization refers to scheming a scheme for accumulating a complete part/product family. Convertibility means that it is likely to shift rapidly after the assembly of a certain product to the assembly of diverse products in the product family. Scalability means the facility to alter system output in a moderately short period to both customer request and demand.

With the distinct model assembly line design, two significant parameters must be enhanced: line arrangement and product sequence [8]. In an active manufacturing atmosphere, these inconstant might necessarily to be reviewed whenever the production state deviates to retain the line well organized and efficient. Though, regular or thorough reconfiguration of the line might be expensive. Firms/industries might choose not to re-configure their outlines, positions and run uneconomically. So far, the unproductive assembly system can affect substantial operative worth. The choice to retain or to modify the complete or portion of the line arrangement is a dominant obstruction that has to be resolved after the line is planned. Hence, a RAS is required to prevent the price of reconfiguration whereas sustain the line effectively.

This paper aims to develop a model of a reconfigurable assembly system design that is reconfigurable to pact with reconfiguring the cost and time as one of the objectives.

The rest of this paper is systematized as given here. Section 2 of the paper addressed about the relevant literature survey in the area of reconfigurable assembly system. The proposed model of RAS is discussed in Sect. 3. Conclusion and references are discussed in the last section.

2 Literature Survey

In this section, some of the applicable literature survey regarding reconfigurable assembly system and line has been briefed. Numerous researcher points out that, to become really marketable, RASs need to have appropriate reconfiguration time and cost. By the arguments of Zhang et al. [9], the reconfiguration time is the period of time taken by system. Therefore, the reconfiguration time can be apprehended as a cost (i.e. opportunity costs) since it touches the service level due to the fatalities linked to the reconfiguration period.

Substantial work has been carried out in the field of modelling of RAS since the beginning of this new paradigm. The major strategy of modelling the RAS is to handle the fluctuating functionality and capacity demand over the forecasting cycle consisting of multiple time horizons. In most of the optimization approaches, the reconfiguration effort and assembly line balancing have been considered as an objective.

Fianfeng et al. [10] recommended an information centred timed coloured object-focused Petri net as an exhibiting technique for RAS and compared the traditional flexible assembly systems with it. Ho [11] proposed a tactic designed for reconfiguration of flexible assembly line systems through motion genetics. He smears motion genes to answer the difficulties of reconfiguring flexible assembly line systems to handle with ever-changing production prerequisite. Jain et al. [12] gave a modelling and simulated performance investigation of reconfigurable assembly line. The RAL model measured in his work has a set of inline movable robots, which can move on a directed path to serve various workstations. Each robot in the RAL can access and assist more than one workstation by moving to their position while satisfying the accessibility constraints. Meng et al. [13] planned an RAS, which comprises automobile line and assembly segments. This self-analogous administrative configuration has many innovative characters, for example capacity, flexibility, robustness, scalability, and so on. With CPM-PERT, a scalability procedure is anticipated and a scheduling method is likewise accessible to understand sturdiness of the assembly system. Bi et al. [3] presented a review on the development of RAL. He suggested the needs and drivers in evolving RAS and recognized both academic and real-world matters critical to the advance of RAS. Hu et al. [14] reviewed the advanced research in the zones of assembly system design, planning, and operations in the occurrence of product variety. Damayanti and Toha [8] planned a prototype of a reconfigurable distinct archetypal assembly line strategy in a lively manufacturing atmosphere. The archetypal is expressed in a nonlinear numeral program writing construction that decreases entire price of assembly line reconfiguration for several period.

Gyulai et al. [5] presented a simulation-based method that defines the borders and components of an RAS conferring to historical order-streams. The proposed method disconnected the low- and high-capacity products and product families vigorously, by conveying to them the appropriate reconfigurable or devoted production lines, respectively. Eguia et al. [15] proposed a procedure for designing and sequencing product families in reconfigurable disassembly system (RDS). His methodology is established in a two-phase approach, where products are first assembled into families and then families are sequenced through the RDS, calculating the required machines and modules configuration for the respective family. Gyulai et al. [16] introduced a novel method that supports the long-term choice to reposition the assembly of a product with declining demand from a dedicated to a reconfigurable line, based on the calculated investment and operational costs. Delorme et al. [17] proposed the use of multi-objective approaches in the field of assembly line design.

Zulfiqar et al. [18] proposed the finding of RAS as a justification for operative production ramp-up planning. Other core matters of production systems are automated and manual assembly elements in order to manage an operative and fast production ramp-up to respond quickly to the niche market variations in the demand cycles. Pattanaik and Jena [19] identified research problems in design and operation of RAS. These research problems were the identification of product family for RAS, design of integrated products, design of optimal configuration for RAS, optimal design for assembly, identifying and enumerating design necessities, use of proper recreation tool for RAS, identification of quality complications, etc. Challenges and possibilities in each area of these issues are identified.

Yuan et al. [20] presented a multi-objective scheduling problem for RAS with costs of assembly line renovation, production load equalization, and minimization of the delayed workload as the three objective functions. A particle swarm optimization metaheuristic was applied to discover the optimal solutions.

3 Proposed Model

3.1 Mathematical Formulation

A multi-objective optimization model is presented where the objectives are minimizing the total reconfiguration cost and time of delivery schedule while minimizing the work content imbalance.

Assumptions

Proposed model of reconfigurable assembly system is characterized by some conditions as follows:

(1) The assembly line is unpaced diverse model assembly line.
(2) The assembly line is premeditated for some periods of t, where $t = 1, 2, \ldots T$.
(3) Variety and volume of various products to be assembled can differ for every period.
(4) Volume of products in each period is recognized at design phase and is deterministic.
(5) Each product has a known and fixed process routing.
(6) The machining time includes the setup time and the reconfiguration time is considered separately.
(7) Reconfiguration is performed only when the product type is to be changed.
(8) Reconfiguration time is primarily the sum of module change times.

Based on the features and characteristics of a typical RAS as discussed in the previous section, three objective functions are developed and formulated here. The following notations and variables are used in the formulation of objective functions:

Notations and Variables

i	Index of product type, $i = 1, \ldots, P$
c	Index of components, $c = 1, 2, \ldots, C$
j	Index of operation type, $j = 1, 2, \ldots, O$
k	Index of workstation, $k = 1, 2, \ldots, M$
t	Index of time periods, $t = 1, 2, \ldots, T$
q_i	Demand quantity of product i
$S_{ii'}$	Commonality index between products i and i', $0 \leq S_{ii'} \leq 1$
$S_{i'}^{i}$	Reconfiguration efficiency from product i to i', $0 \leq S_{i'}^{i} \leq 1$
t_{ci}	Assembly setup time for component c on product i
$T_{ii'}$	Assembly reconfiguration/changeover time from product i to i'
$C_{ii'}$	Number of common components between products i and i'
$(\mathrm{TRC})_t$	Total reconfiguration cost in period t
D_i	Time available (based on due date) for product i
W_i	Work content of product i
N_i	Number of components in product i
A_i	Assembly station with index i
P_i	Priority rank of product i based on delivery schedule
R_i	Rank of product i in a schedule plan.

The first objective is related to the fact that RAS follows a product family-based design where some commonality exists among the products. Further, the ease or complexity of assembling operation for a particular component can vary from product to product. Depending on these two aspects, the commonality among products can be established. In industrial practice, it is relatively easier to changeover the assembly setup and to assemble the components between two products with higher commonality and vice-*a*-versa.

The first objective function also includes another measure of reconfiguration efficiency in terms of the total time or cost associated with each changeover of assembly system. Commonality index between products i and i' is expressed as in (1).

$$S_{ii'} = \frac{C_{ii'}}{\min(N_i, N_{i'})}, \quad 0 \leq S_{ii'} \leq 1 \tag{1}$$

Similarly, the assembly reconfiguration complexity between products i and i' is formulated as in (2).

$$S_{i'}^{i} = \frac{\sum_{c=1}^{N_{i'}} t_{ci'} - \sum_{c=1}^{N_i} t_{ci}}{\sum_{c=1}^{N_{i'}} t_{ci'}}, \quad 0 \leq S_{i'}^{i} \leq 1 \tag{2}$$

Using these two indices, the first objective (F_1) is formulated as a minimization function

$$(\mathrm{TRC})_t \propto \sum_{i=1}^{P} \sum_{i'=1}^{P} \frac{S_{i'}^{i}}{S_{ii'}} \text{ or } (\mathrm{TRC})_t = \text{a constant.} \sum_{i=1}^{P} \sum_{i'=1}^{P} \frac{S_{i'}^{i}}{S_{ii'}}$$

$$\text{In order to minimize } (\mathrm{TRC})_t, \text{ Minimize } F_1 = \sum_{i=1}^{P} \sum_{i'=1}^{P} \frac{S_{i'}^{i}}{S_{ii'}} \tag{3}$$

P is the number of product types to be assembled in the time period t.

The second objective function addresses the issue of fulfilling the delivery date constraints for the products. Using the priority for each product based on their due date information, an ideal sequence for assembly can be developed. Any other sequence for assembling the products which violate the priority rankings is expressed as a penalty function.

$$\text{Minimize } F_2 = \sum_{i=1}^{P} |P_i - R_i| \tag{4}$$

Based on their due date, products are scheduled ideally and rank P_i is assigned to each product. R_i is the sequential rank of product 'i' as generated in a potential solution for assembly sequence.

The third objective function of this model is balancing of assembly load over the time period. Considering the volume of products and their assembly operation times, work content for each product type is established. This objective is expressed as a minimization function which is the sum of the difference between mean work content of all the products and average of each pair of consecutive product in the selected schedule.

$$\text{Minimize } \sum_{i=1}^{K} \left| \overline{W} - \frac{(W_i + W_{i+1})}{2} \right| \tag{5}$$

Further, the practical significance of this objective is that in the absence of an equitable distribution of work contents caused due to product sequence, the delivery schedule can also be affected owing to clustering of high work content products.

In order to solve this multi-objective optimization problem which is NP-hard with conflicting objectives, a self-organizing migratory algorithm (SOMA) is implemented. The three objective functions are numerically expressed as minimization fitness values F_1, F_2 and F_3. The Pareto optimal solutions are plotted in 3D space (Fig. 1) to represent the various optimum sequences of products for the RAS.

The best solution can be selected from the set of Pareto optimal solutions by the decision-maker based on any priority for the objective functions.

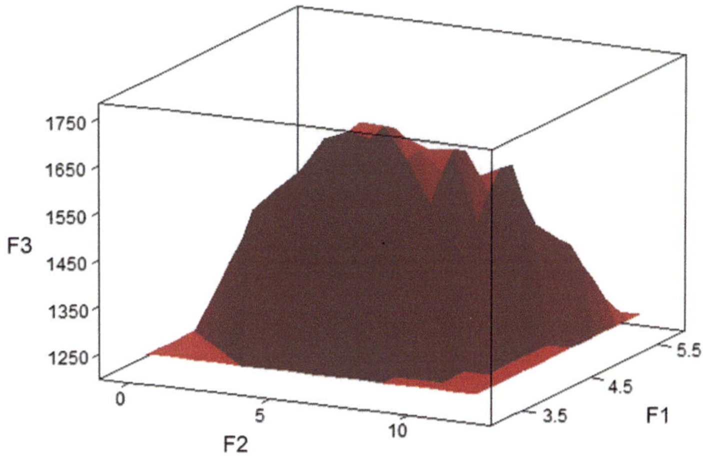

Fig. 1 Pareto fronts for the tri-objective optimization

4 Conclusion

RAS being a relatively newer topic of research needs more emphasis on operational issues like scheduling of products to optimize various cost components. The classical scheduling approaches cannot be applied to RAS owing to the difference in their characteristics and features. In the present paper, the model of a reconfigurable assembly system for finding the optimal schedule plan is proposed by optimizing three objective functions namely minimization of reconfiguration cost and time, minimization of delay in delivery, and minimization of work content imbalance over a time horizon. The optimum solutions are expressed as the sequence of assembling the products in the product family. Mathematical objective functions are developed and solved using a multi-objective self-organizing migratory algorithm.

References

1. Koren Y, Heisel U, Jovane F, Moriwaki T, Pritschow G, Ulsoy G and VanBrussel H (1999) Reconfigurable manufacturing system CIRP Annals – Manufacturing Technology. 527–540.
2. Hoda A. ElMaraghy., (2006) Flexible and reconfigurable manufacturing systems paradigms, Int J Flex Manuf Syst 17:261–276.
3. Bi, Z.M., Wang, L. and Lang, S.Y.T. (2007) Current status of reconfigurable assembly systems, *Int. J. Manufacturing Research*, Vol. 2, No. 3, pp. 303–328.
4. Mehrabi, M.G., Ulsoy, A.G. and Koren, Y. (2000) Reconfigurable manufacturing system and their enabling technologies, International Journal of Manufacturing Technology and Management, Vol. 1, No. 1, pp. 113–130.

5. Gyulai D, Ven Z, Pfeiffer A, Vancza J and Monostori L (2012), Matching demand and system structure in Reconfigurable assembly system. Procedia CIRP (1 ed., Vol.3, pp 579–584).
6. Kalpakjian, S. (2001) Manufacturing Technology Fundamentals, New York: Addisson-Wesley.
7. Bellgran, M. and Johansson, C. (1995) A method for the design of flexible assembly systems, International Journal of Production Economics, Vol. 41, pp. 93–102.
8. D Damayanti and I. S. Toha (2012), reconfigurable mixed model assembly line design in a dynamic production environment. IEEE Int Conf on Industrial engineering and engineering management (IEEM).
9. Zhang, W., Freiheit, T. and Yang, H. (2005) Dynamic scheduling in flexible assembly system based on timed Petri nets model, Robotics and Computer-Integrated Manufacturing, Vol. 21, pp. 550–558.
10. Fianfeng Yu, Yuehong Yin, Xinjun Shing and Zhaoneng Chen (2003), Modelling strategies for Reconfigurable assembly systems, Assembly Automation, Vol.23 issue:3, pp 266–272.
11. Ho JKL (2005) A proposed approach for reconfiguration of flexible assembly line systems by motion genes. Int J Prod Res4 3(9):1729–1749.
12. Jain P K, Fukuda Y, Komma V R and Reddy KVS (2006), Performance modelling of reconfigurable assembly line. Int journal of simulation modelling, 5 :16–24.
13. Meng F, Huang X, Tan D, Wang Y (2004) Study on the architecture of reconfigurable assembly system. In: Yan XT., Jiang CY., Juster N.P. (eds) Perspectives from Europe and Asia on Engineering Design and Manufacture. Springer, Dordrecht.
14. Hu SJ, Ko J, Weyand L, Elmaraghy HA, Lien TK, Koren Y, Bley H, Chryssolouris G, Nasr N, Shpitalni M (2011) Assembly system design and operations for product variety. CIRP Ann 60(2):715–733.
15. Eguia I, Lozano S, Racero J and Guerrero F (2011), A methodological approach for designing and sequencing product families in reconfigurable di-assembly systems. Journal of industrial engineering and management JIEM-4(3):418–435.
16. Gyulai D, Kadar B and Monostori L (2014), Capacity management for assembly system with dedicated and reconfigurable resources. CIRP Annals-Manufacturing Technology 63(1).
17. Delorme X, Battaia O and Dolgui A (2014), Multi-objective approach for design of assembly lines. In: L. Beyoncef et al (eds) Application of multi-criteria and game theory approaches, springer series in advanced manufacturing.
18. Zulfiqar Ali-Qureshi, Wagnili H and ElMaraghy (2014), Procurement of Reconfigurable assembly system a justification for effective production ramp-up planning. Product services systems and value creation. Proceedings of the 6th CIRP conference on Industrial Product service systems, 164–169.
19. L. N. Pattanaik and Abinash Jena (2016), Research Issues in Design and Operation of Reconfigurable Assembly Systems, International conference on Evolutions in manufacturing: Technologies and Business Strategies for Global Competiveness, BIT Mesra, Ranchi.
20. Yuan, Minghai, Kun Deng, W. A. Chaovalitwongse, and Shuo Cheng. 2016. Multi-objective optimal scheduling of reconfigurable assembly line for cloud manufacturing. Optimization Methods and Software 32, (3): 581–593. https://doi.org/10.1080/10556788.2016.1230210.
21. Goyal, Kapil Kumar; Jain, Pramod Kumar and Jain, Madhu, (2011) A Novel Approach to Measure Machine Reconfigurability in Reconfigurable Manufacturing System, Annals of DAAAM for & Proceedings of the 22nd International DAAAM Symposium, Volume 22, No. 1, ISSN 1726-9679.
22. Z. M. Bi, S. Y. T. Lang, W. Shen and L. Wang (2008): Reconfigurable manufacturing systems: the state of the art, International Journal of Production Research, 46:4, 967–992.
23. Gyulai D and Ven Z (2012), Order-Stream oriented system design for reconfigurable assembly systems. In: proceeding of the factory automation 2012 conference. University of Pannonia, pp. 138–143.

Avinash Kumar is PhD Research Scholar at the Department of Production Engineering of Birla Institute of Technology, Mesra, Ranchi, India. He has done his M.E. in Automated Manufacturing System at the Department of Production Engineering from Birla Institute of Technology, Mesra, Ranchi His main Research interests include Modelling and Optimization of Manufacturing System, Assembly System, Reconfigurable Manufacturing, Quality Control. He has also served as reviewer of various Journal and Conferences.

Dr. L. N. Pattanaik is working as an Associate Professor at Production Engineering department in BIT Mesra, Ranchi. He has numerous publications within the specialty and published in reputed national and international peer-reviewed journals. He is actively associated with different national and international societies and academies. His major research interest is in studies related to Design and Optimization of Manufacturing Systems, Lean and Reconfigurable manufacturing, Cellular Manufacturing, Advanced Quality Engineering, Soft Computing, Applications to Manufacturing, Modelling and Simulation of manufacturing systems.

Dr. Rajeev Agrawal is currently working as an Associate Professor in the Department of Mechanical Engineering at MNIT, Jaipur. He has presented many papers at conferences, published articles and papers in various reputed journals. His research interests include Industrial Engineering, Lean manufacturing, Operations Management, Six Sigma, Sustainable Manufacturing.

An Economic Lot-Size Inventory Model for Deteriorating Items with Time-Sensitive Consumption and Reduced Deterioration Rate

D. K. Singh, K. Prasad and S. Mahto

1 Introduction

Among all the classical inventory models, the researchers assumed the rate of demand and deterioration to be static for the last half-century. All these models implicitly dealing with more or less the same assumption that both the rate of demand as well as deterioration are independent and no any external parameters are influenced by the internal policy of any business organization. The decision as to when and how much to order or to manufacture as well as how to control and decide for maintenance of inventories, in case of deteriorating items, in order to optimize the total average cost or profit related with the inventory, are of the most important concerns of the management. We can reduce the deterioration rate significantly in as much as possible by encouraging the efforts of investment in preservation. Preservation techniques were already been used since the vanguard of civilization and will remain so in future also due to rapid social changes and their increased demand for quality resources corresponding to the tremendous population growth. The importance of this model increases significantly for the situations where the inventory deteriorates in their normal storage period. In broad sense, the annihilation of original utility of an item with span of time refers to as deterioration. In such condition, demand of customers is not fulfilled within their desired constraints. Hence, the losses for such deterioration in an inventory are inevitable and of great concerns.

D. K. Singh (✉) · S. Mahto
University Department of Mathematics, Ranchi University, Ranchi 834001, Jharkhand, India
e-mail: dhrubkumarsingh3@gmail.com

S. Mahto
e-mail: drsahdeomahto10@gmail.com

K. Prasad
Department of Mathematics, Amity University, Ranchi 834001, Jharkhand, India
e-mail: krishna9304@gmail.com

J. Chattopadhyay et al. (eds.), *Innovations in Soft Computing and Information Technology*, https://doi.org/10.1007/978-981-13-3185-5_19

Generally, the demand rate is kept in view as static in most of the conventional model related with inventory but it is not practically true as the demand is neither known with certainty nor it is uniform. If there are slight fluctuations, the formula is practically valid. But, if the fluctuation is considerable, the formula loses its validity. In EOQ models, it is also assumed that the annual demand can be estimated in advance, but annual demand cannot be simply estimated with accuracy and in such cases again the formula loses its importance. Several EOQ models have been formulated by different researchers in the past few decades and thus a day-by-day growth is anticipated in electronic inventory data processing along with advancement in scientific methods and techniques of inventory management. Recently, You [1] proposed a model of inventory items in which consumption rate is a function of price and time. Burwell et al. [2] formulated an economic lot-size model for price-sensitive demand under quality and freight charge discounts.

It is believed that wastage of food items shows immorality, so in present era, every large- or small-scale business organization facing highly increased market competitiveness while the losses due to the deterioration of consumable items are responsible for the significant reduction in profit. Ghare and Schrader [3] initially applied the constant rate of deterioration in his model. Covert and Philip [4] introduced a model assuming two-parameter Weibull distribution rate, and this was further enhanced by Pal et al. [5]. Mishra [6] formulated an instantaneous deteriorating inventory model under time-dependent consumption rate with reduced deterioration. Prasad and Mukherjee [7] have discussed the production inventory model with production disruption and shortage. Lee and Dye [8] introduced a preservation technique in an inventory model for deteriorating items under stock-sensitive consumption rate with replacement policy. Parmar and Gothi [9] investigated a deteriorating inventory model in which rate of deterioration follows exponential distribution and demand is linked with time later, and they extended the above deterministic model by using two- and three- parameter Weibull distributions taking different decision variables. Benkherouf et al. [10] worked on a finite horizon inventory model for two substitutable items with time-varying consumption rate.

This study proposes a deteriorating inventory model under quadratic rate of consumption. We have used the concept of preservation technique to control the rate of deterioration that results in significant reduction in total cost of the system. An illustration of the model has been carried out with the help of numerical problem, followed by conclusion and extension to further research.

2 Postulations and Documentations of the Model

The proposed model is formulated using the following postulations and documentations:

2.1 Postulations

1. A single type of item is assumed.
2. Lead time is taken as zero.
3. A finite inventory cycle is taken.
4. The demand function is considered as $D(t) = \lambda t^2, \lambda$ is a non-negative constant.
5. The deterioration is assumed as constant and is controlled by preservation techniques.
6. It is assumed that there is no replenishment for the deteriorated items.
7. In each cycle, the inventory is replenished at once only.

2.2 Documentations

$D(t)$	Time quadratic demand rate.
C_h	The storage cost per unit per unit time.
θ	The rate of deterioration.
C_θ	Deterioration cost per unit per unit time.
C_p	Purchase cost per unit of item.
ATC	Total cost per unit time.
M	Ordering cost of inventory per order.
$I(t)$	The inventory on-hand at time t.
Q	Highest level of inventory.
r	The reduction produced in the deterioration rate on applying preservation techniques.
ρ	The difference in the deterioration rates due to reduction in deterioration and is given by $\rho = (\theta - r)$.
t_θ	The time when deterioration begins.
t_c	The length of one cycle time which is a decision variable.

3 Mathematical Model Formulation

With the above documentation and postulations, we derived the optimal ordering cycle and order quantity such that the total average cost can be made as minimum as tenable on applying the scientific techniques of preservation. The pictorial representation of the inventory cycle is given in Fig. 1. In the present study, demand is time-dependent and in each cycle inventory level depleted due to both demand and deterioration. When replenishment occurred at $t = 0$, the inventory level attains its highest level Q which decrease during the interval $[0, t_\theta]$ and finally at time t_c the inventory diminishes to zero level as shown in Fig. 1.

Fig. 1 Geometry of the Problem

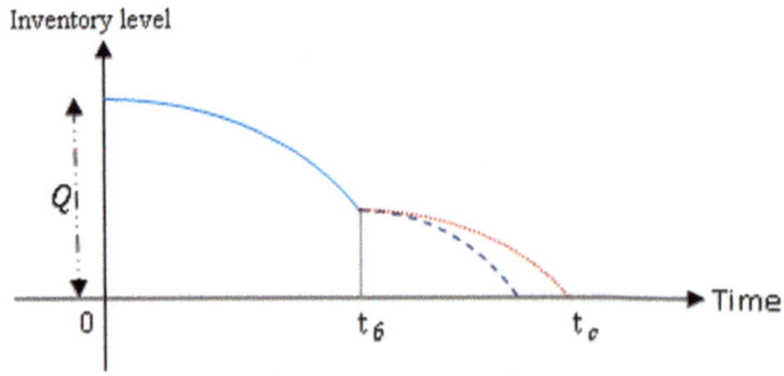

With the above assumptions of the proposed model, the differential equations describing the stock level $I(t)$ at time t are governed by Eqs. (1) and (2) as follows:

$$\frac{\mathrm{d}I(t)}{\mathrm{d}t} = -\lambda t^2, \quad 0 \le t \le t_\theta \tag{1}$$

$$\frac{\mathrm{d}I(t)}{\mathrm{d}t} + \rho I(t) = -\lambda t^2, \quad t_\theta \le t \le t_c \tag{2}$$

With the boundary conditions, $I(0) = Q$ and $I(t_c) = 0$. Applying the boundary condition, $I(0) = Q$, the solution of Eq. (1) is given by

$$I(t) = Q - \frac{\lambda t^3}{3}, \quad 0 \le t \le t_\theta \tag{3}$$

Similarly, on solving Eq. (2) we get,

$$\begin{aligned} e^{\rho t} I(t) &= -\int \lambda t^2 e^{\rho t} \mathrm{d}t + c, \quad t_\theta \le t \le t_c \\ I(t) e^{\rho t} &= -\frac{\lambda t^2 e^{\rho t}}{\rho} + \frac{2\lambda t e^{\rho t}}{\rho^2} - \frac{2\lambda e^{\rho t}}{\rho^3} + c \end{aligned} \tag{4}$$

Now applying the boundary condition $I(t_c) = 0$,
we get, $c = \frac{\lambda t^2 e^{\rho t_c}}{\rho} - \frac{2\lambda t_c e^{\rho t_c}}{\rho^2} + \frac{2\lambda e^{\rho t_c}}{\rho^3}$, then from 'Eq' (4), we get

$$\begin{aligned} I(t) e^{\rho t} &= -\frac{\lambda t^2 e^{\rho t}}{\rho} + \frac{2\lambda t e^{\rho t}}{\rho^2} - \frac{2\lambda e^{\rho t}}{\rho^3} + \frac{\lambda t^2 e^{\rho t_c}}{\rho} - \frac{2\lambda t_c e^{\rho t_c}}{\rho^2} + \frac{2\lambda e^{\rho t_c}}{\rho^3} \\ I(t) &= \frac{\lambda}{\rho}\left(t_c^2 e^{\rho(t_c - t)} - t^2\right) - \frac{2\lambda}{\rho^2}\left(t_c e^{\rho(t_c - t)} - t\right) + \frac{2\lambda}{\rho^3}\left(e^{\rho(t_c - t)} - 1\right), \quad t_\theta \le t \le t_c \end{aligned} \tag{5}$$

In Eqs. (3) and (5), values of $I(t)$ will be the same at $t = t_\theta$, which means that

$$Q - \frac{\lambda t_\theta^3}{3} = \frac{\lambda}{\rho}\left(t_c^2 e^{\rho(t_c - t_\theta)} - t_\theta^2\right) - \frac{2\lambda}{\rho^2}\left(t_c e^{\rho(t_c - t_\theta)} - t_\theta\right) + \frac{2\lambda}{\rho^3}\left(e^{\rho(t_c - t_\theta)} - 1\right)$$

Hence,

$$Q = \frac{\lambda t_\theta^3}{3} + \frac{\lambda}{\rho}\left(t_c^2 e^{\rho(t_c - t_\theta)} - t_\theta^2\right) - \frac{2\lambda}{\rho^2}\left(t_c e^{\rho(t_c - t_\theta)} - t_\theta\right) + \frac{2\lambda}{\rho^3}\left(e^{\rho(t_c - t_\theta)} - 1\right) \quad (6)$$

Hence, the total cost is expressed as:

$$\begin{aligned}\text{Total cost (TC)} &= \text{Ordering cost (OC)} + \text{Deterioration cost (DC)}\\ &\quad + \text{Inventory holding cost (HC)} + \text{Purchase cost (PC)}\end{aligned}$$

(a) Let the ordering cost OC $= M$
(b) The deterioration cost during the period $[t_\theta, t_c]$ given by DC $= c_\theta\{Q - \int_0^{t_c} D(t)\mathrm{d}t\}$

$$\therefore \mathrm{DC} = c_\theta\left\{Q - \frac{\lambda t_c^3}{3}\right\} \quad (7)$$

(c) The inventory holding cost for the period $[0, t_c]$ is given by

$$\mathrm{HC} = \int_0^{t_\theta} c_h I(t)\mathrm{d}t + \int_{t_\theta}^{t_c} c_h I(t)\mathrm{d}t$$

Therefore, by using Eqs. (3) and (5) we get

$$\begin{aligned}\mathrm{HC} = c_h\Bigg[Q t_\theta - \frac{\lambda t_\theta^4}{12} + \frac{\lambda}{\rho}\left\{\frac{t_c^2}{\rho}\left(e^{\rho(t_c - t_\theta)} - 1\right) + \frac{t_\theta^3 - t_c^3}{3}\right\} - \frac{2\lambda}{\rho^2}\left\{\frac{t_c}{\rho}\left(e^{\rho(t_c - t_\theta)} - 1\right) + \frac{t_\theta^2 - t_c^2}{2}\right\}\\ + \frac{2\lambda}{\rho^3}\left\{\frac{1}{\rho}\left(e^{\rho(t_c - t_\theta)} - 1\right) + t_\theta - t_c\right\}\Bigg]\end{aligned} \quad (8)$$

(d) The purchase cost is defined as PC $= C_p Q$

$$\begin{aligned}\mathrm{PC} = c_p\Bigg[\frac{\lambda t_\theta^3}{3} + \frac{\lambda}{\rho}\left(t_c^2 e^{\rho(t_c - t_\theta)} - t_\theta^2\right) - \frac{2\lambda}{\rho^2}\left(t_c e^{\rho(t_c - t_\theta)} - t_\theta\right)\\ + \frac{2\lambda}{\rho^3}\left(e^{\rho(t_c - t_\theta)} - 1\right)\Bigg]\end{aligned} \quad (9)$$

Hence, the total average cost for one complete cycle is expressed as:

$$\mathrm{ATC} = (\mathrm{OC}+\mathrm{DC}+\mathrm{HC}+\mathrm{PC})/t_c$$

$$\begin{aligned}
\mathrm{ATC} = \frac{1}{t_c}\Bigg[& M + c_\theta\left\{Q - \frac{\lambda t_c^3}{3}\right\} + c_h\Bigg\{Qt_\theta - \frac{\lambda t_\theta^4}{12} + \frac{\lambda}{\rho}\left(\frac{t_c^2}{\rho}\left(e^{\rho(t_c-t_\theta)}-1\right) + \frac{t_\theta^3 - t_c^3}{3}\right) \\
& - \frac{2\lambda}{\rho^2}\left(\frac{t_c}{\rho}\left(e^{\rho(t_c-t_\theta)}-1\right) + \frac{t_\theta^2 - t_c^2}{2}\right) \\
& + \frac{2\lambda}{\rho^3}\left(\frac{1}{\rho}\left(e^{\rho(t_c-t_\theta)}-1\right) + t_\theta - t_c\right)\Bigg\} + c_p\Bigg\{\frac{\lambda t_\theta^3}{3} + \frac{\lambda}{\rho}\left(t_c^2 e^{\rho(t_c-t_\theta)} - t_\theta^2\right) \\
& - \frac{2\lambda}{\rho^2}\left(t_c e^{\rho(t_c-t_\theta)} - t_\theta\right) + \frac{2\lambda}{\rho^3}\left(e^{\rho(t_c-t_\theta)}-1\right)\Bigg\}\Bigg]
\end{aligned} \tag{10}$$

Now, to compute the optimum value of t_c and to minimize ATC, we are left to find the value of t_c for which $\frac{\partial \mathrm{ATC}}{\partial t_c} = 0$, satisfying the condition $\frac{\partial^2 \mathrm{ATC}}{\partial t_c^2} > 0$.

We illustrated the result for optimal solution of Eq. 10 using the codes of "Mathematica software" with the help of following numerical example.

4 Numerical Example and Its Graphical Validation

We considered the numerical values of the parameters as follows:

$$M = 30, \quad \lambda = 1, \quad \theta = 0.7, \quad c_\theta = 2, \quad h = 5.0, \quad r = 0.004, \quad t_\theta = 0.4, \quad c_p = 5$$

We get, the optimal value of t_c i.e $t_c^* = 1.2537$ units, $Q^* = 0.9711$ and minimized total average cost $(\mathrm{ATC})^* = 31.6459$ (Fig. 2).

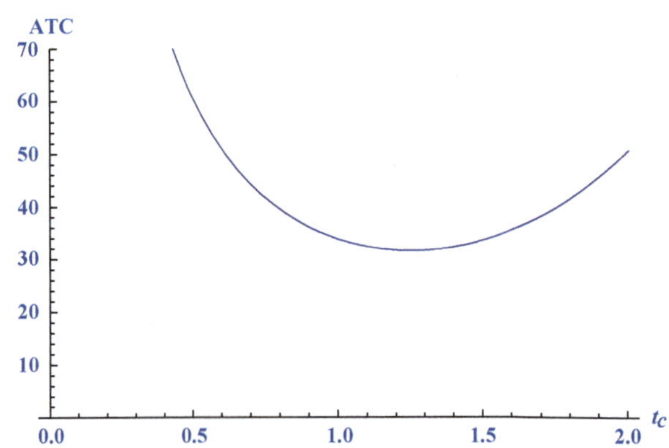

Fig. 2 Variation of total average cost ATC against time t_c

5 Summary and Conclusion

The present study has been made for deteriorating inventory model in which single type of item is considered and deterioration rate of the inventory items considered to be reducing during the decay period with the use of scientific preservation techniques. A procedure of solution is derived to compute an optimal ordering quantity and optimal cycle time to reduce the total relevant average cost of the system. It is quite vivid from the graph that the minimum average cost is 31.6459 and the optimal cycle time t_c is 1.2537 which validates our numerical result. There are still rooms to extend this model in the context of some more realistic assumptions, for example, allowing shortages or by taking any stochastic demand rate.

References

1. S.P. You, Inventory policy for products with price and time-dependent demands, *Journal of the Operation Research Society*. 56(2005) 870–873.
2. T.H. Burwell, D.S. Dave K.E. Fitzpatrick and M.R. Roy, Economic lot-size model for price demand under quantity and freight discounts, *International Journal of Production Economics*. 48(2) (1997) 141–155.
3. P.M. Ghare and G.F. Schrader, A Model for an Exponentially Decaying Inventory, *Journal of Industrial Engineering*. 14(1963) 238–243.
4. R.P. Covert and G.C Philip, An EOQ model for items with Weibull distribution deterioration, *AIIE Transactions*. 5(4) (1973) 323–326.
5. S. Pal, A. Goswami and K. S. Choudhuri, A deterministic inventory model for deteriorating items with stock-dependent demand rate, *International Journal of Production Economics*. 32 (1993) 291–299.
6. V. K. Mishra, An inventory model with Weibull distribution deteriorating items with controllable deterioration rate for item dependent demand and time varying holding cost, *Yugoslav Journal of Operation research*. 24 (2014) 67–98.
7. K. Prasad and B. Mukherjee, Production Inventory model for deteriorating item with production disruption and shortage, *International Journal of Applied Mathematical Sciences*. 6(1) (2013) 117–126.
8. Y.P. Lee and C.Y. Dye, An inventory model for deteriorating items under stock-dependent demand and controllable deterioration rate, *Computers and Industrial Engineering*. 3(2) (2012) 474–482.
9. K. Parmar and U.B. Gothi, Order level inventory model for deteriorating items under quadratic demand with time dependent IHC2014, *Sankhaya Vigyan NSV*. 10(2) (2014) 1–12.
10. L. Benkherouf, K. Skouri and I. Konstantaras, Inventory decisions for a finite horizon problem with product substitution options and time varying demand, *Applied mathematical modeling*. 51 (2017) 669–685.

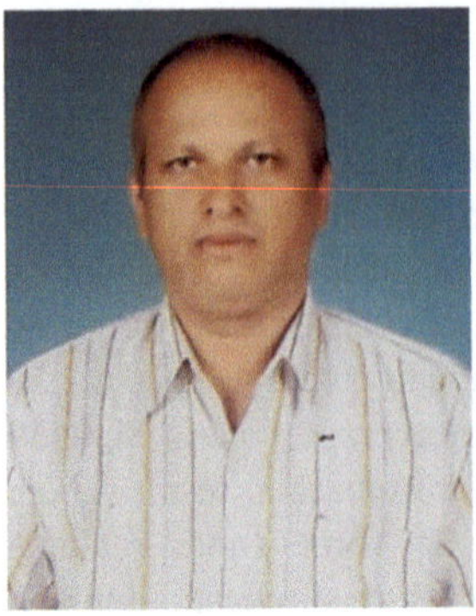

D. K. Singh is an Assistant Professor in the department of Applied Mathematics in an Engineering College, RTCIT, Ormanjhi, Ranchi, Jharkhand. Earlier he worked in the same capacity in another Engineering college namely CIT, Tatisilwai, Ranchi for about 9 years. He earned his M.Sc. & M. Phil. in Mathematics from Ranchi University, Ranchi and he is now a Research Scholar in the University department of Mathematics, Ranchi University, Ranchi, registered in the year 2015. His research interest lies in the fields of deterministic and stochastic Inventory models for deteriorating inventory items assuming different key parameters such as stock dependent, price dependent and time dependent cases. He has publications in "Opsearch" and "Journal of Computer and Mathematical sciences (An international Research Journal)". E-mail: dhrubkumarsingh3@gmail.com

Dr. K. Prasad Assistant Professor, Amity University Jharkhand. He received his M.Sc in Mathematics and Computing and Ph.D in applied mathematics from ISM Dhanbad. His research interests are Inventory control and Optimization. E-mail: krishna9304@gmail.com

Prof. (Dr.) S. Mahto is presently working as an Head of the department, University department of Mathematics, Ranchi University Ranchi. He earned his M.Sc. & Ph.D. degree from Ranchi University, Ranchi. His title of Ph. D. Thesis was "A study of certain Integral Transforms of generalized functions". He has 34 years of post Ph.D experience in teaching in undergraduate and postgraduate students in the department of mathematics under Ranchi University, Ranchi. One Research scholar has been awarded Ph.D.degree under his supervision in 2009 and subsequently six Research Scholars have been registered under his generous supervision. He has got specialisation in Operation Research, Mathematical Statistics and Integral Transforms. Email: drsahdeomahto10@gmail.com

A Statistical Analysis on Climatic Temperature Using Exponential Moving Average

Saurav Bhaumik and Abhishek Mukherjee

1 Introduction

Science had tried to define the elements of nature with all sorts of effects and causes from the earlier times. Temperature is one of them which we are going to deal in this paper. As per the definition of temperature, it is the comparative study of hotness and coldness of an object (in this case air) measured by thermometer in the scale of degree with the unit as Fahrenheit or Celsius. Except some geophysical reasons, greenhouse gases ejected by the peoples cause the climatic temperature to increase. Since it has found that if the people continuously add greenhouse gases in the atmosphere at the present rate, then the temperature will rise by 4–12 °F by 2100. This might be less up to 2–5 °F, if renewable resources instead of fossil fuels are used. Heat waves start with the increase of temperature. So more the temperature, more the heat waves which effects the people in the form of illnesses such as heat cramp and heat stroke or even death. Increasing air temperature also affects people and the environment such as the oceans, weather patterns, snow and ice, and plants and animals. An analysis has shown in [1] that the increase in the minimum monthly temperatures in the Antarctic Peninsula is approximately 6.78 °C over 53 years (1951–2003). This fluctuation of temperature in climate is very unpredictable and unstable. So in our paper, we are going to analyse the temperature data of a certain period of time to predict its nature.

To succeed in this mission, a theoretical approach has been sketched in this paper. Theory of **exponential moving average** in the Sect. 3.1 is been used to analyse the temperature with respect to the time series, in the pattern of **Fibonacci**

S. Bhaumik (✉) · A. Mukherjee
Department of Computer Science Engineering, Bengal College of Engineering and Technology, Durgapur, West Bengal, India
e-mail: saurav.bil@gmail.com

A. Mukherjee
e-mail: abhishekmukherjee009@gmail.com

J. Chattopadhyay et al. (eds.), *Innovations in Soft Computing and Information Technology*, https://doi.org/10.1007/978-981-13-3185-5_20

sequence discussed in the Sect. 3.2. To prove the effectiveness of this theoretical approach to a certain extent, the time series data of temperature is been collected over Kolkata in the year 2016 in a pattern of Fibonacci sequence and then the strategy of exponential moving average is applied on that collected data, which will prove that the exponential moving average (EMA) can not only be applicable to the stock market evaluation prediction but can also be applicable to analyse the dogmatic nature of one of the natural properties, i.e. temperature.

2 Problem Definition

From the prospect of problem definition, let us define the analysis of upcoming temperature in the form that prediction of climatic temperature by the application of exponential moving average algorithm and time series analysis in the pattern of Fibonacci on the collected temperature data from www.accuweather.com over a certain specified area and for a short while of time.

3 Proposed Method

3.1 Exponential Moving Average

Exponential moving average (**EMA**) is also known as **weighted exponential moving average (WEMA)**. After the theory of exponential moving average (EMA) has been proved to be very efficient for the prediction of stock market evaluation, it is one of the goals to prove the same theory to be effective and efficient for the analysis of climatic temperature behaviour. Exponential moving average is been defined as a type of average where it reduces the lag caused by the **simple moving average**, by applying more weight to the recent data [2]. The calculation of EMA consists of three steps:

a. Calculate the simple moving average which is the sum of the data of a specified period by that specified period.
b. Then, calculate the multiplier, which is calculated as 2 divided by that specified period plus 1.
c. Then, calculate the EMA by

$$\mathrm{EMA} = \alpha \cdot Y_{\mathrm{t}} + (1 - \alpha) \cdot S_{\mathrm{t}-1} \tag{1}$$

where $S_1 = Y_1$ and for $t > 1$.

The coefficient α is the multiplier whose value is always between 0 and 1. Y_{t} is the data at the time period t. S_{t} is the data of the EMA at any time period t. EMA algorithm is vastly used in the field of stock market.

3.2 *Fibonacci Sequence*

Fibonacci sequence/series is one of the greatest achievements in the field of mathematics, which is invented during twelfth and thirteenth centuries by Leonardo (born: c1170AD), a great mathematician of Pisa who was named Fibonacci after his father name Gugliemo Bonaccio. In this sequence, each term is the sum of the previous two terms. The sequence is as follows:

$$1, 1, 2, 3, 5, 8, 13, 21\ldots$$

The modern mathematics had extended the Fibonacci sequence terms by adding one zero at the initial position. Its sequence is in the following form:

$$0, 1, 1, 2, 3, 5, 8, 13, 21\ldots$$

After a long research, some of the scientist through their research work had tried to prove that Fibonacci sequence was very close to the events of nature. That is the sequence that had been found among the arrangement of leaf in plants, the pattern of florets of a flower, the bracts of a pine core or the scales of a pineapple and even in the shell of the chambered Nautilus [3, 4]. This fact has enthusiast us to consider the time series interval in the Fibonacci sequence.

4 The Methodology

In analysing temperature, each day temperature data for eleven months of Kolkata, India, is measured. The high temperature and low temperature of each day are analysed using exponential moving average method. In this process, each day is considered as one unit and exponential moving average is calculated for both high and low temperatures of the day. To calculate the smoothing factor, Fibonacci numbers are chosen as time intervals.

We followed the following algorithm:

a. Inputs: Temperature data (T)
b. Step 1: Choose time interval "N", which is a Fibonacci number.
c. Step 2: Calculate $\alpha = 2/(1 + N)$
d. Step 3: For time interval 1 to N, calculate simple moving average, $S = (T_1 + T_2 + T_3 + \cdots + T_N)/N$
e. Step 4: For time interval $N + 1$, calculate exponential moving average $E_{N+1} = \alpha * T_{N+1} + (1 - \alpha) * S$.
f. Step 5: For time interval $N + 2$ to T, calculate exponential moving average $E_i = \alpha * T_i + (1 - \alpha) * E_i - 1$, where $N + 2 \leq i \leq T$
g. Step 6: Continue step 1–5 for another time interval.
h. Step 7: Exit

After calculating exponential moving average for all inputs, we can generate a graph for all the input data and their corresponding EMA to continue the analysing process.

5 Result and Discussion

In the process of analysing the climatic temperature over a selected area of India, i.e. Kolkata from 1 January 2016 to 31 December 2016, a continuous sequence of temperature data is been collected on a daily basis over a period of 12 months from www.accuweather.com. Everyday's maximum temperature and minimum temperature are been recorded and being used at the time of temperature analysis by the exponential moving average method. Each day is considered as one unit. On the basis of each unit, EMA is calculated for both maximum and minimum temperatures. To calculate the smoothing factor or the multiplier (α), three numbers are being chosen among the Fibonacci sequence as a time intervals

$$(\alpha = 2/(1+N), \quad \text{where } N \text{ is the interval and } N = 5, 8, 13).$$

The temperature analysis process consists of three EMAs which are calculated at an interval of 5, 8 and 13 days.

Following graph is plotted by measuring high and low-temperature details from 1 January 2016 to 31 December 2016 of Kolkata, India, by considering the number of days as x-axis and the degree of temperature as y-axis.

In this graph, the "EMA_{Low1}" and "EMA_{High1}" are calculated with $N = 5$, "EMA_{Low2}" and "EMA_{High2}" are plotted with $N = 8$, and the "EMA_{Low3}" and "EMA_{High3}" are plotted with $N = 13$, i.e. the time interval of 5, 8 and 13 days. In Fig. 1, the original low- and high-temperature data ($TEMP_{Low}$ and $TEMP_{High}$) is measured throughout the year which is fluctuating over the lines of "EMA_{Low}" and "EMA_{High}", respectively. According to the theory of exponential moving average, when the original line of "$TEMP_{High}$" or "$TEMP_{Low}$" remains above its "EMA_{High}" line or "EMA_{Low}" line, the tendency of increasing in temperature will be high and when it remains below the EMA lines, the tendency changes. In Fig. 1, we have plotted three EMA values with respect to a temperature data, which makes the analysing process more prominent. Now, if we analyse only the low-temperature data in respect to the "EMA_{Low1}", "EMA_{Low2}" and "EMA_{Low3}", we observe that when the "$TEMP_{Low}$" line always remains close to the "EMA_{Low1}" line, rise and fall of the low temperature are very much close to it. When "$TEMP_{Low}$" line crosses the "EMA_{Low1}" line in any directions, there is a rise or fall in the temperature but that was not clear with one "EMA_{Low}" line. To analyse the low-temperature data more sharply, we use other two "EMA_{Low}" lines. When the low line in the graph crosses the "EMA_{Low2}", the tendencies of rise or fall become more than

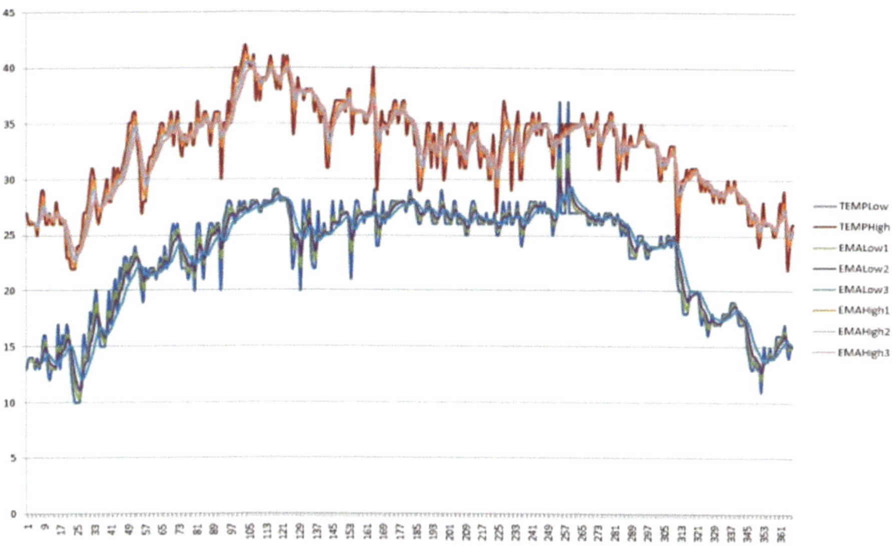

Fig. 1 EMA graph over Kolkata from 1 January 2016 to 30 November 2016

"EMA_{Low1}". In case of "EMA_{Low3}", whenever the original "$TEMP_{Low}$" line tends to cross the "EMA_{Low3}" line, the tendency of rise or fall will be more than other two EMA lines. Now, if we observe the "$TEMP_{High}$" data line, it also moves like the way "$TEMP_{Low}$" data line moves, according to its EMA lines. So, the tendency of rise and fall of "$TEMP_{High}$" is also same as that of the "$TEMP_{Low}$".

In the graphical representation of low temperature, high temperature and the EMAs, if we notice the 121st to 145th day's fluctuation, it is clearly understood that the EMAs can be used to analyse the temperature data. On 121st day in the graph, the "$TEMP_{High}$" line crossed all the three "EMA_{High}" lines in an upward direction, so the chances of increase in temperature remain high until it cuts "EMA_{High1}" in a downward direction. Now, again when it crosses the "EMA_{High1}" line downwardly, the tendency of temperature drop increases. So, when it crosses all the "EMA_{High}" lines, the tendency of drop in temperature becomes clear.

We have also applied EMA in every month's temperature data which is shown in Figs. 2, 3, 4, 5, 6 and 7.

But in this case, we have done some changes in the calculation of EMA. We have calculated the EMA on 3 and 5 days of time interval, and we got a clear view on temperature fluctuation over the EMA lines. In this paper, we have shown graphical representation of six months temperature data with its respective EMA and in Tables 1, 2, the EMA for both high and low temperature is calculated.

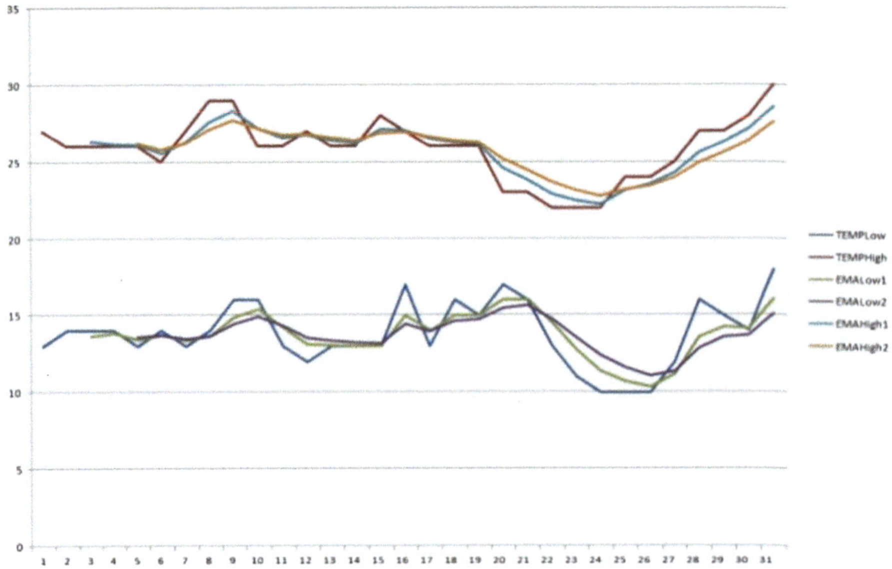

Fig. 2 January 2016

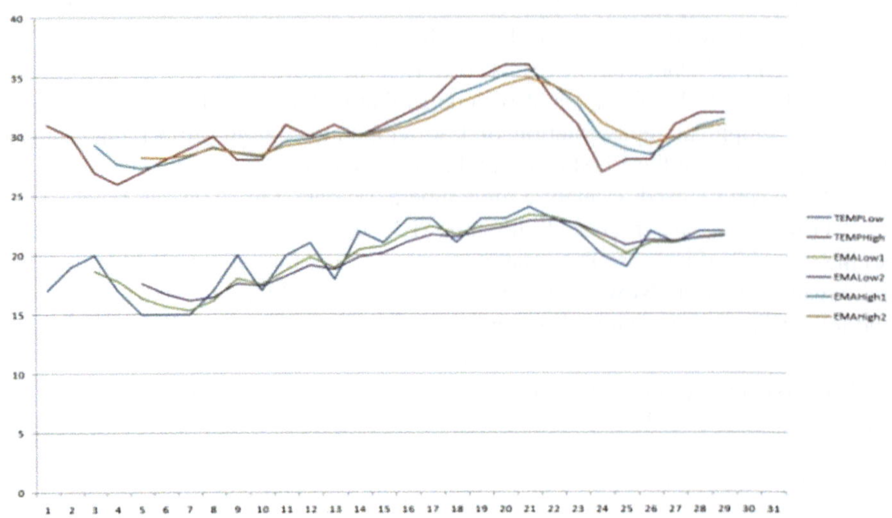

Fig. 3 February 2016

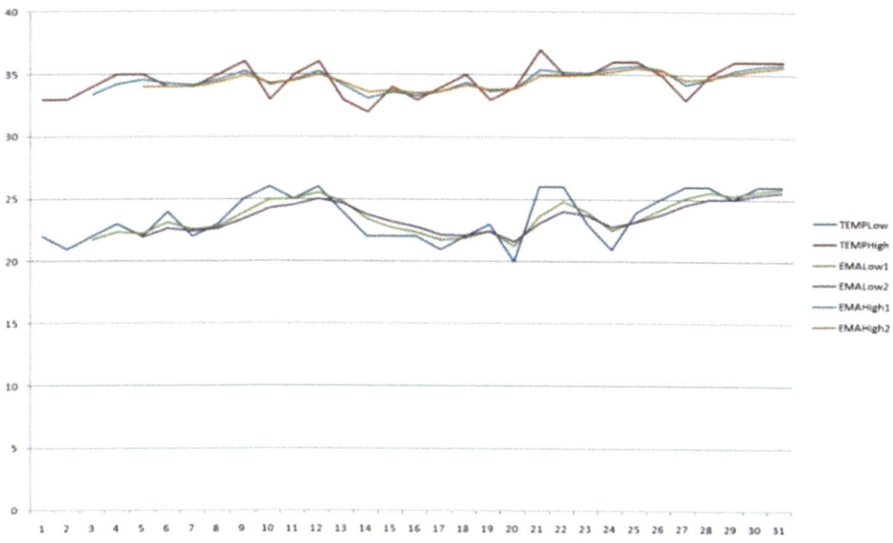

Fig. 4 March 2016

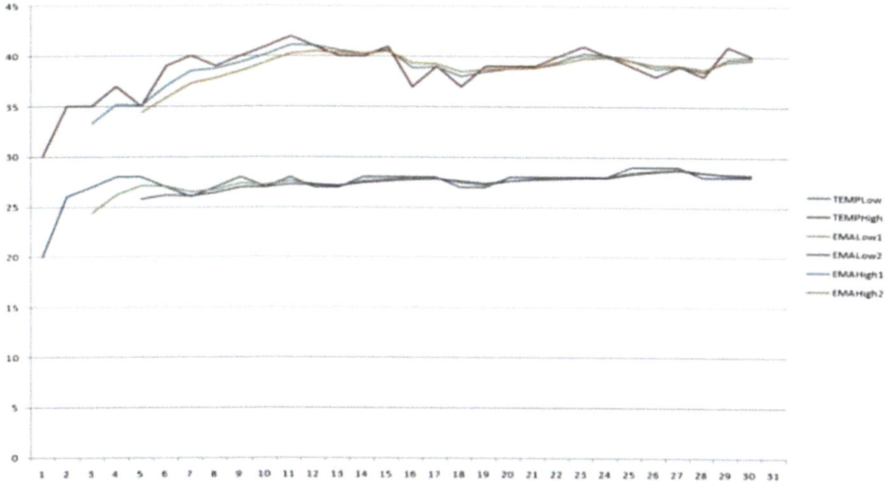

Fig. 5 April 2016

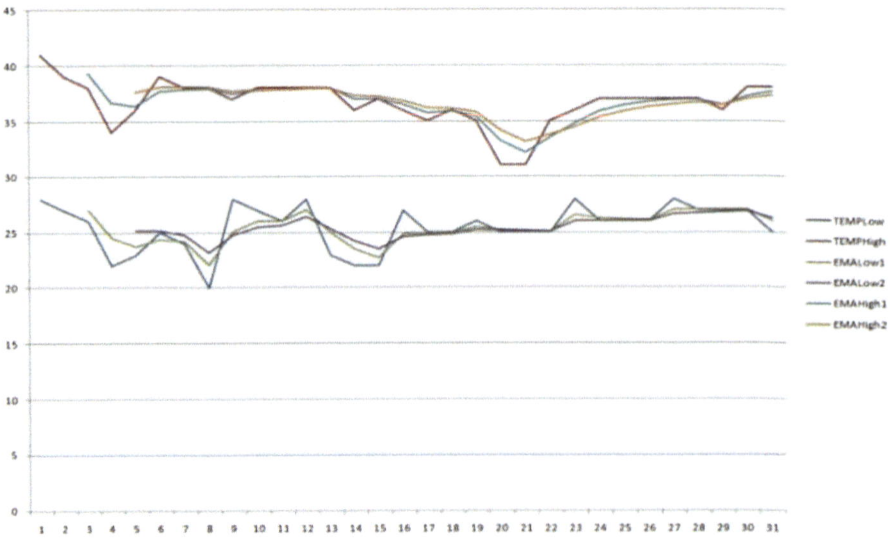

Fig. 6 May 2016

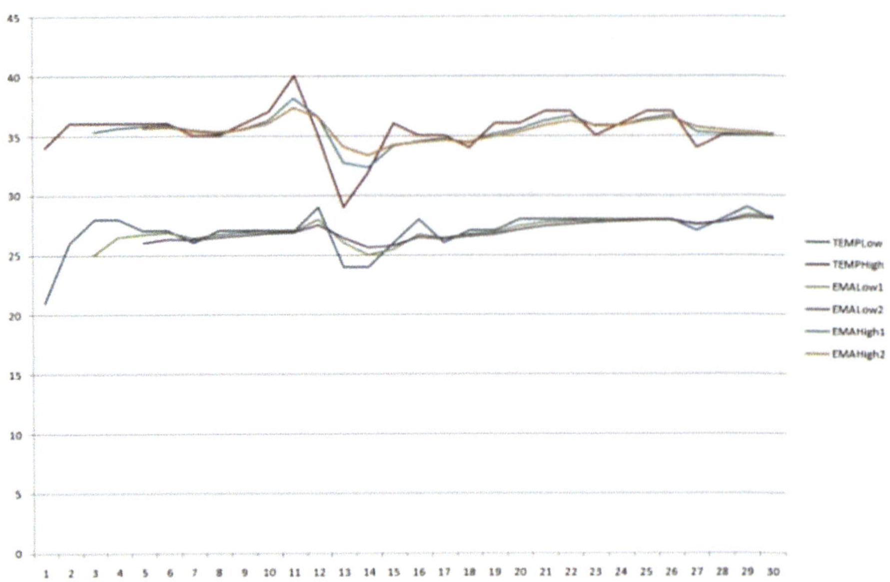

Fig. 7 June 2016

Table 1 Calculation of EMA_{lows} data for January 2016

Date	Temp	Temp	SMA	Multiplier	EMA	SMA	Multiplier	EMA	SMA	Multiplier	EMA
	Low	High	Low1		Low1	Low2		Low2	Low3		Low3
1	13	27									
2	14	26									
3	14	26	13.667	0.5	13.667						
4	14	26	14	0.5	13.833						
5	13	26	13.667	0.5	13.417	13.6	0.33	13.6			
6	14	25	13.667	0.5	13.708	13.8	0.33	13.732			
7	13	27	13.333	0.5	13.354	13.6	0.33	13.49			
8	14	29	13.667	0.5	13.677	13.6	0.33	13.659	13.625	0.22	13.625
9	16	29	14.333	0.5	14.839	14	0.33	14.431	14	0.22	13.781
10	16	26	15.333	0.5	15.419	14.6	0.33	14.949	14.25	0.22	14.122
11	13	26	15	0.5	14.21	14.4	0.33	14.306	14.125	0.22	14.315
12	12	27	13.667	0.5	13.105	14.2	0.33	13.545	13.875	0.22	14.173
13	13	26	12.667	0.5	13.052	14	0.33	13.365	13.875	0.22	13.841
14	13	26	12.667	0.5	13.026	13.4	0.33	13.245	13.75	0.22	13.583
15	13	28	13	0.5	13.013	12.8	0.33	13.164	13.75	0.22	13.455
16	17	27	14.333	0.5	15.007	13.6	0.33	14.43	14.125	0.22	13.648
17	13	26	14.333	0.5	14.003	13.8	0.33	13.958	13.75	0.22	13.799
18	16	26	15.333	0.5	15.002	14.4	0.33	14.632	13.75	0.22	14.136
19	15	26	14.667	0.5	15.001	14.8	0.33	14.753	14	0.22	14.253
20	17	23	16	0.5	16	15.6	0.33	15.495	14.625	0.22	14.637
21	16	23	16	0.5	16	15.4	0.33	15.661	15	0.22	14.937
22	13	22	15.333	0.5	14.5	15.4	0.33	14.783	15	0.22	15.024
23	11	22	13.333	0.5	12.75	14.4	0.33	13.535	14.75	0.22	14.652

(continued)

Table 1 (continued)

Date	Temp	Temp	SMA	Multiplier	EMA	SMA	Multiplier	EMA	SMA	Multiplier	EMA
	Low	High	Low1		Low1	Low2		Low2	Low3		Low3
24	10	22	11.333	0.5	11.375	13.4	0.33	12.368	13.875	0.22	13.922
25	10	24	10.333	0.5	10.688	12	0.33	11.587	13.5	0.22	13.133
26	10	24	10	0.5	10.344	10.8	0.33	11.063	12.75	0.22	12.443
27	12	25	10.667	0.5	11.172	10.6	0.33	11.372	12.375	0.22	12.053
28	16	27	12.667	0.5	13.586	11.6	0.33	12.899	12.25	0.22	12.188
29	15	27	14.333	0.5	14.293	12.6	0.33	13.593	12.125	0.22	12.66
30	14	28	15	0.5	14.146	13.4	0.33	13.727	12.25	0.22	13.175
31	18	30	15.667	0.5	16.073	15	0.33	15.137	13.125	0.22	13.723

SMA Simple moving average

Table 2 Calculation of EMA_{Highs} data for January 2016

Date	Temp	Temp	SMA	Multiplier	EMA	SMA	Multiplier	EMA	SMA	Multiplier	EMA
	Low	High	High		High1	High2		High2	High3		High3
1	13	27									
2	14	26									
3	14	26	26.333	0.5	26.333						
4	14	26	26	0.5	26.167						
5	13	26	26	0.5	26.083	26.2	0.33	26.2			
6	14	25	25.667	0.5	25.542	25.8	0.33	25.804			
7	13	27	26	0.5	26.271	26	0.33	26.199			
8	14	29	27	0.5	27.635	26.6	0.33	27.123	26.5	0.22	26.5
9	16	29	28.333	0.5	28.318	27.2	0.33	27.742	26.75	0.22	27.05
10	16	26	28	0.5	27.159	27.2	0.33	27.167	26.75	0.22	26.819
11	13	26	27	0.5	26.579	27.4	0.33	26.782	26.75	0.22	26.639
12	12	27	26.333	0.5	26.79	27.4	0.33	26.854	26.875	0.22	26.718
13	13	26	26.333	0.5	26.395	26.8	0.33	26.572	26.875	0.22	26.56
14	13	26	26.333	0.5	26.197	26.2	0.33	26.383	27	0.22	26.437
15	13	28	26.667	0.5	27.099	26.6	0.33	26.917	27.125	0.22	26.781
16	17	27	27	0.5	27.049	26.8	0.33	26.944	26.875	0.22	26.829
17	13	26	27	0.5	26.525	26.6	0.33	26.633	26.5	0.22	26.647
18	16	26	26.333	0.5	26.262	26.6	0.33	26.424	26.5	0.22	26.504
19	15	26	26	0.5	26.131	26.6	0.33	26.284	26.5	0.22	26.393
20	17	23	25	0.5	24.566	25.6	0.33	25.2	26	0.22	25.647
21	16	23	24	0.5	23.783	24.8	0.33	24.474	25.625	0.22	25.065
22	13	22	22.667	0.5	22.891	24	0.33	23.658	25.125	0.22	24.39
23	11	22	22.333	0.5	22.446	23.2	0.33	23.111	24.375	0.22	23.864

(continued)

Table 2 (continued)

Date	Temp	Temp	SMA	Multiplier	EMA	SMA	Multiplier	EMA	SMA	Multiplier	EMA
	Low	High	High		High1	High2		High2	High3		High3
24	10	22	22	0.5	22.223	22.4	0.33	22.744	23.75	0.22	23.454
25	10	24	22.667	0.5	23.111	22.6	0.33	23.159	23.5	0.22	23.574
26	10	24	23.333	0.5	23.556	22.8	0.33	23.436	23.25	0.22	23.668
27	12	25	24.333	0.5	24.278	23.4	0.33	23.952	23.125	0.22	23.961
28	16	27	25.333	0.5	25.639	24.4	0.33	24.958	23.625	0.22	24.63
29	15	27	26.333	0.5	26.319	25.4	0.33	25.632	24.125	0.22	25.151
30	14	28	27.333	0.5	27.16	26.2	0.33	26.413	24.875	0.22	25.778
31	18	30	28.333	0.5	28.58	27.4	0.33	27.597	25.875	0.22	26.707

SMA Simple moving average

6 Conclusions

From the earlier days, the analysis of nature's behaviour seems to be a great challenge. Through this paper, we had tried to statistically analyse the probability of the increase and decrease of one of the natural properties, i.e. temperature, with respect to some specific area. In this theoretical model, exponential moving average (EMA) and Fibonacci series are the tools we have with us to prove our approach to be effective. After a long discussion on our graphically represented result, we come to a conclusion that though exponential moving average is one of the efficient mathematical calculations in the field of stock market exchange, still it can be applicable in the field of atmospheric science. This approach can also be taken into consideration to warn the human race from natural disasters that cause due to drastic increase or decrease of climatic temperature. Our approach to find the probability of rising and falling of climatic temperature was applicable to a certain extent, but fails to find the exact or approximate range of temperature variation through the same theoretical model.

References

1. GILLIAN L. HUGHES, SUHASINI SUBBA RAO AND TATA SUBBA RAO, School of Mathematics, University of Manchester, Manchester M60 1QD, UK 2 Texas A&M University, College Station, TX 77843-3143, USA, School of Mathematics, University of Manchester, Manchester M60 1QD, UK 2 Texas A&M University, College Station, TX 77843-3143, USA.
2. Stanley Dash, *A Comparative study of moving averages: Simple Weighted and Exponential, Trade Station Labs Analysis Concepts,* Issue 38 Wednesday, May 9,2012.
3. Lynn D. Newton, Fibonacci and Nature: *Mathematics Investigations for Schools, The Mathematical Association*, Vol. 16, No. 5 (Nov., 1987), pp. 2–8.
4. Omotehinwa T. O, Ramon S.O, Fibonacci Numbers and Golden Ratio in Mathematics and Science, International Journal of Computer and Information Technology (ISSN: 2279 – 0764) Volume 02–Issue 04, July 2013.

Saurav Bhaumik is an Assistant Professor of Computer science & Engineering Department at Bengal College of engineering & Technology, Durgapur. He started his Career in the year 2014 as Assistant Professor of Computer Science & Engineering Department at Govt. Engineering College Ramgarh, Jharkhand. He received his B.Tech and M.Tech degree from West Bengal University of Technology, Kolkata. His current research interests are Data Science & Open Source hardware.

Abhishek Mukherjee is an Assistant Professor of Computer science & Engineering Department at Ramgarh Engineering College, Ramgarh. He started his Career in the year 2016 as Assistant Professor of Computer science & Engineering Department at Bengal College of engineering & Technology, Durgapur. He received his B.Tech and M.Tech degree from West Bengal University of Technology, Kolkata. His current research interests are Data Science & Multi Agent Based systems.

Determining All Possible Candidate Keys for Relational Database Design

Kunal Kumar and Sachindra Kumar Azad

1 Introduction

Back-end programming language that enables us to insert, update and delete the data from the tables of the relational database is referred as database management system. Database management system is divided into different categories: small database system executes on personal computers and large server database system for huge storage of data executes on supercomputer or mainframe computers [1]. Relational database system is mainly developed to solve the problem that arises during the file processing system [2]. Earlier, when we are conventionally operating systems, the file processing system was proposed, and where the data is stored in scattered fashion as single data that is stored in more than one file, which increased the data redundancy problems. When data is stored in many places, it may also increase the searching time and very difficult to update the particular entity [3].

Database is a collection of more than one table. In tables, data is stored in rows known as tuple of the relation and column referred to the attribute(s) of the database relation schema [4]. Cardinality of the table defined as the number of tuples in a relation schema and degree of table represents the number of column in a relation schema. Tuples in database tables always share common property to each other [5]. In file processing system, data is not atomic, i.e. every data is stored in more than one place in a different file. In relational database, data in table should be atomic at any cost. Database system also ensures that database relation may not hold attribute having multi-value, composite attribute and set of almost identical type of value. Before designing the database table, we should design the pictorial diagram of the

K. Kumar (✉) · S. K. Azad
University Department of Statistics and Computer Applications,
T.M. Bhagalpur University, Bhagalpur 812007, India
e-mail: kunal.mishra.mca@hotmail.com

S. K. Azad
e-mail: skazad@rediffmail.com

J. Chattopadhyay et al. (eds.), *Innovations in Soft Computing and Information Technology*, https://doi.org/10.1007/978-981-13-3185-5_21

table that is known as entity–relationship diagram. ER diagram always views the real world as object and relationship shared between these objects [6]. ER diagram represents the conceptual structure of the relational database. It is also assumed as the requirement analysis phase of the database designing [7]. The client and the database developer are communicating between the requirement analysis phases and getting all necessary information from the end-user, and then database designer designs the entity–relationship model. Relational database is always considered as set of tuples. Tuples refer the row of database relation schema and the column of database relation schema is known as attribute. Row of the table always has atomic entry in the table [8]. The row of the table always shared the common property with each other. To avoid the multiple entries of the row, the attribute of the table uses a feature known as key attribute. An attribute which is set as the key attribute will never contain the repeated value in the corresponding row [9]. Therefore, the attribute should be determined during the requirement analysis phase. Each table is a combination of attributes and tuples. Key is always defined as minimum set of attributes that is used to uniquely differentiate the tuple from a table. If there is any multi-valued dependency that is defined in the ER diagram, then database designer who converts the ER diagram to table always make a separate table for multi-valued attribute with foreign key attribute [10]. Key of the table is a combination of single attribute or more than one attribute known as composite attribute. The superkey always considered as the superset of the key and the minimal superkey known as candidate of the table. Candidate key of the table is a collection of all possible primary keys in it. It is up to the database designer who chooses the primary key from the given set of candidate key [11]. On an attribute or set of attribute that is able to form primary key of the table, rest of that remains in the candidate key which is known as alternate key of the table. Determining the key of a relational database table is a challenging work. Key of relation should be determined from the functional dependency proposed by the database designer [12]. There are various methods that are proposed by researcher to determine the key of a table. In this, we are also proposed a method to determine the key of the table. Once we know the key of the relation, it helps the database designer to reach the table at highest level of normal form. Normalization is the concept in relational database for the standardization of the table. Normalization may contain single relation or more than one relation. Sometimes, it is impossible to store all the information into a single table. So the database designer decomposed the main table into small tables, and these decomposed must be connected to each other with the help of foreign key [13]. Decomposition may be categorized into two types: lossless decomposition and lossy decomposition. Decomposition is said to be lossless, if after joining the decomposed tables, the original table should be formed without any redundancy. In lossy decomposition after joining the decomposed table, if the original table is formed, then it contains redundant tuple [14]. Decomposition should be lossless, and functional dependency should be preserved in the decomposed table. When a table is decomposed into different small tables, the information can retrieve using the various join operations [15].

Once a table reached the highest level of the normalization, multiple transaction operations are submitted by the remote site to the server computer to access the stored information at the same time. Consistency state should be maintained by the database [16]. When more than one transaction is trying to access the database, either each operation defined in the transaction should be executed in full or nothing, it should not stop in the middle and produced the result. Secondly, when more than one operation is accessed by the remote site at the same time, it should be managed the concurrency control mechanisms [17]. The result of one transaction cannot be seen by the other transaction till the compilation of the first transaction operations. Once the transaction is complete, it should be durable; at the end of every transaction operation, the commit command is executed. After the execution of commit command, database is updated and new information is added to the table. Transaction operations execute under one principle known as *all or nothing* [18].

2 Related Work

Relational data model was introduced by Edgar F. Codd, IBM in 1970. A relational database has two-dimensional structure, i.e. it is divided into rows and columns, where data is stored. Database management system is software that acts as a gatekeeper for the database. The concept of functional dependency and the concept of normal are also introduced by Codd. In 1977, Fagin extends the concept of functional dependency to concept of multifunctional dependency. This concept increases the conceptual design structure of the database [19]. This concept of fourth normal form is based on the concept of multi-valued dependencies. In 2008, Hongbo Li and Li Zhou have proposed the method for finding the all possible candidate keys of the relational database. The set of functional dependency and the canonical cover sets of functional dependency are used to construct the data structure called MAC. The MAC is defined as algorithm to find out the entire candidate key, and it is the fastest algorithm to compute the candidate key of the relation [11, 20]. In year 2009, Zhang Yi-Shun has discussed a method to find set of all candidate keys from given set of functional dependency using Karnaugh map. The K-map is concept that is used to minimize the Boolean algebra expression. Zhang Yi-Shun used the concept of K-map to find all possible candidate keys of the relation. Convert the set of functional dependency into counterpart one in logical algebra and then on applying Karnaugh map all possible candidate keys can be generated. The set of functional dependency is converted into sum of product and connected to OR gate, and then filled the Karnaugh map by one where the sum of product value is given in [12, 21]. Once the Karnaugh map cells are filled and then find the minimum set of the input variables. Finding all possible candidate keys of the relations is a fundamental and complex problem, but solving through Karnaugh map in logic algebra, it can find all possible candidate key, solved simple and complete. Warshall's algorithm for finding the transitive closure of the graph is procedure to find all paths from one node to another node directly or indirectly.

In Warshall's algorithm, the given graph is converted into adjacency matrix of the graph and then applying the algorithm to find the all possible paths. Warshall's algorithm works when the path is directed from one node to other node, and the graph should be not weighted graph. If there is a directed path from one node to another node, then put one in the adjacency matrix column or if there is no directed path from one node to another node, then put zero in the respected column. After determining the matrix from the graph [22], if the value in row of the matrix and the column of the matrix is one in $R(k-1)$, then the value remains same in the respective row and column as $R(k)$ [23]. If the value in row of the matrix and the column of the matrix is zero in $R(k-1)$, the value is updated to one in the respective row and column is $R(k)$. If and only if the values in the row of the matrix and the column of the matrix and the values in the column of the matrix and the row of the matrix are one in $R(k-1)$ [24].

3 Proposed Model for Determining Key

The proposed methodology is used to find out the set of candidate key of any relation of the database based on the given sets of the functional dependency. Functional dependency of a relation shows the various dependencies of the attribute among them; in the proposed method, we will use dependency directed graph representation of the functional dependency and directed graph matrix representation of the directed graph.

A. ***Dependency Directed Graph Diagram***

From the given semantics of the database, we generally get the functional dependency. It is generally referred as the mapping between the two or more attributes of the same relation in the database. Designing a dependency directed graph, we should follow the following rules.

i. Node of the dependency directed graph contains the left-hand side part of the functional dependency. Left-hand side part of the functional dependency should be any number of attribute.
ii. Each node of the dependency directed graph must have self-loop edge because functional dependency in left side has to be a subset of attribute and right side has to be subset of attribute.
iii. Draw the edge from one node of the dependency directed graph to another node according to the functional dependency given for the relation.

Example 1 A relation $R(A, B, C, D)$ has given set of functional dependency $AB \rightarrow C, C \rightarrow D$ and $D \rightarrow AB$.

Figure 1 is the dependency directed graph of the relation R.

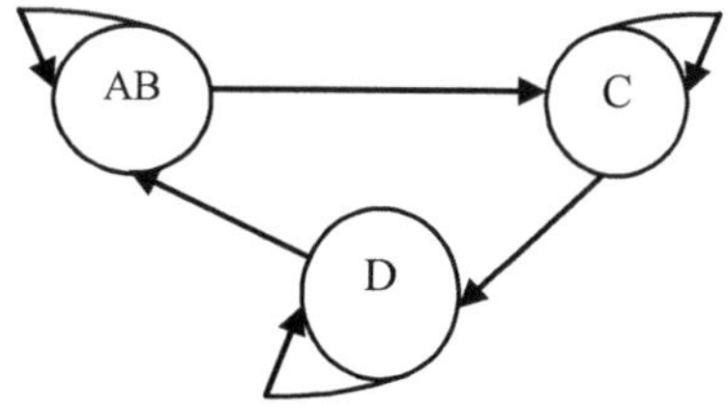

Fig. 1 Dependency directed graph for functional dependency

B. ***Dependency Directed Graph Matrix***

After designing the dependency directed graph from the given set of the functional dependency, now formed the dependency directed matrix with following rules.

i. The matrix is the combinations of rows and column. In the dependency directed matrix, the rows of the matrix represent the left-hand side part of the functional dependency.
ii. The column of the dependency directed matrix represents the list of all attributes of the database relation.
iii. After executing the above two steps, the number of row and column of the matrix should be same. It is possible that the number of row of the matrix is lesser than column. In their case, add dummy entry in row.
iv. If there is edge between two nodes or there is self-loop in the dependency directed graph, put 1 in dependency directed matrix, and put 0 if there is no path between two nodes.

A relation $R(A, B, C, D)$ has given set of functional dependency $AB \to C, C \to D$ and $D \to AB$, $D^{(0)}$ is a dependency directed matrix of Fig. 2.

B' is the dummy entry in the row of the Matrix $D^{(0)}$. There is no functional dependency which is defined for B.

C. ***Algorithm for Finding Key***

The algorithm that we are going to implement is Warshall's algorithm for computing the transitive closure of the dependency directed matrix. Applying Warshall's algorithm on dependency directed matrix and select first row and first column in $D^{(0)}$ matrix (Figs. 3, 4, 5 and 6).

Fig. 2 Dependency directed matrix of the relation

$$D^{(0)} = \begin{array}{c} \\ AB \\ B' \\ C \\ D \end{array} \begin{array}{c} \begin{array}{cccc} A & B & C & D \end{array} \\ \begin{bmatrix} 1 & 1 & 1 & 0 \\ 0 & 1 & 0 & 0 \\ 0 & 0 & 1 & 1 \\ 1 & 1 & 0 & 1 \end{bmatrix} \end{array}$$

Fig. 3 After first iteration, $D^{(1)}$ is formed

$$D^{(1)} = \begin{array}{c} \\ AB \\ B' \\ C \\ D \end{array} \begin{array}{c} \begin{array}{cccc} A & B & C & D \end{array} \\ \begin{bmatrix} 1 & 1 & 1 & 0 \\ 0 & 1 & 0 & 0 \\ 0 & 0 & 1 & 1 \\ 1 & 1 & 1 & 1 \end{bmatrix} \end{array}$$

Fig. 4 After second iteration, $D^{(2)}$ is formed

$$D^{(2)} = \begin{array}{c} \\ AB \\ B' \\ C \\ D \end{array} \begin{array}{c} \begin{array}{cccc} A & B & C & D \end{array} \\ \begin{bmatrix} 1 & 1 & 1 & 0 \\ 0 & 1 & 0 & 0 \\ 0 & 0 & 1 & 1 \\ 1 & 1 & 1 & 1 \end{bmatrix} \end{array}$$

Fig. 5 After third iteration, $D^{(3)}$ is formed

$$D^{(3)} = \begin{array}{c} \\ AB \\ B' \\ C \\ D \end{array} \begin{array}{c} \begin{array}{cccc} A & B & C & D \end{array} \\ \begin{bmatrix} 1 & 1 & 1 & 1 \\ 0 & 1 & 0 & 0 \\ 0 & 0 & 1 & 1 \\ 1 & 1 & 1 & 1 \end{bmatrix} \end{array}$$

Fig. 6 After fourth iteration, $D^{(4)}$ is formed

$$D^{(4)} = \begin{array}{c} \\ AB \\ B' \\ C \\ D \end{array} \begin{array}{c} \begin{array}{cccc} A & B & C & D \end{array} \\ \begin{bmatrix} 1 & 1 & 1 & 1 \\ 0 & 1 & 0 & 0 \\ 1 & 1 & 1 & 1 \\ 1 & 1 & 1 & 1 \end{bmatrix} \end{array}$$

The operations on four rows and columns are completed using Warshall's algorithm, and the $D^{(4)}$ matrix shows that all possible paths from one node to all other nodes. The final matrix $D^{(4)}$ shows that there is an indirect path form *AB, C and D* to all other nodes in the dependency directed graph. It means that the through *AB, C and D* node, we can reach all other nodes i.e. *AB, C and D* is set of all candidate key through which we can find all other attribute of the relation.

4 Conclusions

The algorithmic method in this paper is used to determine all possible candidate keys on any database relation. The set of functional dependency that has complex attributes or if there is no information about the attributes in the list of functional dependency sets can be easily solved by this proposed method. Once generating the dependency graph matrix of the functional dependency, after that it is very simple to determine the candidate key on the computer. We used Java environment to realize. The row of the matrix having all one indicates that there is path form this attribute to all other attributes. It means from this particular attribute, we can reach all their attributes. The candidate key attribute of the relation is an attribute or set of attributes from which we can determine all other attributes of the relation. The row of the matrix having all one is doing the same work. Therefore, we can consider a row of the dependency matrix having all one as a candidate key of the database relation.

References

1. Bahmani, Amir Hassan, Naghibzadeh, Mahmoud, Bahmani, Behnam, "Automatic database normalization and primary key generation", Canadian Conference on Electrical and Computer Engineering, Page No 11–16, 2008
2. Puangsaijai, Wittawat, Puntheeranurak, Sutheera, "A comparative study of relational database and key-value database for big data applications", 2017 International Electrical Engineering Congress, iEECON 2017
3. Demba, Moussa "Algorithm for Relational Database Normalization Up To 3Nf", International Journal of Database Management Systems (IJDMS), Vol 5, Issue 3, Page No 39-1, 2013
4. Abraham Silberschatz, Henry F. Korth, S. Sudarshan. Database system concepts [M]. Higher Education Press, 2006
5. Darwen, Hugh, Date, C. J., Fagin, Ronald, "A normal form for preventing redundant tuples in relational databases", Proceedings of the 15th International Conference on Database Theory - ICDT'12, Page 114, 2012
6. Vimala, S., Khanna Nehemiah, H., Saranya, G. and Kannan, A. "Applying Game Theory to Restructure PL/SQL Code", International Journal of Soft Computing, Vol. 7, No. 6, pp. 264–270, 2012
7. Zichen Xu, Yi-Cheng Tu, and Xiaorui Wang, Online Energy Estimation of Relational Operations in Database Systems, IEEE transactions on computers, vol. 64, no. 11, November 2015
8. Amir Hassan Bahmani, Mahmoud Naghibzadeh, Behnam Bahmani "Automatic database normalization and primary key generation" *IEEE CCECE/CCGEI May 5-7*, Niagara Falls, Canada, 2008
9. C. J. Date, A. Kannan, S. Swaminathan, "An Introduction to Database Systems", 8th Edition, Pearson Education (Dorling Kindersley (India) Pvt. Ltd.), 2008
10. Silberschatz, Abraham, Korth, H.F., Sudarshan, S., "Database System Concepts - 6th. ed.", Publication Database, Volume 4, pages 1376, 2011
11. Ali Muhammad Rushdi, Muhammad Ali Rushdi, "Switching-algebraic algorithmic derivation of candidate keys in relational databases", Emerging Trends in Communication Technologies (ETCT) International Conference on, pp. 1–5, 2016
12. Garcia-Molina, Hector, Ullman, Jeffrey D, Widom, Jennifer, Özsu, MT, Valduriez, P Connolly, Thomas, "Database Systems: A Practical Approach to Design, Implementation, and Management", International Journal of Computer Applications, Vol 49, Issue 4, Page 90–107, 2010
13. Chilson, David W., Kudlac, Michael E., "Database design", ACM SIGMIS Database, Vol 15, Issue 1, Pages 11–19, 1983

14. Djordjević-Kajan, Slobodanka, "Fundamentals of database systems", Microelectronics Journal, Vol 28, Issue 5, Pages 603–604, 1997
15. Letkowski, Jerzy, "Doing database design with MySQL", Western New England University, Vol 6, Issue 1,Pages 1–15,2014
16. Omiecinski, Edward R., "A Parallel Algorithm for Relational Database Normalization", IEEE Transactions on Parallel and Distributed Systems, Vol 1, Issue 4, Pages 415–423, 1990
17. Churcher, Clare, "Beginning database design: From novice to professional", Beginning Database Design: From Novice to Professional, Vol 9781430242109, Pages 1–225, 2012
18. Wesley, D, "Relational database design", Journal of insurance medicine (New York, N.Y.), Vol 32, Issue 2, Pages 63–70, 2000
19. Teorey, Toby J. Fry, James P., "The Logical Record Access Approach to Database Design", ACM Computing Surveys, Vol 12, Issue 4, Pages 465, 1980
20. Hellerstein, Joseph M. Stonebraker, Michael Hamilton, James, "Architecture of a Database System", Foundations and Trends® in Databases, Vol 1, Issue 2, Pages 141–259,2007
21. Qian, Li LeFevre, K Jagadish, Hv, "CRIUS: user-friendly database design", Proceedings of the VLDB Endowment, Pages 81–92, 2010
22. Lee, Heeseok, "Justifying database normalization: a cost/benefit model", Information Processing and Management, Vol 31, Issue 1, Pages 59–67, 1995
23. Beaubouef, Theresa Petry, Frederick E. Ladner, Roy, "Normalization in a rough relational database", Lecture Notes in Computer Science (including subseries Lecture Notes in Artificial Intelligence and Lecture Notes in Bioinformatics), Vol 3641 LNAI, Pages 275–282, 2005
24. Al-Absi, Ahmed Abdulhakim Kang, Dae Ki, "Relational database normalization algorithm: A TAM database analyst technique", Information (Japan), Vol 17, Issue 17A, Pages 3223–3228, 2014

Mr. Kunal Kumar is currently working as Research Scholar toward the PhD degree at the University Department of Statistics and Computer Applications. Tilka Manjhi Bhagalpur University, Bhagalpur, India. He received the Master of computer applications and Bachelor of computer applications degree from the Birla Institute of Technology, Mesra, Ranchi, India in year 2012 and 2008. His area of research interests includes Database theory, Database Design and Big Data analytics.

Dr. Sachindra Kumar Azad is an Associate Professor and Head, in University Department of Statistics and Computer Applications at Tilka Manjhi Bhagalpur University, Bhagalpur, India. His area of research interests includes Database theory, Database Design and Big Data analytics.

PID Controller Tuning in Smith Predictor Configuration for Stable Processes with Large Time Delay Using IMC Scheme

Md Nishat Anwar, Somnath Pan and Ashraf Raza

1 Introduction

The PID controller is having long history and is widely implemented in process industry either in its basic form or with some modification. The problem arises when the time delay of the process is significantly more in comparison with its dominant time constant. In such cases, the closed-loop system performs poorly and it is hard to control through traditional controller in simple control architecture [1, 2]. The Smith predictor control scheme [3] is effective and accepted for long dead-time processes where the controller is designed with consideration of the delay-free part of the process. The set-point performance of Smith predictor is good, but load disturbance rejection is poor especially when the processes having very slow dynamics [4, 5]. Astrom et al. [6] proposed a "modified Smith predictor scheme with two degrees of freedom" to consider for both the set-point and the load disturbance performances separately. Numerous modified Smith predictor schemes have been proposed in the literature to achieve improved closed-loop performance [7–11].

The internal model control (IMC) scheme [12, 13] is a simple, attractive and powerful design method which offers only one tuning parameter to adjust the tradeoff between the performance and the robustness of the system. An equivalent unity negative feedback controller from the IMC scheme is obtained and further converted into a PID controller by various methods [14–16].

M. N. Anwar · A. Raza (✉)
Department of Electrical Engineering, National Institute of Technology,
Ashok Rajpath, Patna 800005, Bihar, India
e-mail: ashrafraza510@gmail.com

S. Pan
Department of Electrical Engineering, Indian School of Mines,
Dhanbad 826004, India

J. Chattopadhyay et al. (eds.), *Innovations in Soft Computing and Information Technology*, https://doi.org/10.1007/978-981-13-3185-5_22

A modified Smith predictor scheme is proposed by Kaya in [17] to control the processes with long dead time to achieve enhanced responses for set-point as well as load disturbance. In [18], an automatic tuning method for PID controller through IMC technique in Smith predictor configuration has been proposed. The method is proposed for the first and second orders with dead-time processes and for higher-order processes, and the method is applied after reducing the process into low-order process.

An IMC scheme in the Smith predictor configuration is used in this paper to design PID controller for large dead-time processes. The IMC controller is designed by obtaining the IMC equivalent of the Smith predictor configuration which gives the Smith predictor controller and is further simplified to PID controller by approximate frequency response matching. The simulation of various examples demonstrates the effectiveness of the method.

The remaining paper is organized as follows. "The design method is discussed in Sect. 2 and simulation of various examples in Sect. 3." Section 4 gives the conclusion for the proposed work.

2 The Design Method

The control scheme of Smith predictor is used to control the processes that are having very large dead-time dynamics compared to their time constants. The Smith predictor scheme is shown in Fig. 1, where, $G_P(s)$ is the plant, $G_C(s)$ is the controller, l is the time delay of the process, $r(t)$ is the input and $y(t)$ is the output. $F(s)$ is used to improve the set-point performance. The process is considered as:

$$G_P(s) = g(s)e^{-ls} \tag{1}$$

where, $g(s)$ is the part of the process free from delay. The model of the process is considered as

$$G_M(s) = g_m(s)e^{-ls} \tag{2}$$

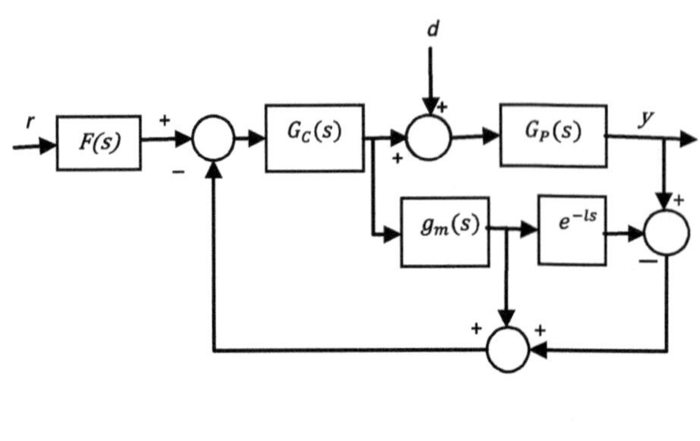

Fig. 1 Control scheme of Smith predictor

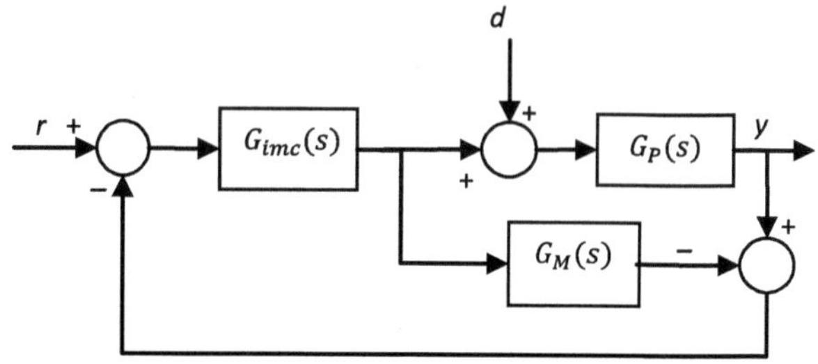

Fig. 2 IMC control scheme

where, $g_m(s)$ is the process model delay-free part. The transfer function from r to y may be obtained as:

$$G_{r,y,\text{smith}}(s) = \frac{F(s)G_C(s)G_P(s)}{1 + G_C(s)[g_m(s) + g(s)e^{-ls} - g_m(s)e^{-ls}]} \tag{3}$$

For perfect modelling of the process, i.e., $G_M(s) = G_P(s)$, the denominator of the closed-loop transfer function (CLTF) in the Smith predictor scheme becomes free of the delay term as shown below:

$$G_{r,y,\text{smith}}(s) = \frac{G_C(s)G_P(s)}{1 + G_C(s)g_m(s)} = \frac{G_C(s)g_m(s)e^{-ls}}{1 + G_C(s)g_m(s)} \tag{4}$$

The IMC scheme is shown in Fig. 2 where, $G_{\text{imc}}(s)$ is the IMC controller, $G_P(s)$ is the process and $G_M(s)$ is the model of the process. In case of perfect modelling of the process, the IMC scheme becomes open loop. At this condition, there is no stability problem for a stable process if the controller is stable and the controller may be designed relatively easily for the stable open-loop system. The disturbances or model mismatch can be taken care of by the feedback of the control scheme.

The overall transfer function in the IMC scheme from r to y is given by

$$G_{r,y,\text{imc}}(s) = \frac{G_{\text{imc}}(s)G_P(s)}{1 + [G_P(s) - G_M(s)]G_{\text{imc}}(s)} \tag{5}$$

In case of perfect modelling of the process, the CLTF may be written as Eq. (6).

$$G_{r,y,\text{imc}}(s) = G_{\text{imc}}(s)G_P(s) = G_{\text{imc}}(s)g_m(s)e^{-ls} \tag{6}$$

According to the IMC scheme, $G_M(s)$ is splitted into two parts as:

$$G_M(s) = g_m^+(s)g_m^-(s) \tag{7}$$

where, $g_{\mathrm{m}}^{-}(s)$ and $g_{\mathrm{m}}^{+}(s)$ are the invertible (i.e., left-hand side poles and zeros with process gain) and non-invertible (non-minimum phase part including the time delay term) parts, respectively.

The IMC controller is designed by choosing

$$G_{\mathrm{imc}}(s) = \frac{1}{g_{\mathrm{m}}^{-}(s)} f(s) = \frac{1}{g_{\mathrm{m}}^{-}(s)(\lambda s+1)^{n}} \quad (8)$$

and thus the CLTF becomes (at the condition of perfect modelling of the process):

$$G_{r,y,\mathrm{imc}}(s) = \frac{g_{\mathrm{m}}^{+}(s)}{(\lambda s+1)^{n}}$$

where, $f(s) = \frac{1}{(\lambda s+1)^{n}}$ is a filter in the IMC scheme with two objectives: (*i*) to make $G_{\mathrm{imc}}(s)$ proper and realizable by appropriate selection of the positive integer *n;* and (*ii*) the system response is governed by the filter time constant λ. Here, the only tuning parameter λ is tuned to adjust the speed of the response against the robustness of the closed-loop performance.

Now, with block-algebra manipulation, the Smith predictor scheme in Fig. 1 may be rearranged as shown in Fig. 3. For the same input–output relationship, i.e., for $G_{r,y,\mathrm{smith}}(s) = G_{r,y,\mathrm{imc}}(s)$. Fig. 3 is compared with Fig. 2 to give rise to the relation that $G_C(s)$ along with the feedback element $G_{\mathrm{m}}(s)$ becomes equal to $G_{\mathrm{imc}}(s)$.

Hence,

$$G_{\mathrm{imc}}(s) = \frac{G_{\mathrm{C}}(s)}{1+G_{\mathrm{C}}(s)g_{\mathrm{m}}(s)} \quad \text{or} \quad G_{\mathrm{C}}(s) = \frac{G_{\mathrm{imc}}(s)}{1-G_{\mathrm{imc}}(s)g_{\mathrm{m}}(s)} = \frac{1}{g_{\mathrm{m}}^{-}(s)(\lambda s+1)^{n} - g_{\mathrm{m}}(s)}.$$

The order of $G_{\mathrm{C}}(s)$ would be high and usually may not be preferred for practical implementation. Hence, it is intended to convert the Smith controller $G_{\mathrm{C}}(s)$ into a PID controller $G_{\mathrm{C}}^{\mathrm{PID}}(s)$ of the following structure.

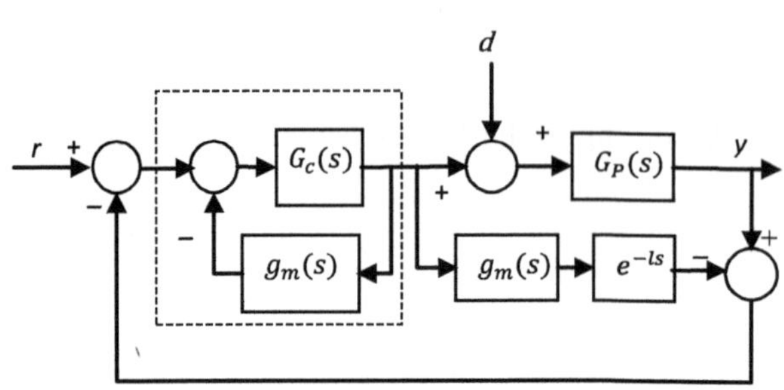

Fig. 3 Scheme of Smith predictor in Fig. 1 is rearranged

$$G_{\mathrm{C}}^{\mathrm{PID}}(s) = K_{\mathrm{P}} + \frac{K_{\mathrm{I}}}{s} + K_{\mathrm{D}}s$$

where, "K_{P}, K_{I} and K_{D} are the proportional, the integral and the derivative gains, respectively." For designing a PID controller from $G_{\mathrm{C}}(s)$, the frequency response matching method [19, 20] has been considered. Hence, the following expression may be written:

$$G_{\mathrm{C}}^{\mathrm{PID}}(s)\big|_{s=j\omega} \cong G_{\mathrm{C}}(s)\big|_{s=j\omega} \tag{9}$$

where, the above expression (Eq. 9) is equivalent in terms of frequency response.

Equation (9) may also be written in terms of real and imaginary parts as:

$$G_{\mathrm{CR}}^{\mathrm{PID}}(\omega) + jG_{\mathrm{CI}}^{\mathrm{PID}}(\omega) \cong G_{\mathrm{CR}}(\omega) + jG_{\mathrm{CI}}(\omega) \tag{10}$$

where,

$$G_{\mathrm{C}}^{\mathrm{PID}}(s)\big|_{s=j\omega} = G_{\mathrm{CR}}^{\mathrm{PID}}(\omega) + jG_{\mathrm{CI}}^{\mathrm{PID}}(\omega)$$

and $G_{\mathrm{C}}(s)\big|_{s=j\omega} = G_{\mathrm{CR}}(\omega) + jG_{\mathrm{CI}}(\omega)$.

Here, $G_{\mathrm{CR}}^{\mathrm{PID}}(\omega), G_{\mathrm{CI}}^{\mathrm{PID}}(\omega), G_{\mathrm{CR}}(\omega)$ and $G_{CI}(\omega)$ are all real functions of ω. Separating the real and the imaginary parts, Eq. (10) may be written as:

$$\begin{aligned} G_{\mathrm{CR}}^{\mathrm{PID}}(\omega) &\cong G_{\mathrm{CR}}(\omega) \\ \text{and,}\quad G_{\mathrm{CI}}^{\mathrm{PID}}(\omega) &\cong G_{\mathrm{CI}}(\omega) \end{aligned} \tag{11}$$

Expanding both the expressions of Eq. (11) by Taylor series expansion about $\omega = 0$ and equating the initial N derivatives of the corresponding functions, the following equations may be obtained.

$$\frac{\mathrm{d}^p}{\mathrm{d}\omega^p}\left[G_{\mathrm{CR}}^{\mathrm{PID}}(\omega)\right]\bigg|_{\omega=0} = \frac{\mathrm{d}^p}{\mathrm{d}\omega^p}\left[G_{\mathrm{CR}}(\omega)\right]\bigg|_{\omega=0} \tag{12}$$

$$\frac{\mathrm{d}^p}{\mathrm{d}\omega^p}\left[G_{\mathrm{CI}}^{\mathrm{PID}}(\omega)\right]\bigg|_{\omega=0} = \frac{\mathrm{d}^p}{\mathrm{d}\omega^p}\left[G_{\mathrm{CI}}(\omega)\right]\bigg|_{\omega=0} \tag{13}$$

where, where, $p \in [0, N-1]$

Now, using the results of the divided difference calculus [21, 22], $G_{\mathrm{CR}}^{\mathrm{PID}}(\omega)$ approximately matches $G_{\mathrm{CR}}(\omega)$, if

$$G_{\mathrm{CR}}^{\mathrm{PID}}(\omega)\big|_{\omega=\omega_p} = G_{\mathrm{CR}}(\omega)\big|_{\omega=\omega_p}; \quad p \in [0, N-1] \tag{14}$$

and

$$G_{\mathrm{CI}}^{\mathrm{PID}}(\omega)\big|_{\omega=\omega_p} = G_{\mathrm{CI}}(\omega)|_{\omega=\omega_p}; \quad p \in [0, N-1] \tag{15}$$

where ω_p are very small positive values near zero. From Eqs. (14) and (15) at least two frequency points are needed for evaluating the three unknown parameter of the PID controller. Therefore for frequency points ω_0 and ω_1 around $\omega = 0$, the following equation is obtained.

$$\boldsymbol{A}\bar{x} = \bar{b} \tag{16}$$

where,

$$\boldsymbol{A} = \begin{bmatrix} 1 & 0 & 0 \\ 0 & -\frac{1}{\omega_0} & \omega_0 \\ 1 & 0 & 0 \\ 0 & -\frac{1}{\omega_1} & \omega_1 \end{bmatrix}; \quad \bar{x} = [K_P \quad K_I \quad K_D]^{\mathrm{T}};$$

and, and, $\bar{b} = [G_{CR}(\omega_0) \quad G_{CI}(\omega_0) \quad G_{CR}(\omega_1) \quad G_{CI}(\omega_1)]^T$

The solution of Eq. (16) will give the parameters of the PID controller and thus the PID controller is evaluated.

- **Selection of low-frequency points**

As from Eqs. (12) and (13), the matching of frequency response at $\omega \to 0$ will yield better controller. In this regard, frequency points for matching have been selected around 0.001 times of the bandwidth frequency. The effectiveness of the approximation is shown in Fig. 4 by considering the process $G_1(s)$ and the close matching of the Smith predictor controller $G_C(s)$ and the PID controller $G_C^{\mathrm{PID}}(s)$ in the low-frequency region is observed.

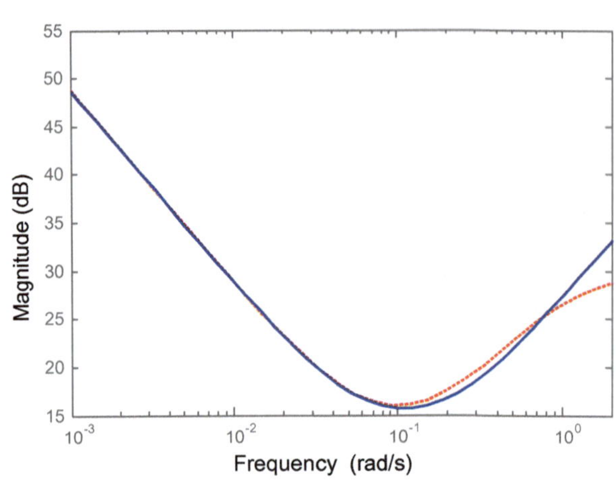

Fig. 4 Bode magnitude plots of the controllers for the process $G_1(s)$: *Solid line*—the PID controller, *Dashed line*—the Smith predictor controller

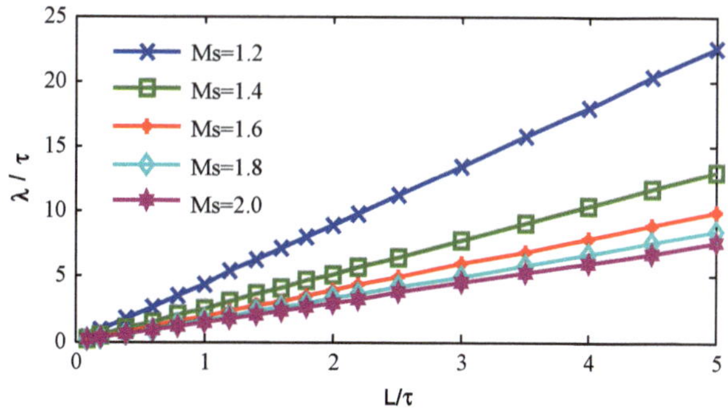

Fig. 5 λ guidelines for FOPDT process

- **Selection of filter time constant λ**

The tuning parameter λ in the proposed method is related to the performance and the robustness of the system. A smaller value of λ results in fast response with poor robustness and vice versa. For analysing the effect of λ on the robustness of the system, an FOPDT model is considered as below:

$$G_P(s) = \frac{1e^{-ls}}{\tau s + 1}, \quad \tau = 1 \tag{17}$$

The value of λ has been obtained for different time delay of the system l for a fixed value of robustness level (M_s). Figure 5 shows the guideline for selection of λ for the FOPDT model for different M_s values where each curve is a plot of λ/τ versus l/τ ratios.

3 Simulation Results

The performance of the proposed method is illustrated through various examples from the literature [18, 23] having dead-time dominant dynamics. The processes considered along with the designed IMC controllers and the selected frequency points for matching are shown in Table 1. The method has been compared with the

Table 1 Processes and IMC controllers

Process, $G_P(s)$	IMC controller, $G_{imc}(s)$	Frequency points ω_0, ω_1
$G_1(s) = \frac{e^{-10s}}{(17s+1)(6s+1)}$ [18]	$\frac{(17s+1)(6s+1)}{(1.8s+1)^2}$	0.003, 0.006
$G_2(s) = \frac{e^{-10s}}{(s+1)^2}$ [18]	$\frac{(s+1)^2}{(2.4s+1)^2}$	0.002, 0.004
$G_3(s) = \frac{e^{-20s}}{(3s+1)(2s+1)(s+1)(0.5s+1)}$ [18]	$\frac{(3s+1)(2s+1)(s+1)(0.5s+1)}{(1.19s+1)^4}$	0.003, 0.006
$G_4(s) = \frac{e^{-8s}}{(2s+1)^3(s+1)^2}$ [23]	$\frac{(2s+1)^3(s+1)^2}{(0.96s+1)^5}$	0.004, 0.008

well-known methods. The proposed method may be evaluated by considering the following parameters:

- Maximum sensitivity (M_s): It is considered for the robustness of the control system and is defined as: $M_s = \max_{0 \le \omega \le \infty} |1/1 + C(j\omega)G_P(j\omega)|$ where, $C(s)$ is the controller in unity negative feedback configuration. In Smith predictor configuration, the M_s is evaluated from its unity negative feedback configuration equivalent. Smaller M_s value in the range of 1.2–2.0 [24] is preferred.
- Integral absolute error (*IAE*): It is an indication of the performance measure of the step-response and is given by:"$\text{IAE} = \int_0^\infty |e(t)|\text{d}t$ where, $e(t)$ is the error signal."
- Total variation (*TV*): It is the measure of the manipulated variable $u(t)$. "It is defined as $\text{TV} = \sum_{i=1}^{\infty} |u_{i+1} - u_i|$" where, u_i is the discretized manipulated variable. Lower is the value of *TV* better is the smoothness of the controller output [25].

A unit step input in both the set-point and the load disturbance is applied for evaluation of the control system performance and also compared with some other well-known methods. The PID parameters, the performance and the robustness measures are listed in Table 2. The $F(s)$ is considered for the improved set-point response. The simulation results of processes G_1 and G_4 are shown in Figs. 6, 7, 8

Table 2 PID parameters and performance comparison

System	Method	K_P	K_I	K_D	M_s	Set-point response		Load disturbance response	
						IAE	TV	IAE	TV
$G_1(s)$	Proposed[a]	6.13	0.276	22.73	1.6	14.58	34.09	13.59	2.02
	Kaya [18]	2.3	0.10	10.18	5.0	23.14	3.59	19.97	2.00
$G_2(s)$	Proposed	0.1619	0.206	0.009	1.46	29.72	3.0	17.85	1.9
	Kaya [18]	0.40	0.20	0.20	1.46	30.0	3.0	14.85	2.0
	Benouarets and Atherton [26]	0.47	0.453	0.11	23.15	24.80	3.21	12.34	2.06
$G_3(s)$	Proposed[b]	0.98	0.20	0.86	1.60	25.87	2.3	24.78	2.02
	Kaya [18]	0.527	0.0951	0.729	1.60	30.51	2.0	30.41	2.0
	Benouarets and Atherton [26]	1.873	0.35	2.51	2.17	24.66	4.26	22.85	2.06
$G_4(s)$	Proposed[c]	1.25	0.21	2.35	1.60	16.24	2.27	15.50	2.02
	Yang et al. [23]	0.63	0.065	1.74	1.95	18.70	1.71	15.51	1.24
	Maclaurin-PID [15]	0.58	0.065	1.59	1.84	18.69	1.52	15.58	1.13

[a] $F(s) = 1/(s+1)$
[b] $F(s) = 1/(s+1)$
[c] $F(s) = 1/(s+1)$

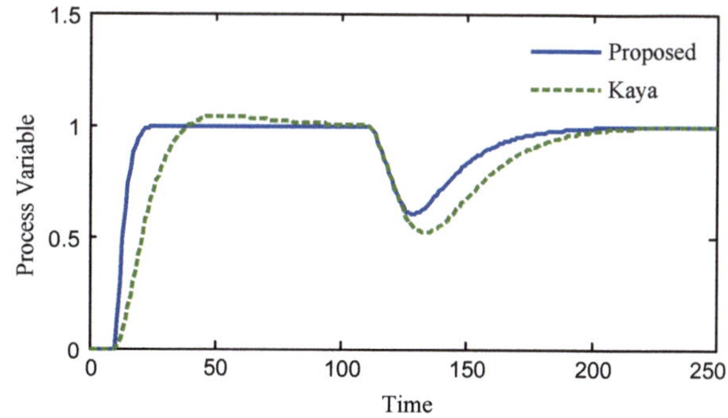

Fig. 6 Response of the system $G_1(s)$, set-point input at $t = 0$ s and disturbance input at $t = 100$ s

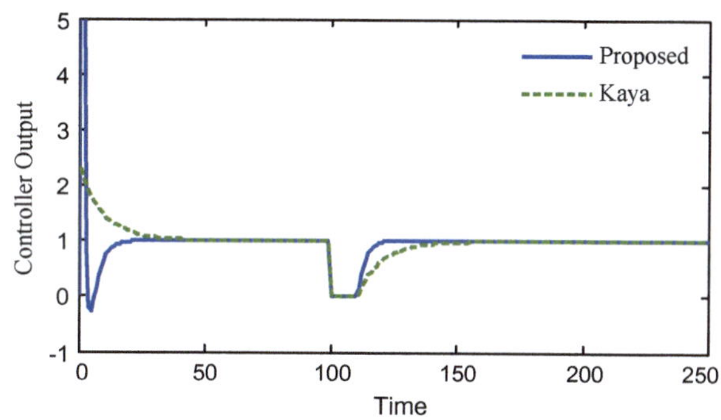

Fig. 7 Controller output of the nominal process $G_1(s)$

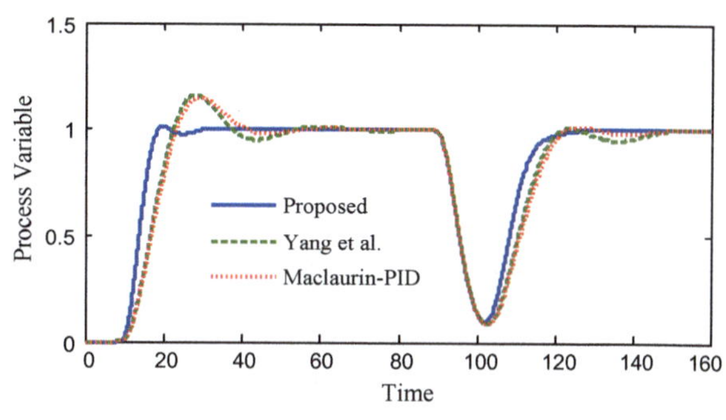

Fig. 8 Response of the system $G_4(s)$, set-point input at $t = 0$ s and disturbance input at $t = 80$ s

and 9, and similar observations have been found for other examples also. It shows that the proposed method gives the better response in most of the cases.

The proposed controller has been analysed for robustness by inserting uncertainty (10% perturbations) in the dead time of the processes, and the simulation result for the process G_1 is shown in Fig. 10.

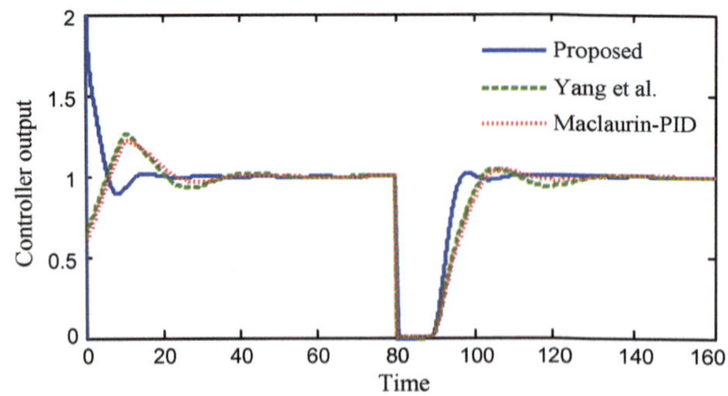

Fig. 9 Controller output of the system $G_4(s)$

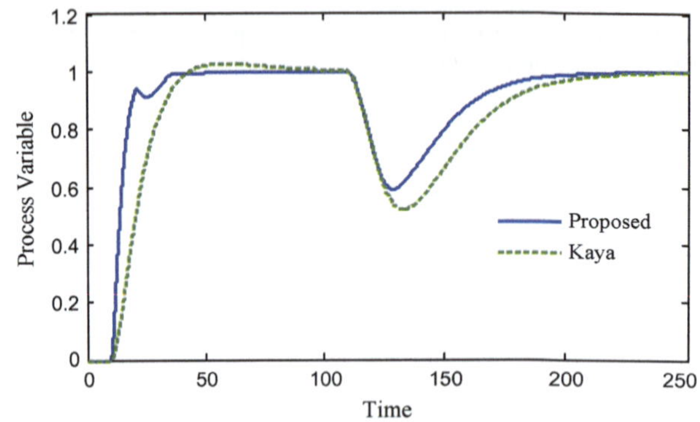

Fig. 10 Response of the perturbed system $G_1(s)$, set-point input at $t = 0$ s and disturbance input at $t = 100$ s

4 Conclusions

A PID controller design method in Smith predictor configuration through IMC scheme has been presented for processes with long dead time. Here from, an equivalent Smith predictor controller is obtained which is further approximated to a PID one by frequency response matching. The method is simple, and only two low-frequency points are needed for matching purpose and is applicable to high-order as well as low-order processes with large dead time without reduction of the process model. Efficacy of the method against mathematical simplicity along with very small computational burden has been shown through some examples and comparing favourably with some prevalent methods.

Acknowledgements The authors acknowledge the support from Department of Science and Technology, India (under the project grant no: ECR/16/001547) at NIT Patna, India.

References

1. Meyer C, Seborg DE (1976) Chem Eng Scie 31:775–778.
2. Hagglund T (1992) IEEE Control System Magazine 12(1):57–60.
3. Smith OJM (1959) ISA J 6(2):28–33.
4. Hang CC, Wong FS (1979) Proc ISA Annual Conf:33–44.
5. Watanabe K, Ito M (1981) IEEE trans Auto Contr 26(6):1261–1266.
6. Astrom KJ, Hang CC, Lim BC (1994) IEEE Trans Auto Contr 39(2):343–345.
7. Matausek MR, Micic AD (1996) IEEE Trans Auto Cont 41(8):1199–1203.
8. Kwak HJ, Whan S, Lee IB (2001) Ind Eng Chem Res 40:1500–1506.
9. Hang CC, Wang QG, Yang XP (2003) Ind Eng Chem Res 42:484–489.
10. Uma S, Chidambaram M, Rao AS (2010) Chem Eng Res and Des 88:592–601.
11. Padhan DG, Majhi S (2011) Electronics Letters 47(17): 959–961.
12. Gracia CE, Morari M (1982) Ind Eng Chem Process Des Dev 21:308–323.
13. Morari M, Zafiriou E (1989) Robust process control. New Jersey: Prentice-Hall Inc.
14. Wang QG, Hang CC, Yang XP (2001) Automatica 37:2041–2048.
15. Li Y, Park S, Lee M, Brosilow C (1998) AIChE Journal 44(1):106–115.
16. Panda RC (2008) Ind Eng Chem Res 47:8684–8692.
17. Kaya I (2003) ISA Trans 42:101–110.
18. Kaya I (2004) Computer and Cheminical Engineering 28:281–290.
19. Anwar MN, Pan S (2015) ISA Transactions 55:175–187.
20. Anwar MN, Shamsuzzoha M, Pan S (2015) Arabian Journal of Science and Engineering 40:995–1004.
21. Pal J (1986) International Journal of Control 43(1):257–269.
22. Pan S, Pal J (1995) Appl Math Modelling 19:133–138.
23. Yang X, Xu B, Chiu MS (2011) Ind Eng Chem Res 50:1352–1359.
24. Astrom KJ, Hagglund T (1995) PID Controllers Theory Design and Tuning, 2nd ed. North Carolina: Instrument Society of America, Research Triangle Park.
25. Skogestad S (2003) Journal of Process Control 13:291–309.
26. Benouarets M, Atherton DP (1994) UKACC International Conference on Control: 795–800.

Md Nishat Anwar He obtained his bachelor and master degree in Electrical Engineering from AMU, Aligarh, India. He obtained his Ph.D. degree in Electrical Engineering from IIT (ISM), Dhanbad, India in 2015. He has teaching experience of 4 years and currently working as an Assistant Professor in the department of Electrical Engineering, NIT Patna, India. His research interests include industrial control and automation, PID controller, model predictive control etc. and has several masters and Ph.D. students working under him. His publication includes 4 papers in referred journals, 12 papers in international conferences and 5 papers in national conferences.

Somnath Pan He received his B.E. and M.E. degree in Electrical Engineering from Jadavpur University, Kolkata, India and Ph. D. degree from IIT, Kharagpur, India. He has almost 23 years of teaching and research experience and currently working as Professor in the department of Electrical Engineering, IIT, Dhanbad, India. His research interest includes reduced order modelling & controller design, process control and industrial automation etc. He has several papers published in referred journals and conferences and guided several masters and Ph.D. students.

Ashraf Raza Received bachelor degree in Electrical Engineering from AMU, Aligarh, India, in 2010 and master degree from IIT, Roorkee, India, in 2012 in the specialization Electric drives and power electronics. He has 3 years teaching experience and currently pursuing Ph.D. from NIT Patna in Electrical Engineering Department. The research interest includes process control, PID control for time delay system, control of integrating and unstable process etc.

A Simple, Secure, and Time-Efficient Bit-Plane Operated Bit-Level Image Encryption Scheme Using 1-D Chaotic Maps

K. Abhimanyu Kumar Patro and Bibhudendra Acharya

1 Introduction

Due to the advancements in Internet technology, large numbers of images are transmitted over the internet. As a result, privacy and security of images have become a major issue in the present scenario. Encryption is one of the well-known techniques to secure images against various attacks. Though various traditional encryption techniques [1, 2] are present, they are not much suitable to encrypt images because of some image in-built characteristics such as high-correlation of adjacent pixels, bulky data, high redundancy, etc [3–6].

Chaos-based image encryption provides a better security to the images. Chaotic maps have certain good characteristics such as sensitivity of their parameters, non-periodicity, ergodicity, nonlinearity [7, 8]. Matthews [9] first proposed chaos-based encryption technique, since then, many encryption techniques are used chaotic maps in image encryption. Chaotic maps are classified into high-dimensional and one-dimensional (1-D) [10]. High-dimensional chaotic maps have the disadvantage of complex structure and therefore requires high amount of hardware resources for implementation. Complexity can be reduced using 1-D chaotic maps which are simple in structure with low hardware resource requirement and high efficiency [11]. Hence, the proposed algorithm uses 1-D chaotic maps to perform image encryption. A number of researches have been done using 1-D chaotic maps. Multiple 1-D chaotic mapbased image encryption is proposed by Ahmed et al. in [12]. A variable-size hash function is developed using multiple chaotic maps by Musheer Ahmad et al. in [13].

K. A. K. Patro (✉) · B. Acharya
Department of Electronics and Telecommunication Engineering, National Institute of Technology Raipur, Raipur 492010, Chhattisgarh, India
e-mail: abhimanyu.patro@gmail.com

B. Acharya
e-mail: bacharya.etc@nitrr.ac.in

J. Chattopadhyay et al. (eds.), *Innovations in Soft Computing and Information Technology*, https://doi.org/10.1007/978-981-13-3185-5_23

This paper contributes the followings.

1. To reduce the total simulation time and also to make the algorithm simple, two stages of bit-plane diffusion operation and bit-level column–row shuffling operation are presented.
2. The Beta map along with the Piece-wise Linear Chaotic Map (PWLCM) system is used in this algorithm to obtain large key space and high key sensitivity.
3. Hash-based key operations are performed to prevent the algorithm against chosen-plaintext attack (CPA) and known-plaintext attack (KPA).

The remaining sections are as follows. Section 2 describes the 1-D chaotic maps. Section 3 presents the proposed key generation and encryption algorithm. The security analyses and the computer simulations are presented in Sect. 4. At last, the conclusion is reached in Sect. 5.

2 One-Dimensional Chaotic Maps

2.1 Beta Map

The Beta function [14] is defined as,

$$\text{Beta}(x;\alpha,\beta,x_1,x_2)=\begin{cases}\left(\frac{x-x_1}{x_0-x_1}\right)^{\alpha}\left(\frac{x_2-x}{x_2-x_0}\right)^{\beta} & \text{if } x\in\,]x1,x2[\\ 0 & \text{else}\end{cases}\tag{1}$$

where α,β,x_1 and $x_2\in R, x_1<x_2$ and x_0,

$$x_0=\frac{(\alpha x_2+\beta x_1)}{(\alpha+\beta)}\tag{2}$$

By using the Beta function, the Beta map is defined as,

$$x_{n+1}=k\times\text{Beta}(x_n;x_1,x_2,\alpha,\beta)\tag{3}$$

where

$$\alpha=p_1+q_1\times r\tag{4}$$

and

$$\beta=p_2+q_2\times r\tag{5}$$

where k is the control parameter, r is the bifurcation parameter, and (p_1,q_1,p_2,q_2) are the constants.

2.2 PWLCM System

Recently, in image encryptions, PWLCM systems are widely used because of its good characteristics such as simpler representation, efficient implementation, excellent ergodicity, and good dynamical behavior [15, 16].

The PWLCM system is defined as,

$$x_{n+1} = \begin{cases} \frac{(x_n)}{(\mu)} & \text{if } 0 \le x_n < \mu \\ \frac{(x_n - \mu)}{(0.5-\mu)} & \text{if } \mu \le x_n < 0.5 \\ (1 - x_n) & \text{if } 0.5 \le x_n < 1 \end{cases} \tag{6}$$

where μ lies in the range (0, 0.5).

3 Proposed Methodology

3.1 Key Generation Operation

The steps for key generation are

Step 1: Input an original grayscale image, *I*.
Step 2: Generate 32-decimal values by converting 256-bit hash values of original image. The generated hash values are represented as,

$$\text{hash} = h1, h2, h3, \ldots, h31, h32 \tag{7}$$

Step 3: Using the hash values, the secret keys are generated as,

$$\begin{cases} \text{mue} = \text{mue1} + (\text{mod}(\text{sum}(h1 : h2), 256)/2^9) \times 0.1 \\ x1(1) = x(1) + (\text{mod}(\text{sum}(h3 : h5), 256)/2^9) \times 0.1 \end{cases} \tag{8}$$

where $(\text{mue1}, \text{mue})$ are the original and generated system parameters, $(x(1), x1(1))$ are the original and generated initial values of PWLCM system only.

$$\begin{cases} xxx1(1) = xxx(1) + (\text{mod}(\text{sum}(h6 : h8), 256)/2^9) \times 0.1 \\ k = kk + (\text{mod}(\text{sum}(h9 : h11), 256)/2^9) \times 0.1 \\ x1 = xx1 + (\text{mod}(\text{sum}(h12 : h14), 256)/2^9) \times 0.1 \\ x2 = xx2 + (\text{mod}(\text{sum}(h15 : h17), 256)/2^9) \times 0.1 \\ p1 = pp1 + (\text{mod}(\text{sum}(h18 : h20), 256)/2^9) \times 0.1 \\ p2 = pp2 + (\text{mod}(\text{sum}(h21 : h23), 256)/2^9) \times 0.1 \\ q1 = qq1 + (\text{mod}(\text{sum}(h24 : h26), 256)/2^9) \times 0.1 \\ q2 = qq2 + (\text{mod}(\text{sum}(h27 : h29), 256)/2^9) \times 0.1 \\ r = rr + (\text{mod}(\text{sum}(h30 : h32), 256)/2^9) \times 0.1 \end{cases} \tag{9}$$

where $(xxx(1), xxx1(1))$ are the original and generated initial values of Beta map only. $(kk, xx1, xx2, pp1, pp2, qq1, qq2, rr)$ and $(k, x1, x2, p1, p2, q1, q2, r)$ are the original and generated system parameters of Beta map only.

3.2 Encryption Operation

Figure 1 shows the block diagram of the encryption operation of the proposed cryptosystem. The encryption operation is as follows:

Step 1: Input an $M \times N$ sized grayscale image, I.
Step 2: Generate bit-planes of the image, I. The generated 8 bit-planes are,

$$IBBP1, IBBP2, IBBP3, IBBP4, IBBP5, IBBP6, IBBP7, IBBP8$$

Step 3: Generate secret keys $(\text{mue}, x1(1))$ of PWLCM system as Step 3 of Sect. 3.1.
Step 4: Iterate the PWLCM system of Eq. (6) for $200 + (M \times N \times 8)$ times. To avoid transit effect, remove 200 iterations from the first position. Therefore, the generated PWLCM sequence is,

$$x1 = (x1(1), x1(2), \ldots, x1(M \times N \times 8)) \tag{10}$$

Step 5: Using the sequence $x1$ of PWLCM system, generate a key image of size $M \times N$. The key image generation process is as follows:

$$\begin{cases} \textit{for } i = 1:\ (M \times N \times 8) \\ \quad \textit{if } x1(i) < 0.5 \\ \qquad x1(i) = 0; \\ \quad \textit{else} \\ \qquad x1(i) = 1; \\ \quad \textit{end} \\ \textit{end} \\ K = \textit{reshape}(x1, [(M \times N), 8]); \\ C = \textit{bi2de}(K); \\ D = \textit{reshape}(C, [M, N]); \end{cases} \tag{11}$$

where D is the generated key image. Here, the threshold value is taken as 0.5.
Step 6: Generate bit-planes of the key image, D. The generated 8 bit-planes are, $KBBP1, KBBP2, KBBP3, KBBP4, KBBP5, KBBP6, KBBP7, KBBP8$
Step 7: Perform Stage-1 of bit-plane diffusion operation between bit-planes of the original image and the key image. Let the diffused bit-plane outputs are, $IKBP1, IKBP2, IKBP3, IKBP4, IKBP5, IKBP6, IKBP7, IKBP8$

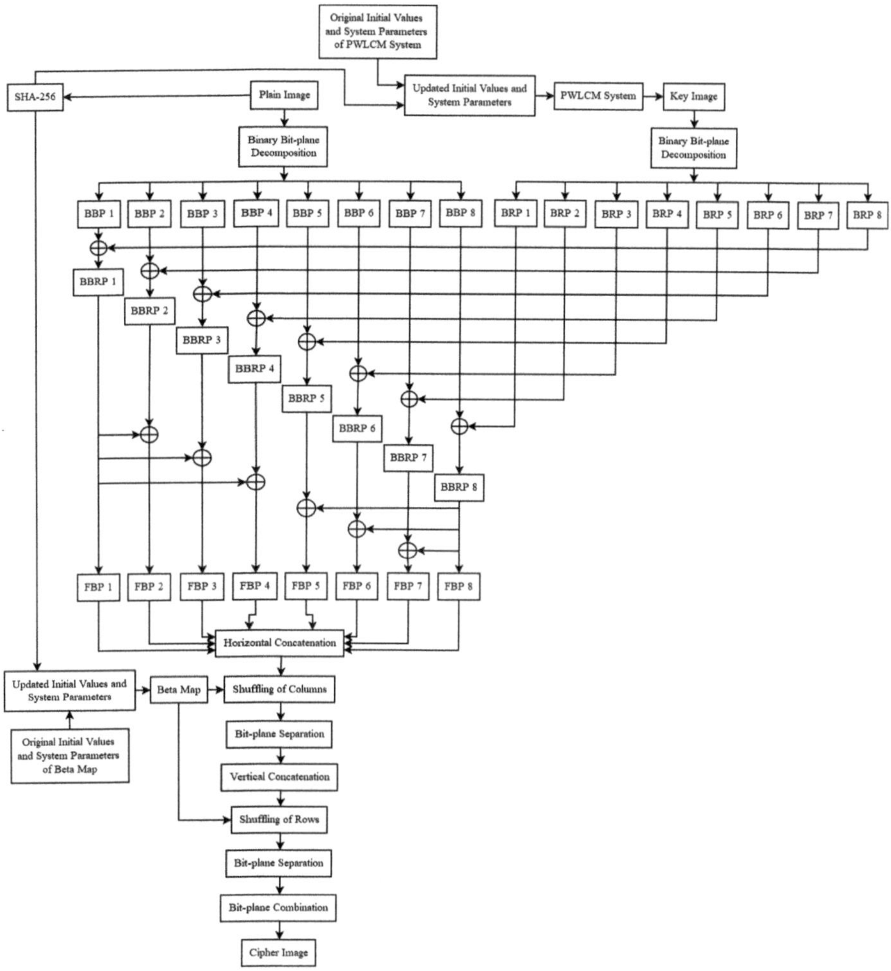

Fig. 1 Encryption system block diagram

The diffusion operation is as follows:

$$\begin{cases} IKBP1 = IBBP1 \oplus KBBP8 \\ IKBP2 = IBBP2 \oplus KBBP7 \\ IKBP3 = IBBP3 \oplus KBBP6 \\ IKBP4 = IBBP4 \oplus KBBP5 \\ IKBP5 = IBBP5 \oplus KBBP4 \\ IKBP6 = IBBP6 \oplus KBBP3 \\ IKBP7 = IBBP7 \oplus KBBP2 \\ IKBP8 = IBBP8 \oplus KBBP1 \end{cases} \tag{12}$$

Step 8: Perform Stage-2 of bit-plane diffusion operation between the bit-plane outputs of **Step 7**. Let the diffused bit-plane outputs are denoted as:

$$FDBP1, FDBP2, FDBP3, FDBP4, FDBP5, FDBP6, FDBP7, FDBP8$$

The diffusion operation is as follows:

$$\begin{cases} FDBP1 = IKBP1 \\ FDBP2 = IKBP1 \oplus IKBP2 \\ FDBP3 = IKBP1 \oplus IKBP3 \\ FDBP4 = IKBP1 \oplus IKBP4 \\ FDBP5 = IKBP5 \oplus IKBP8 \\ FDBP6 = IKBP6 \oplus IKBP8 \\ FDBP7 = IKBP7 \oplus IKBP8 \\ FDBP8 = IKBP8 \end{cases} \tag{13}$$

Step 9: Generate the secret keys $(xxx1(1), k, x1, x2, p1, p2, q1, q2, r)$ of Beta map as **Step 3** of Sect. 3.1.
Step 10: Iterate the Beta map of Eq. 1–5 for $200 + ((N \times 8) + (M \times 8))$ times. To avoid transit effect, remove 200 iterations from the first position. Therefore, the generated Beta map sequence is,

$$xxx1 = \begin{pmatrix} xxx1(1), xxx1(2), xxx1(3), \ldots, \\ xxx1((N \times 8) + (M \times 8)) \end{pmatrix} \tag{14}$$

Step 11: Separate the Beta map sequence into two parts which are represented as:

$$\begin{cases} xxx11 = \begin{pmatrix} xxx1(1), xxx1(2), xxx1(3), \ldots, \\ xxx1(N \times 8) \end{pmatrix} \\ xxx12 = \begin{pmatrix} xxx1((N \times 8) + 1), xxx1((N \times 8) + 2), \\ xxx1((N \times 8) + 3), \ldots, \\ xxx1((N \times 8) + (M \times 8)) \end{pmatrix} \end{cases} \tag{15}$$

Step 12: Sort the Beta map sequences as,

$$\begin{cases} [xxxC\text{sort}, xxxC\text{index}] = \text{sort}(xxx11) \\ [xxxR\text{sort}, xxxR\text{index}] = \text{sort}(xxx12) \end{cases} \tag{16}$$

where $xxxC$sort and $xxxR$sort are the sort sequences of $xxx11$ and $xxx12$, respectively. $xxxC$index and $xxxR$index are the index values of $xxxC$sort and $xxxR$sort, respectively.
Step 13: Horizontal concatenate the bit-plane outputs of **Step 8**. Perform column shuffling operation by using the index value, $xxxC$index.
Step 14: Separate the column shuffled bit-plane outputs in **Step 13**. Vertical concatenate the bit-planes to perform row-shuffling operation by using the index value, $xxxR$index.

Step 15: Separate the bit-planes in **Step 14** and then combine all the bit-planes to generate an encrypted image.

The decryption operation is performed in the reverse way of performing encryption operation.

4 Security Analyses and Computer Simulations

Simulation results are shown in Fig. 2. In this algorithm, the keys are taken as $x(1) = 0.2, \text{mue1} = 0.3, xxx(1) = -0.2, kk = 0.9, xx1 = -0.7, xx2 = 1, pp1 = 8, pp2 = 3, qq1 = 1, qq2 = -1, rr = -0.2$. In Fig. 2, we can see that the proposed

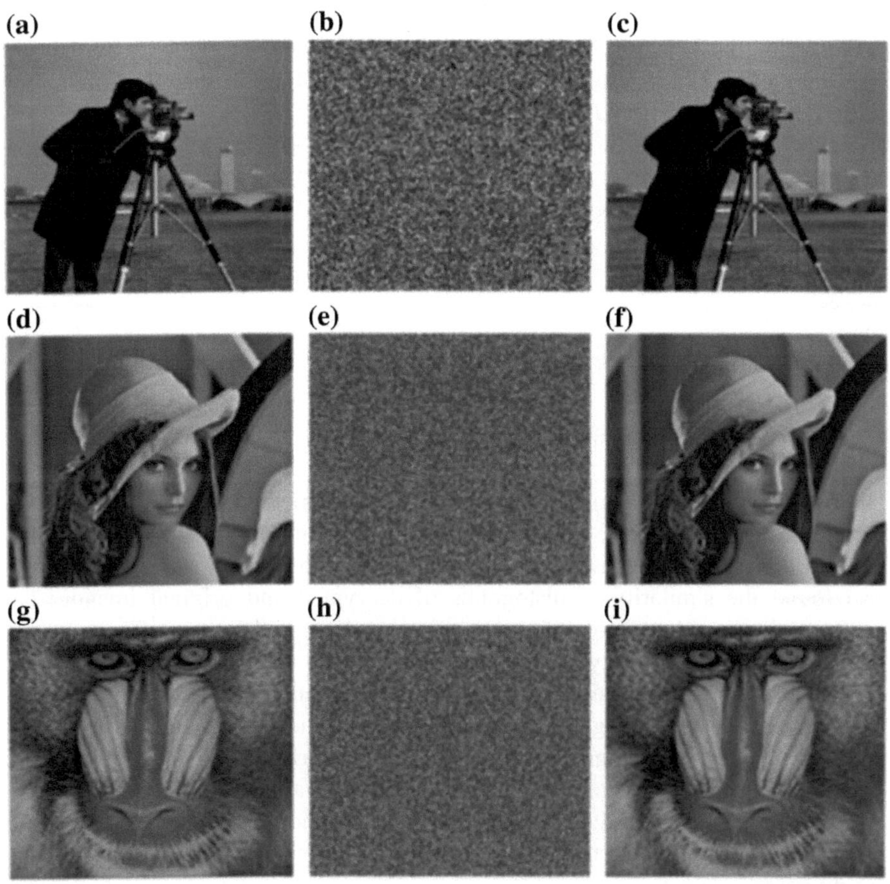

Fig. 2 Computer simulation outputs: original images of **a** "Cameraman (256 × 256)," **d** "Lena (512 × 512)," **g** "Baboon (512 × 512)"; encrypted images of **b** "Cameraman (256 × 256)," **e** "Lena (512 × 512)," **h** "Baboon (512 × 512)"; decrypted images of **c** "Cameraman (256 × 256)," **f** "Lena (512 × 512)," **i** "Baboon (512 × 512)"

algorithm encrypts and decrypts images properly using the right keys. This shows that the proposed method is suitable to encrypt and decrypt grayscale images.

The security analyses are as follows.

4.1 Key Space Analysis

Key space is the number of different keys which are used in an encryption–decryption operation. The key space should be larger than 2^{128} to protect brute-force attack [17]. The secret keys used in this scheme are,

a. system parameters and initial values of chaotic maps
b. hash value of 256-bits.

To define the keys of chaotic maps, a precision value of 10^{-15} is used in this algorithm and for hash value, a key space of 2^{128} is taken in this algorithm.

So the key space in total is,

$\left(10^{15} \times 10^{15}\right) \times \left(10^{15} \times 10^{15} \times 10^{15} \times 10^{15} \times 10^{15} \times 10^{15} \times 10^{15} \times 10^{15} \times 10^{15}\right) \times 2^{128} = 10^{165} \times 2^{128} = 1.0853 \times 2^{676}$ which is larger than 2^{128} [17].

This proves that the proposed scheme strongly protects brute-force attack.

4.2 Statistical Attack Analysis

4.2.1 Histogram Analysis

Figure 3 shows the histogram results of the proposed method. In the results, we see the uniformity of grayscale values in the histograms of encrypted images and also see the differences in the histograms of encrypted and original images. In Fig. 3, we also found the similarity of histograms of decrypted and original images. This confirms the zero losses of information in the decrypted images. These in turn proves the strong resistivity of statistical attack in the encryption scheme.

The histogram uniformity can be evaluated quantitatively by calculating the variance of encrypted images. The lower is the variance, higher the uniformity of grayscale values in histogram images [18]. The variance results are shown in Table 1. In Table 1, we can see the lesser value of variance of encrypted images using the proposed method as compared to the encryption methods in [19, 20]. This proves that a greater uniformity occurs in the histograms of encrypted images using the proposed scheme.

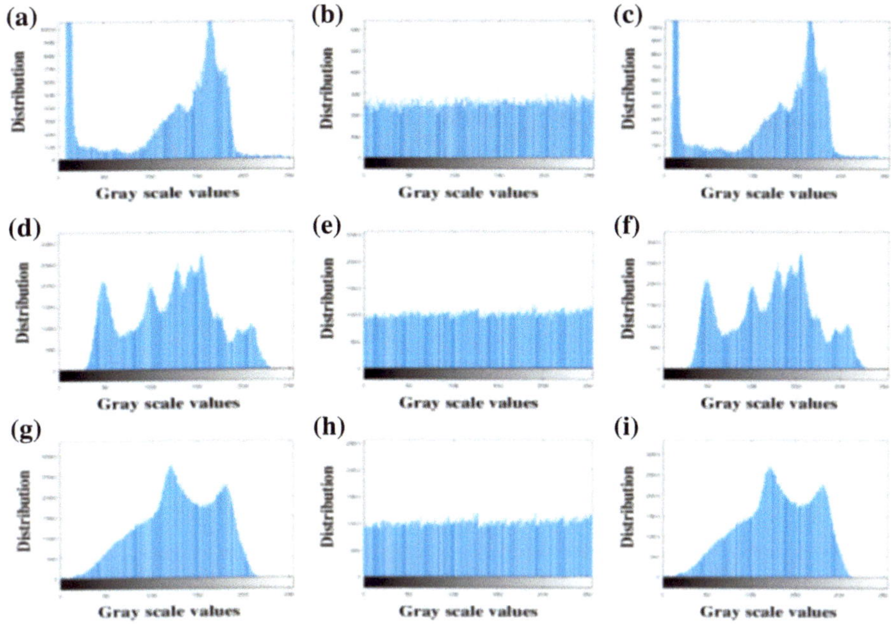

Fig. 3 Histogram outputs: original images of **a** "Cameraman," **d** "Lena," **g** "Baboon"; encrypted images of **b** "Cameraman," **e** "Lena," **h** "Baboon"; decrypted images of **c** "Cameraman," **f** "Lena," **i** "Baboon"

Table 1 Results of variance

Algorithms	Images	Original images	Encrypted images
Ours	Cameraman.tiff	110,970	334.5391
	Lena.tiff	633,400	3480.7
	Baboon.tiff	749,430	3857.3
Ref. [19]	Lena	–	5554.8293
Ref. [20]	Lena	–	5335.8309

4.2.2 Correlation Analysis

Table 2 shows the correlation results of the proposed scheme. In this result, we can see that, in all the three directions, the original images have correlation value close to 1 and the encrypted images have correlation value close to 0. The correlation plot of "Lena" image is shown in Fig. 4. These plots clearly show the high adjacent pixel correlation of original images and weak adjacent pixel correlation of encrypted images. This proves that the proposed scheme strongly resists statistical attack.

Table 2 Correlation coefficient results

Image	Org. images			Enc. images		
	Diag.	Vert.	Horz.	Diag.	Vert.	Horz.
Cameraman.tiff (256 × 256)	0.9082	0.9556	0.9345	-0.0070	0.0140	0.0122
Lena.tiff (512 × 512)	0.9626	0.9853	0.9702	0.0183	0.0017	0.0200
Baboon.tiff (512 × 512)	0.7297	0.7607	0.8681	0.0022	−0.0146	−0.0093

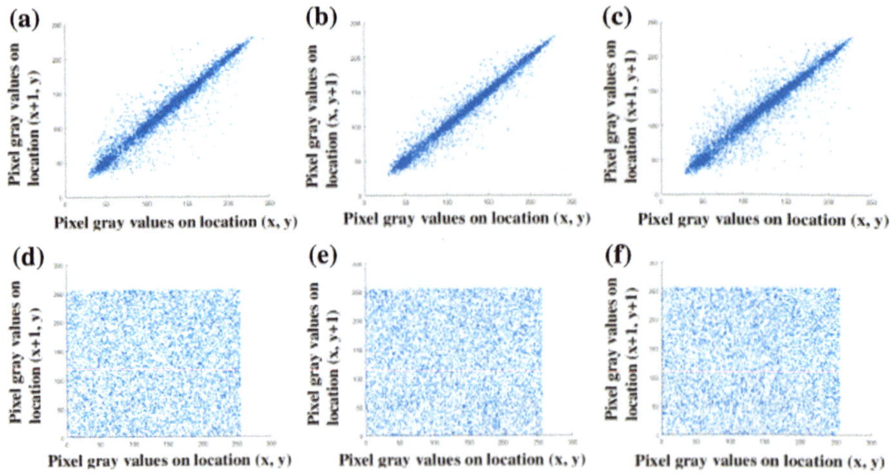

Fig. 4 Correlation outputs of original "Lena" image: **a** horz., **b** vert., and **c** diag.; correlation outputs of encrypted "Lena" image: **d** horz., **e** vert., **f** diag

4.3 Differential Attack Analysis

Two measures Unified Average Changing Intensity (UACI) and Numbers of Pixel Changing Rate (NPCR) are mostly used to analyze the differential attack in an encryption scheme. The expected value of UACI and NPCR of 256-level grayscale image are 33.4635% and 99.6094%, respectively [5]. Table 3 shows the min., max., and avg. UACI and NPCR results (100 values) of the proposed scheme. In Table 3, we can see that the average UACI and NPCR results for all the images are very close to the expected UACI and NPCR values, respectively. This proves the high resistivity of differential attack in the proposed scheme.

4.4 Information Entropy

It measures the degree of randomness of pixels in images. For an ideal encryption system, the entropy of encrypted images falls to 8. The higher is the closeness to 8,

Table 3 UACI and NPCR results

Image	UACI (%)			NPCR (%)		
	Min.	Max.	Avg.	Min.	Max.	Avg.
Cameraman.tiff (256 × 256)	33.2526	33.6811	33.4827	99.5514	99.6536	99.6060
Lena.tiff (512 × 512)	33.3694	33.6682	33.4941	99.5819	99.6414	99.6077
Baboon.tiff (512 × 512)	33.3695	33.6025	33.4727	99.5762	99.6460	99.6079

Table 4 Information entropy results

Images	Original images	Encrypted images
Cameraman.tiff	7.0097	7.9963
Lena.tiff	7.4451	7.9976
Baboon.tiff	7.3583	7.9974

the stronger is the entropy. The entropy results are shown in Table 4. The results show that the entropies of all the encrypted images are very close to 8. This confirms the high randomness of pixels in the encrypted images by using the proposed scheme.

4.5 Key Sensitivity Analysis

Key sensitivity measures the sensitivity of keys in decryption and encryption of images. That means a small change in the key will lead to large changes in the output. A cryptosystem should have high key sensitivity to protect brute-force attack [18, 21, 22]. The key sensitivity in both decryption and encryption process is as follows.

4.5.1 Encryption Level Key Sensitivity Analysis

In this process, the encryption operation is executed by changing a single bit in 10^{-15} position of one key and the remaining keys are unchanged and finally analyze the difference image of two encrypted images. The encryption level key sensitivity results of "Lena" image are shown in Fig. 5. The UACI and NPCR results are shown in Table 5. The difference images of Fig. 5 and the NPCR, UACI results of Table 5 shows that more than 99% pixels are changed in changing of one key. This proves the high encryption level key sensitivity of the proposed algorithm.

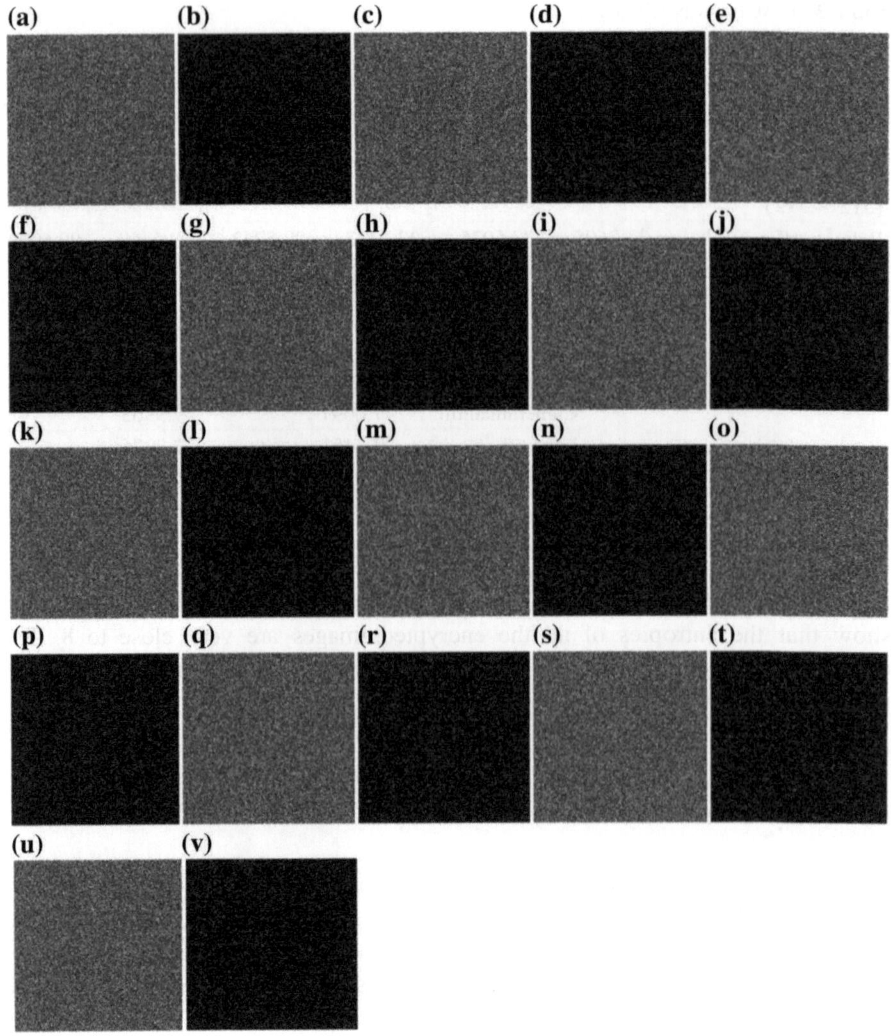

Fig. 5 Encryption level key sensitivity results of "Lena" image: encrypted images when changing of key from **a** mue1 to mue1 $+10^{-15}$, **c** $x(1)$ to $x(1)+10^{-15}$, **e** $xxx(1)$ to $xxx(1)+10^{-15}$, **g** kk to $kk+10^{-15}$, **i** $xx1$ to $xx1+10^{-15}$, **k** $xx2$ to $xx2+10^{-15}$, **m** $pp1$ to $pp1+10^{-15}$, **o** $pp2$ to $pp2+10^{-15}$, **q** $qq1$ to $qq1+10^{-15}$, **s** $qq2$ to $qq2+10^{-15}$, **u** rr to $rr+10^{-15}$; Difference of **b** Figs. 2e and 5a, **d** Figs. 2e and 5c, **f** Figs. 2e and 5e, **h** Figs. 2e and 5g, **j** Figs. 2e and 5i, **l** Figs. 2e and 5k, **n** Figs. 2e and 5m, **p** Figs. 2e and 5o, **r** Figs. 2e and 5q, **t** Figs. 2e and 5s, **v** Figs. 2e and 5u

Table 5 UACI and NPCR results of "Lena" image

Encryption keys	$\text{mue1} + 10^{-15}$	$x(1) + 10^{-15}$	$xxx(1) + 10^{-15}$	$kk + 10^{-15}$	$xx1 + 10^{-15}$	$xx2 + 10^{-15}$
NPCR (%)	99.4892	99.4579	99.6166	99.6235	99.5949	99.5892
UACI (%)	32.7018	32.6263	33.4090	33.4851	33.4971	33.4666
Encryption keys	$pp1 + 10^{-15}$	$pp2 + 10^{-15}$	$qq1 + 10^{-15}$	$qq2 + 10^{-15}$	$rr + 10^{-15}$	
NPCR (%)	99.6090	99.6235	99.6006	99.6220	99.6151	
UACI (%)	33.4944	33.5019	33.5604	33.5239	33.5390	

4.5.2 Decryption Level Key Sensitivity Analysis

In this process, the decryption operation is executed by changing a single bit in 10^{-15} position of one decryption key and the remaining decryption keys are unchanged and finally analyze the key changed decrypted images. The key sensitivity results of "Lena" image are shown in Fig. 6. The NPCR and UACI results are shown in Table 6. Figure 5 and Table 6 results show that more than 99% pixels are changed in changing of one decryption key. This shows the high decryption level key sensitivity of the algorithm.

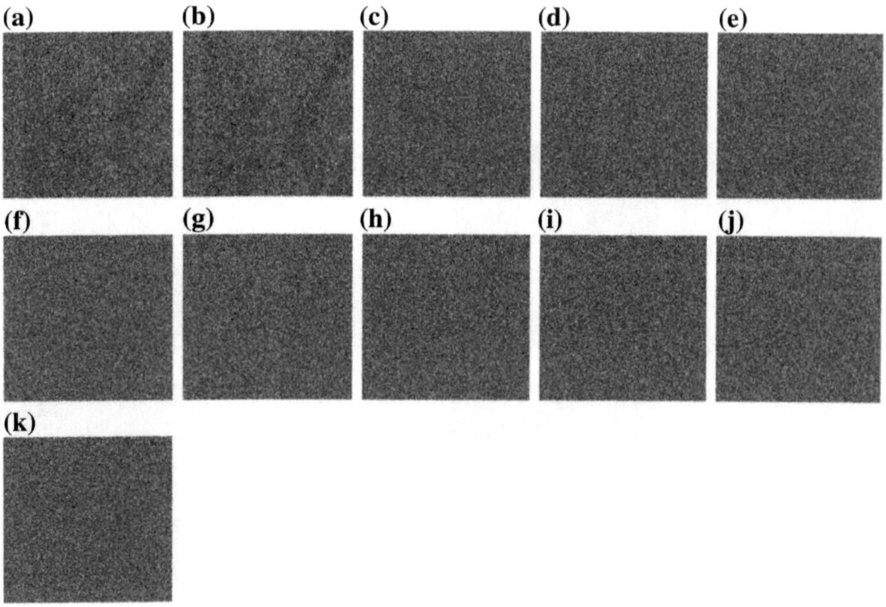

Fig. 6 Decryption level key sensitivity results of "Lena" image: Decrypted images when changing of key from **a** mue1 to $\text{mue1} + 10^{-15}$, **b** $x(1)$ to $x(1) + 10^{-15}$, **c** $xxx(1)$ to $xxx(1) + 10^{-15}$, **d** kk to $kk + 10^{-15}$, **e** $xx1$ to $xx1 + 10^{-15}$, **f** $xx2$ to $xx2 + 10^{-15}$, **g** $pp1$ to $pp1 + 10^{-15}$, **h** $pp2$ to $pp2 + 10^{-15}$, **i** $qq1$ to $qq1 + 10^{-15}$, **j** $qq2$ to $qq2 + 10^{-15}$, **k** rr to $rr + 10^{-15}$

Table 6 UACI and NPCR results of "Lena" image

Encryption keys	$mue1 + 10^{-15}$	$x(1) + 10^{-15}$	$xxx(1) + 10^{-15}$	$kk + 10^{-15}$	$xx1 + 10^{-15}$	$xx2 + 10^{-15}$
NPCR (%)	99.0734	99.1058	99.5991	99.6090	99.5789	99.6246
Encryption keys	$pp1 + 10^{-15}$	$pp2 + 10^{-15}$	$qq1 + 10^{-15}$	$qq2 + 10^{-15}$	$rr + 10^{-15}$	
NPCR (%)	99.6128	99.5998	99.5995	99.5998	99.6246	

4.6 Plaintext Sensitivity Analysis

Plaintext sensitivity measures the sensitivity of pixel values in original images. This is done first by randomly changing one-pixel value and constant other pixel values in original image. Then the proposed encryption algorithm is applied to get the difference of two encrypted images (one-pixel value changed encrypted image and original encrypted image). Figure 7 shows the results of plaintext sensitivity at positions (1, 1), (64, 64), (128, 128), and (256, 256). Table 7 shows the corresponding UACI and NPCR results when the pixel values are changed at random positions. The images of Fig. 7b, d, f, h and the UACI, NPCR results of Table 7 shows that more than 99% pixels are changed in encrypted image while changing of one pixel in original image. This proves the high sensitivity of the plaintext of the proposed scheme.

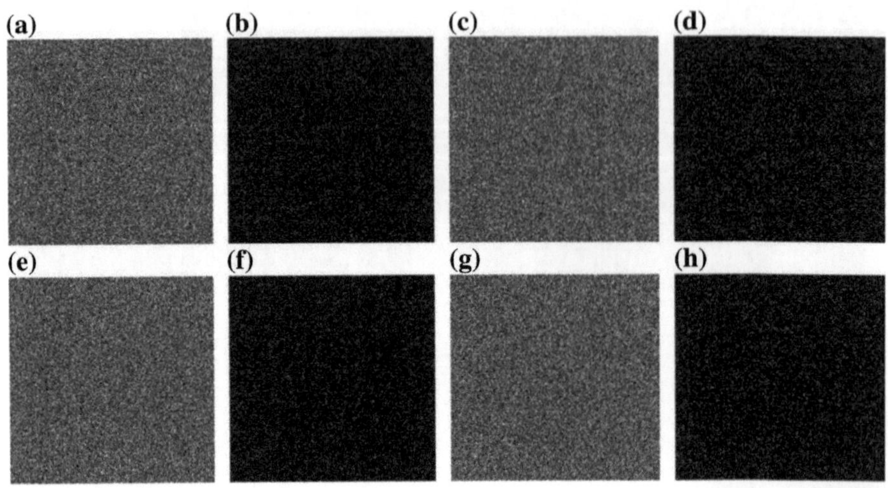

Fig. 7 Plaintext sensitivity results of "Lena" image: encrypted images when pixels changed at **a** (1, 1), **c** (64, 64), **e** (128, 128), **g** (256, 256) positions; Difference of **b** Figs. 2e and 7a, **d** Figs. 2e and 7c, **f** Figs. 2e and 7e, **h** Figs. 2e and 7g

Table 7 UACI and NPCR results of "Lena" image

Pixel positions	(1, 1)	(64, 64)	(128, 128)	(256, 256)
NPCR (%)	99.5926	99.6067	99.5995	99.6132
UACI (%)	33.4845	33.5027	33.4715	33.4923

4.7 Chosen-Plaintext Attack (CPA) and Known-Plaintext Attack (KPA) Analysis

The CPA and KPA are easily attacked by the attacker when the algorithm is not much depended on the original image. The proposed scheme highly depends on the original image in both diffusion and permutation operations. This shows the high resistivity of the proposed algorithm against CPA and KPA.

4.8 Comparison Analysis

The comparative results are shown in Table 8. The comparison is done on "Lena" image. By observing Table 8 comparison results, we found the followings:

- Higher key space of the proposed scheme as compared to the schemes in [19, 20].
- The correlation values of all the schemes are close to 0.
- Better UACI and NPCR results of the proposed scheme as compared to the schemes in [19, 20].
- The entropy values of all the schemes are close to 8.

Table 8 Comparison of "Lena (512 × 512)" image

Algorithms	Key space	Correlation coefficients			NPCR	UACI	Entropy
		Horizontal	Vertical	Diagonal			
Ours (1 round)	1.0853×2^{676}	0.0200	0.0017	0.0183	99.6077	33.3694	7.9976
Ref. [19] (1 round)	$10^{42} \approx 1.4349 \times 2^{139}$	0.0020	−0.0009	0.0016	0.4222	0.1365	7.9993
Ref. [20] (1 round)	More than $10^{120} \approx 2^{400}$	0.0013	−0.0002	−0.0003	99.5838	17.0035	–

4.9 Encryption Time

The proposed algorithm is simulated using MATLAB R2012a in a system having configuration 3.40 GHz processor with 4 GB RAM. The total simulation time required to simulate 256×256 and 512×512 sized images are 0.148 s and 0.733 s, respectively. Due to the requirement of less simulation time, the proposed algorithm is suitable to use in real-time applications.

5 Conclusion

This paper proposes a bit-level image encryption scheme using 1-D chaotic maps. In this proposed method, two-stages of bit-plane diffusion operations using PWLCM system and one-stage of bit-level column–row shuffling operation using Beta map are performed. This scheme yields good encryption outputs, produces higher key space, less simulation time to execute the algorithm, and resists all the widely known security attacks. In this algorithm, simple bit-plane diffusion operations and bit-level column–row shuffling operations are used, and hence, the proposed method provides simplicity in the algorithm. The comparison results show the better security of the proposed scheme as compared to other reported schemes. All these features show that the proposed algorithm is time-efficient, secure, and simple to encrypt grayscale images.

References

1. Coppersmith D (1994) The Data Encryption Standard (DES) and its strengths against attacks, IBM J Res Dev 38:243–250.
2. Pub NF. 197 (2001) Advanced encryption standard (AES). Federal Information Processing Standards Publication. 197, 441-0311.
3. Gao H, Zhang Y, Liang S, Li D (2006) A new chaotic algorithm for image encryption, Chaos Solitons Fractals 29:393–399.
4. Samhita P, Prasad P, Patro KAK, Acharya B (2016) A Secure Chaos-based Image Encryption and Decryption Using Crossover and Mutation Operator, Int J Cont T Appl 9:17–28.
5. Gupta A, Thawait R, Patro KAK, Acharya B (2016) A Novel Image Encryption Based on Bit-Shuffled Improved Tent Map, Int J Cont T Appl 9:1–16.
6. Shadangi V, Choudhary SK, Patro KAK, Acharya B (2017) Novel Arnold Scrambling Based CBC-AES Image Encryption, Int J Cont T Appl 10:93–105.
7. Guesmi R, Amine M, Farah B et al. (2016) Hash key-based image encryption using crossover operator and chaos, Multimed Tools Appl 75:4753–4769.
8. Guesmi R, Farah MAB, Kachouri A, Samet M (2016) A novel chaos-based image encryption using DNA sequence operation and Secure Hash Algorithm SHA-2, Nonlinear Dyn 83:1123–1136.
9. Matthews R (2010) On the derivation of a chaotic encryption algorithm, Cryptologia XIII 37–41.
10. Liu W, Sun K, Zhu C (2016) A fast image encryption algorithm based on chaotic map, Opt Lasers Eng 84:26–36.

11. Wang X, Wang S, Zhang Y, Guo K (2017) A novel image encryption algorithm based on chaotic shuffling method, Inf Secur J A Glob Perspect 0:1–10.
12. Abd El-Latif AA, Li L, Zhang T, Wang N, Song X, Niu X (2012) Digital image encryption scheme based on multiple chaotic systems, Sens Imaging 13:67–88.
13. Ahmad M, Khurana S, Singh S, AlSharari HD (2017) A Simple Secure Hash Function Scheme Using Multiple Chaotic Maps, 3D Res 8.
14. Zahmoul R, Ejbali R, Zaied M (2017) Image encryption based on new Beta chaotic maps, Optics and Lasers in Engineering 96:39–49.
15. Wang XY, Yang L (2012) Design of pseudo-random bit generator based on chaotic maps, Int J Mod Phy B 26:1250208.
16. Li S, Chen G, Mou X (2005) On the dynamical degradation of digital piecewise linear chaotic maps, Int J Bifurcation Chaos 15:3119–3151.
17. Kulsoom A, Xiao D, Abbas SA (2016) An efficient and noise resistive selective image encryption scheme for gray images based on chaotic maps and DNA complementary rules, Multimed Tools Appl 75:1–23.
18. Chai X, Chen Y, Broyde L (2017) A novel chaos-based image encryption algorithm using DNA sequence operations, Opt Laser Engg 88:197–213.
19. Zhu ZL, Zhang W, Wong KW, Yu H (2011) A chaos-based symmetric image encryption scheme using a bit-level permutation, Information Sciences 181:1171–1186.
20. Zhang YQ, Wang XY (2014) A symmetric image encryption algorithm based on mixed linear–nonlinear coupled map lattice, Information Sciences 273:329–351.
21. Patro KAK, Acharya B (2018) Secure multi–level permutation operation based multiple colour image encryption, J Inf Secur Appl 40:111–133.
22. Patro KAK, Banerjee A, Acharya B (2017) A Simple, Secure and Time Efficient Multi-way Rotational Permutation and Diffusion Based Image Encryption by Using Multiple 1-D Chaotic Maps, In International Conference on Next Generation Computing Technologies, Springer, Singapore, 396–418.

K. Abhimanyu Kumar Patro received the M.Tech. degree in Electronics and Telecommunication Engineering from the National Institute of Science and Technology, Berhampur, India.

He is currently pursuing Ph.D. in the Department of Electronics and Telecommunication Engineering, National Institute of Technology, Raipur, India. His research interests include Cryptography and Network Security, Digital Signal Processing, Digital Image Processing. He has more than 15 research publications in National/International Journals and conferences. He also served as a reviewer of several journals, including IEEE, Journal of Electronic Imaging.

Mr. Patro is a member of Cryptology Research Society of India, International Association of Engineers, Universal Association of Computer and Electronics Engineers, International Association of Computer Science and Information Technology, and Internet Society.

Bibhudendra Acharya received Post Graduation and Ph.D. degree in Electronics and Communication Engineering from National Institute of Technology, Rourkela, India.

He is currently Head and Assistant Professor in the Electronics and Telecommunication Engineering Department, National Institute of Technology, Raipur, India. His research interests include Cryptography and Network Security, Microcontroller and Embedded system, Signal Processing, Mobile Communication. He has more than 60 research publications in National/ International Journals and conferences. He also served as a reviewer of several journals, including IEEE.

Dr. Acharya is a member of Computer Society of India, The Institution of Electronics and Telecommunication Engineers (India), Institute of Engineers (India), Indian Society for Technical Education (India), Cryptology Research Society of India, International Association of Engineers, Universal Association of Computer and Electronics Engineers, International Association of Computer Science and Information Technology, and The Indian Science Congress Association.

Optimization of Semi-solid-Forging Parameters of A356–5TiB$_2$ In Situ Composites Using ANSYS and DEFORM Simulation

S. Deepak Kumar, D. Karthik, S. K. Parida and S. K. Jha

1 Introduction

Thixoforming of metallic alloys and composites in the semi-solid state offers certain advantages compared to conventional forming routes. Semi-solid metal forming (SSMF) is a potential route to develop near-net-shape products, especially for aerospace and automotive applications [1]. Among the processes like rolling, extrusion, and forging for SSMF components, forging has substantial potential for improvement in mechanical properties and thereby better performance [2]. Finite element analysis (FEA) is a useful simulation tool used to assist in verifying and approximating how a product will react under various internal and external loading conditions. FEA involves a computer model of a design that is loaded and then analyzed for precise results, such as deformation, deflection, stress–strain, and temperature distributions and simulates the various loads and boundary conditions and determines the design response in such conditions. Koc et al. [3] have employed FEA-based simulation model to analyze the defects and material flow occurring during the semi-solid forging of aluminum alloys. Similar works were reported by Jafari et al. [4] to optimize the design parameters and investigate the semi-solid extrusion behavior of Al–Si alloy.

S. Deepak Kumar (✉) · S. K. Jha
Department of Production Engineering, Birla Institute of Technology Mesra, Ranchi 835215, Jharkhand, India
e-mail: dks10@iitbbs.ac.in

D. Karthik
Department of Manufacturing Engineering, National Institute of Technology Jamshedpur, Jamshedpur 831014, Jharkhand, India

S. K. Parida
Department of Manufacturing Engineering, National Institute of Foundry and Forge (NIFFT), Ranchi 834003, Jharkhand, India

J. Chattopadhyay et al. (eds.), *Innovations in Soft Computing and Information Technology*, https://doi.org/10.1007/978-981-13-3185-5_24

Today, simulation software like Design Environment for Forming (DEFORM-3D) is widely used to forecast the material and metal flow during semi-solid-forming operations. This has reduced the costly shop floor trials and delay of the finished products in the industry [5]. Satheesh et al. [6] modeled the microstructural evolution in the semi-solid condition by carrying out DEFORM simulations. In a recent work, Randhir et al. [7] studied the effect semi-solid forging on the microstructure, hardness, and strength of Al–Si alloys, by performing simulation trials.

Based on the literature survey, it can be concluded that optimization can be achieved efficiently and quickly through FEA-based simulation softwares like ANSYS and DEFORM-3D simulations for industrial applications. Thus, in the present research work, DEFORM-3D and ANSYS simulations were employed to optimize the semi-solid-forging parameters of in situ-formed A356–5TiB_2 composites.

2 Materials and Methods

2.1 *Synthesis of A356–5TiB_2 Composites*

The A356–5TiB_2 in situ composites were fabricated by mixed salt route. At a reaction temperature of 800 °C, the halide salts, K_2TiF_6, and KBF_4, undergo an exothermic reaction with molten Al–Si alloy and form titanium-diboride (TiB_2) particles in the melt. The details of the synthesis of the in situ composites have been described previously [8].

2.2 *Thixoforming of A356–5TiB_2 In Situ Composite*

The semi-solid forging was carried using an 80-ton hydraulic press (Flowmech Engineers Private Limited, India, Model No.-HP 50) on the alloy and composites. The load applied was at a RAM speed 20 mm/s and pressure of 150 kg/cm^2. A silicone spray was used as a lubricant in order to eliminate or decrease the friction between semi-solid billets and dies. The forming dies were preheated up to 200 °C to avoid the forged specimens from cooling. The heating process was monitored using K-type thermocouples to control the temperature fluctuation in the range of ±2 °C. Once the desired temperature is achieved, the thermocouples were extracted and the billets were independently subjected to semi-solid forging in the order of 30–50% deformation. The complete details of thixoforming of composites can be found elsewhere [9, 10].

2.3 Microstructure and Tensile Testing

Microstructural analyses were performed by optical microscopy and scanning electron microscope (SEM, Model: 40, Carl-Zeiss SMT, Germany) fitted with EDX analysis. Vickers hardness (macro-) measurements were carried out at an indentation load of 5 kg. For tensile testing, the alloy and composite's tensile specimens of size 4 × 4 mm cross section and gauge length of 16 mm were tested the following ASTM E-8M standards.

2.4 Finite Element Analysis Using ANSYS and DEFORM

The 3D geometric model of tensile specimens was developed using CATIAV5 software and is then imported in ANSYS R15 software. A 3D structural solid 164 element was used to obtain refined mesh. The required force to bring the semi-solid A356 alloy to yield point is calculated following Eq. (1)

$$F = \sigma_y \times A \tag{1}$$

where A is the cross-sectional area of the tensile specimen which is 40 mm^2 and σ_y is the yield stress of 165 MPa. Thus, to ensure that the test specimen goes well beyond the yield stress, a load of 6600 N was applied. This same force was applied to the fully elastic and elastic–plastic model to determine how ANSYS analyzes these results. The geometry of the tensile specimen and the boundary conditions applied is depicted in Fig. 1.

In the current research work, efforts have made to analyze material and metal flow during the semi-solid forging of A356–5TiB_2 composite by performing DEFORM-3D simulations. The thermo-physical properties and DEFORM 3D simulation inputs of semi-solid A356–5TiB_2 composite are presented in Table 1.

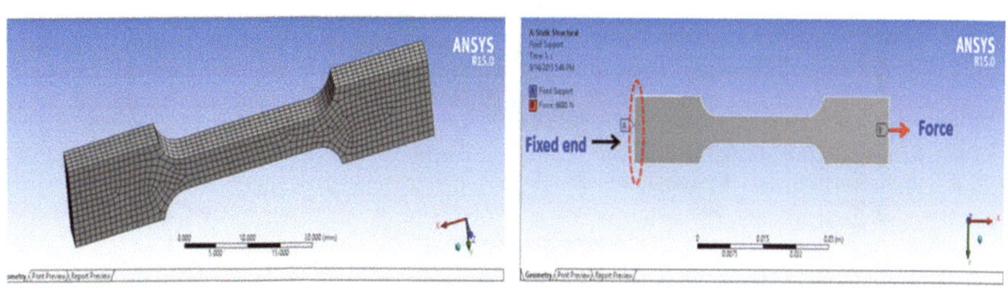

Fig. 1 Meshed model of tensile specimen and boundary conditions in ANSYS

Table 1 Thermo-physical properties of semi-solid A356–5TiB_2 composite

Parameter	Value
Density	2680 kg/m^3
Hardness (H_v5)	99
Yield strength	165 MPa
% elongation	6%
Modulus of elasticity	72.4 GPa
Poisson's ratio	0.33
Coefficient of thermal expansion (CTE)	21.4 μm/m-°C
Thermal conductivity	151 W/m-K
Solidus temperature	555 °C
Liquidus temperature	615 °C

3 Results and Discussion

3.1 *ANSYS Results*

The finite element analysis results comprise of displacements and stress–strain curves including Von-Mises stress criteria in the ANSYS simulation. The analysis was considered for fully elastic–plastic material properties under the tensile loading conditions. It can be seen from Fig. 2 that at the center of the test specimen there is a highest stress and this is due to a notch which is due to the change in section areas. A maximum stress of 270 MPa is recorded and is located toward the fillets and is located within the gauge length zone of the tensile specimen. It is noted that the maximum deflection obtained in the semi-solid A356–5TiB_2 composite at fracture to be 10.6 mm and is presented in Fig. 3.

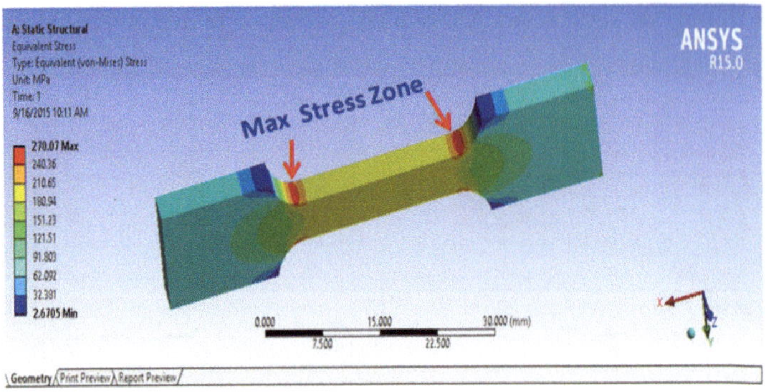

Fig. 2 Von-Mises stress in elastic–plastic zone

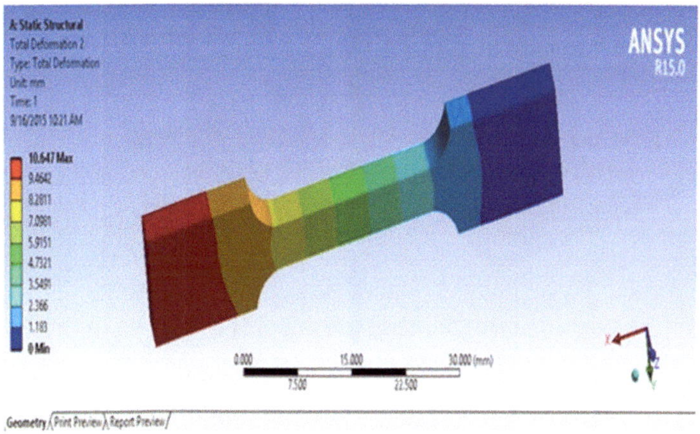

Fig. 3 Deflection in elastic–plastic zone of the tensile specimen

3.2 *Analysis of DEFORM 3D Simulation Results*

The semi-solid forging at 30–50% deformations and the distribution of temperature plots were presented in Fig. 4. In the simulation analysis at 30% deformation, the temperatures reduced from 580 to 576 °C and in the 40% deformed samples, the billet temperatures further reduced to 574 °C. This indicates that, up to 40% deformation, the temperature levels are maintained in the range 580–574 °C

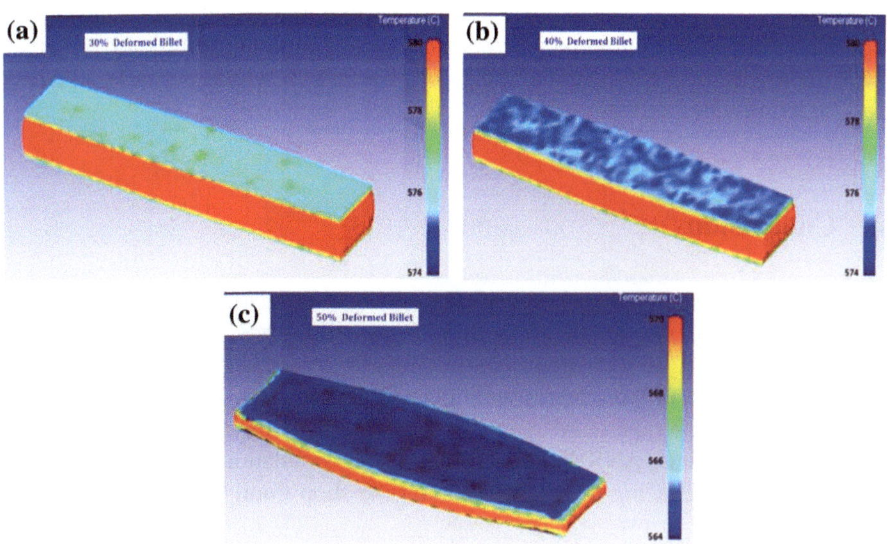

Fig. 4 Prediction of temperature in **a** 30%, **b** 40%, **c** 50% deformed billets

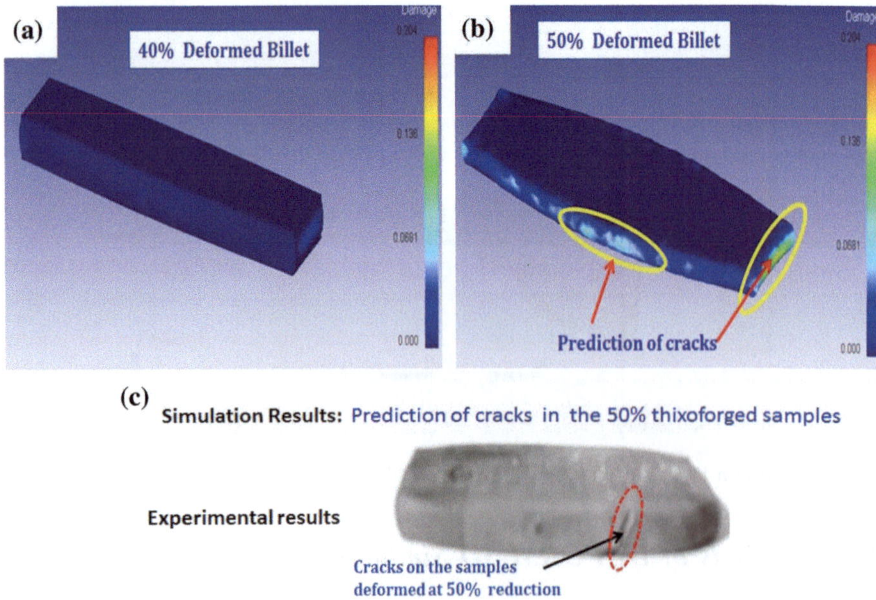

Fig. 5 Analysis of simulation results **a** 40%, **b** 50%, **c** experimental results

corresponding to a solid fraction (f_s) of 0.5–0.55. Moreover, in the 50% deformed samples the temperatures further decreased to 564 °C, which indicated that f_s goes beyond 0.7, which is not desired in semi-solid forging. The simulations in Fig. 5 demonstrate the occurrence of cracks in 50% deformed billets which are in accordance with experimental results. Thus, from the simulation analysis, the optimum semi-solid-forging parameters were identified to be: semi-solid forging at 40% deformation, at a temperature range of 574–580 °C which is nearly equal to a solid fraction value of 0.5.

4 Conclusions

In the current research work, ANSYS and DEFORM simulations were performed to achieve the maximum tensile strength of semi-solid-forged A356–5TiB_2 in situ composites.

Thus, the optimum parameters were identified to be: semi-solid forging at 40% deformation, at a temperature range of 574–580 °C which is nearly at a solid fraction of 0.5. Thus, ANSYS and DEFORM 3D simulation software resulted in saving time and reducing the costly trials on the shop floor.

Acknowledgements The research support provided by Central Research Facility, at Indian Institute of Technology Kharagpur is gratefully acknowledged.

References

1. Flemings MC. Behavior of metal alloys in the semisolid state. Metallurgical Transactions A 1991; 22: 957–981.
2. Fan Z. Semisolid metal processing. International Materials Reviews 2002; 47:49–85.
3. Koc M, Vazquez V, Witulski T, Altan T. Application of the finite element method to predict material flow and defects in the semi-solid forging of A356 aluminum alloys. J Mater. Proc. Tech. 1996; 59:106–112.
4. Jafari MR, Zebarjad SM, Kolahan F. Simulation of thixoformability of A356 aluminum alloy using finite volume method. Mater. Sci. Eng. A. 2007; 454–455: 558–563.
5. DEFORM-3D Software User Manual, Ver. 6.1, Scientific Forming Technologies Corporation, Columbus, Ohio, 2011.
6. Satheesh P, Mahesh NS, Mushib B. Development of forging process for synchronized ring through numerical simulation. Sastech Journal. 2013; 12:1–8.
7. Randhir Kumar, Singh NK, Ohdar RK, In: proc. ATCOM 2011, RDCIS SAIL, Ranchi, July 5–7, 2011.
8. Deepak kumar S, Mandal A, Chakraborty M. Effect of Thixoforming on the Microstructure and Tensile Properties of A356 Alloy and A356-5TiB$_2$ in-situ Composite. Transactions of the Indian Institute of Metals. 2015; 68:123–130.
9. Deepak Kumar S, Vundavilli PR, Mantry S, Mandal A, Chakraborty M. A Taguchi optimization of cooling slope casting process parameters for production of semisolid A356 alloy and A356-5TiB$_2$ in-situ composite feedstock. Procedia Materials Science. 2014; 5: 232–241.
10. Deepak kumar S, Mandal A, Chakraborty M. On the age hardening behavior of thixoformed A356-5TiB$_2$ in-situ composite. Mater. Sci. Eng. A. 2015; 636: 254–262.

Dr. S. Deepak Kumar He earned his Ph.D from School of Mechanical Sciences, Indian Institute of Technology, Bhubaneswar in 2016 and M.Tech (Foundry & Forge Technology) from National Institute of Foundry & Forge Technology (NIFFT) - Ranchi, in the year 2006. He has a First Class Degree in Mechanical & Production Engineering in 2002 from Andhra University, Visakhapatnam. He has an Industry experience of over 3 years at G.E, Bangalore and Bharat Forge Ltd Pune and developed expertise in Finite Element Analysis and Design of various automotive components. He started his profession of Teaching and research as a faculty at Department of Manufacturing Engineering, National Institute of Technology (NIT) Jamshedpur for 3 Years (Jan2015–Dec2017). Currently, he is working as Asst. Professor, since 4Jan -2018 in the Department of Production Engineering, Birla Institute of Technology- Mesra, Ranchi.

Current Research Interests:

- Solidification Processing of Al-Si-Mg based alloys and in-situ composites;
- Semi-solid Metal Processing, (Cooling Slope casting, Thixoforming, Thixocasting, Thixoforging, Thixo extrusion);
- Near-net shape Manufacturing; Manufacturing Process Simulation;
- Casting Process Simulation–Pro-CAST, Fluent and DEFORM Simulation.

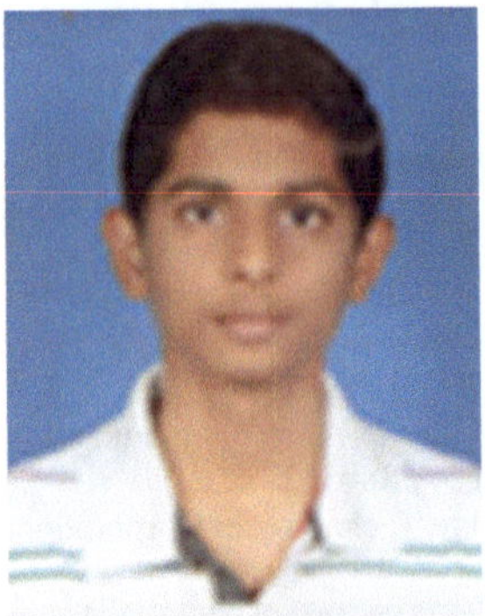

Mr. D. Karthik has pursued B.Tech in the Department of Manufacturing Engineering, National Institute of Technology Jamshedpur during 2013–2017.

His Research Interests include:

- Optimization Techniques using Design of Experiments; Taguchi method;
- Advanced Manufacturing Processes; Casting, Welding & Forming.

Dr. S. K. Parida has earned his Ph.D from School of Mechanical Sciences, Indian Institute of Technology, Bhubaneswar in 2015 and M.Tech (Manufacturing) from Indian Institute of Technology Kharagpur.

His Research Interests include:

- Composite material, Finite element, Fracture mechanics, Design of Machine Element, Manufacturing Systems Engineering;
- Fracture mechanics of adhesive bonded joints.

S. K. Jha

Research Fields:

- Manufacturing technology, Industrial Engineering Green Manufacturing, Productivity Improvement, Friction Stir welding.

Author Index

J. Chattopadhyay et al. (eds.), *Innovations in Soft Computing and Information Technology*, https://doi.org/10.1007/978-981-13-3185-5

MIX
Papier aus verantwortungsvollen Quellen
Paper from responsible sources
FSC® C105338

If you have any concerns about our products,
you can contact us on
ProductSafety@springernature.com

In case Publisher is established outside the EU,
the EU authorized representative is:
Springer Nature Customer Service Center GmbH
Europaplatz 3, 69115 Heidelberg, Germany

Printed by Libri Plureos GmbH
in Hamburg, Germany